BIOTECHNOLOGY IN AGRICULTURE SERIES

General Editor: Gabrielle J. Persley, Biotechnology Manager, Agriculture and Natural Resources Department, The World Bank, Washington DC, USA.

For a number of years, biotechnology has held out the prospect for major advances in agricultural production, but only recently have the results of this new revolution started to reach application in the field. The potential for further rapid developments is, however, immense.

The aim of this new book series is to review advances and current knowledge in key areas of biotechnology as applied to crop and animal production, forestry and food science. Some titles will focus on individual crop species or groups of species, others on specific goals such as plant protection or animal health, with yet others addressing particular methodologies such as tissue culture, transformation or immunoassay. In some cases, relevant molecular and cell biology and genetics will also be covered. Issues of relevance to both industrialized and developing countries will be addressed, and social, economic and legal implications will also be considered. Most titles will be written for research workers in the biological sciences and agriculture, but some will also be useful as textbooks for senior-level students in these disciplines.

Editorial Advisory Board:
P.J. Brumby, formerly of the World Bank, Washington DC, USA.
E.P. Cunningham, FAO, Rome, Italy.
P. Day, Rutgers University, New Jersey, USA.
J.H. Dodds, Michigan State University, USA.
J.J. Doyle, International Laboratory for Research on Animal Diseases, Nairobi, Kenya.
S.L. Krugman, United States Department of Agriculture, Forest Service.
W.J. Peacock, CSIRO, Division of Plant Industry, Australia.

Titles Available:
1: Beyond Mendel's Garden: Biotechnology in the Service of World Agriculture *G.J. Persley*
2: Agricultural Biotechnology: Opportunities for International Development *Edited by G.J. Persley*
3: The Molecular and Cellular Biology of the Potato *Edited by M.E. Vayda and W.D. Park*
4: Advanced Methods in Plant Breeding and Biotechnology *Edited by D.R. Murray*
5: Barley: Genetics, Biochemistry, Molecular Biology and Biotechnology *Edited by P.R. Shewry*
6: Rice Biotechnology *Edited by G.S. Khush and G.H. Toenniessen*
7: Plant Genetic Manipulation for Crop Protection *Edited by A. Gatehouse, V. Hilder and D. Boulter*
8: Biotechnology of Perennial Fruit Crops *Edited by F.A. Hammerschlag and R.E. Litz*
9: Bioconversion of Forest and Agricultural Plant Residues *Edited by J.N. Saddler*
10: Peas: Genetics, Molecular Biology and Biotechnology *Edited by R. Casey and D.R. Davies*

PEAS: GENETICS, MOLECULAR BIOLOGY AND BIOTECHNOLOGY

Edited by

R. CASEY AND D.R. DAVIES

John Innes Institute
Colney Lane
Norwich NR4 7UH
UK

CAB INTERNATIONAL

CAB INTERNATIONAL Tel: Wallingford (0491) 832111
Wallingford   Telex: 847964 (COMAGG G)
Oxon OX10 8DE   Telecom Gold/Dialcom: 84: CAU001
UK   Fax: (0491) 833508

© CAB INTERNATIONAL 1993. All rights reserved. No part of this publication may be reproduced in any form or by any means, electronically, mechanically, by photocopying, recording or otherwise, without the prior permission of the copyright owners.

A catalogue entry for this book is available from the British Library.

ISBN 0 85198 863 6

Typeset by Colset Typesetters Ltd, UK
Printed and bound in Great Britain at the
University Press, Cambridge

Contents

Contributors		vii
Preface		ix
1.	The Pea Crop D.R. Davies	1
2.	The Nuclear Genome T.H.N. Ellis	13
3.	The Plastid Genome T.H.N. Ellis	49
4.	Regulation of Gene Expression P.M. Gilmartin	61
5.	Genetic and Developmental Analysis of the Seed T.L. Wang and C.L. Hedley	83
6.	Biochemistry and Molecular Biology of Seed Products R. Casey, C. Domoney and A.M. Smith	121
7.	Developmental Mutants I.C. Murfet and J.B. Reid	165
8.	Molecular Approaches to the Study of Diseases of Pea A. Vivian	217

9.	Root Nodules, *Rhizobium* and Nitrogen Fixation N.J. BREWIN, M.J. AMBROSE AND J.A. DOWNIE	237
10.	Tissue Culture and Transformation D.R. DAVIES AND P.M. MULLINEAUX	291
11.	Opportunities in Relation to Developments in Plant Biotechnology D.R. DAVIES	303

Index 313
  Appendix A 320
  Appendix B 323

Contributors

M.J. AMBROSE, *John Innes Institute, Colney Lane, Norwich, NR4 7UH, UK.*

N.J. BREWIN, *John Innes Institute, Colney Lane, Norwich, NR4 7UH, UK.*

R. CASEY, *John Innes Institute, Colney Lane, Norwich, NR4 7UH, UK.*

D.R. DAVIES, *John Innes Institute, Colney Lane, Norwich, NR4 7UH, UK.*

C. DOMONEY, *John Innes Institute, Colney Lane, Norwich, NR4 7UH, UK.*

J.A. DOWNIE, *John Innes Institute, Colney Lane, Norwich, NR4 7UH, UK.*

T.H.N. ELLIS, *John Innes Institute, Colney Lane, Norwich, NR4 7UH, UK.*

P.M. GILMARTIN, *Centre of Plant Biochemistry and Biotechnology, University of Leeds, Leeds, LS2 9JT, UK.*

C.L. HEDLEY, *John Innes Institute, Colney Lane, Norwich, NR4 7UH, UK.*

P.M. MULLINEAUX, *John Innes Institute, Colney Lane, Norwich, NR4 7UH, UK.*

I.C. MURFET, *Department of Plant Science, University of Tasmania, GPO Box 252C, Hobart 7001, Australia.*

J.B. REID, *Department of Plant Science, University of Tasmania, GPO Box 252C, Hobart 7001, Australia.*

A.M. SMITH, *John Innes Institute, Colney Lane, Norwich, NR4 7UH, UK.*

A. VIVIAN, *Faculty of Applied Sciences, University of the West of England, Coldharbour Lane, Bristol, BS16 1QY, UK.*

T.L. WANG, *John Innes Institute, Colney Lane, Norwich, NR4 7UH, UK.*

Preface

Peas made an early claim to fame as material for the experimentalist, having been used by Mendel in his seminal experiments over 100 years ago. They are still valuable today as experimental material for plant molecular biologists, biochemists and physiologists; it should also be remembered that they are the fourth most important legume crop in the world. The range of mutations affecting the development of various plant tissues and organs in the pea provides a wealth of material, which will be increasingly exploited by developmental biologists in the future, and at the same time provides valuable germplasm for the plant breeder. The prospect of altering the products stored in the seed is a realistic option, and with pea transformation now possible there are good reasons for being optimistic about future developments in the crop.

It is intended that this book will bring the reader up to date with our present understanding of the genetics, biochemistry and biotechnology of peas. Many of the achievements are of a fundamental nature, but we have tried to couch such studies in the context of new opportunities in, and new uses for, the crop. The book should be of interest to research scientists as well as students in applied and fundamental areas of plant biology, and much of its contents will appeal to those who work with grain legumes other than peas, since many of the approaches adopted, and much of the discussion, are relevant to these other crops.

Those who have contributed to the book deserve our gratitude for keeping close to the schedule we set them and tolerating our promptings and reminders. We are particularly grateful to our secretary, Tarn Dalzell, for her endless patience and good humour in the face of numerous amendments, corrections and rearrangements.

Rod Casey and Roy Davies

The Pea Crop

D.R. DAVIES
John Innes Institute, Colney Lane, NR4 7UH, Norwich, UK

The pea (*Pisum sativum*) has been an important model plant for several generations of plant physiologists, biochemists and geneticists (see Sutcliffe and Pate, 1977) and a crop plant whose utilization can be traced back to neolithic times (Zohary and Hopf, 1988). But the competition is now severe on both fronts. The value of the pea plant as an experimental system remains high for many physiologists and biochemists, but for geneticists the merits of those plant and animal systems which have very short life cycles, or have particularly useful attributes (such as the transposable elements of maize), have become overwhelming. For plant molecular biologists the siren call of *Arabidopsis* has been difficult to resist with its many advantages including a small genome, short life cycle, few repeated sequences in its DNA and ease of transformation.

The future of the pea as a crop plant has to be similarly evaluated. In large areas of the developed world where political factors do not distort the issues, international trade and rapid mass transport have exposed all crops to the economic realities of comparison with any others. Peas as fresh, canned, dehydrated or frozen vegetables have to compete with green beans, corn, brassicas, dry beans, root crops, etc., irrespective of season. Peas as an animal feed are critically assessed by feed compounders on the basis of their nutritional value and price in comparison with any competing feed source of plant or animal origin. The value of peas in a peasant economy is assessed on several criteria – yield, resistance to biotic or abiotic stresses and local acceptability as a food and feed material. Given the above it might be thought that this book is a requiem for the pea as a crop and as an experimental system. For a number of reasons which will become apparent in subsequent chapters, this is not the case.

If we look first at its role as a model crop, the pea has many attributes which make it suitable for particular biochemical investigations. This is well illustrated by the work on pea seeds which have proved to be excellent material for studying the biosynthesis, genetics and molecular biology of starch (Chapters 5 and 6). This in turn has highlighted the opportunities that exist for manipulating the amount and quality of starch and, in doing so, has indicated that the genetic engineering of this product will be among the foremost targets of plant biotechnology in the next decade. Our knowledge of the biosynthesis of starch in peas, and of the DNA sequences that code for the enzymes involved and the availability of a range of mutants with altered seed carbohydrates, together with the recent demonstration that peas can be transformed, have made the pea one of the prime experimental systems for such studies. Undoubtedly, these investigations in turn will open up new opportunities for providing added value to the pea as well as other crops, as new uses emerge for novel starches in the formulation of convenience foods and food products, and in industrial applications. The latter include the use of starch in the paper industry, in the pharmaceutical and cosmetic industries, in the manufacture of adhesives and textiles, and as a flocculant in mineral industries.

There is a body of information on the genetics of flowering, of internode length, of gibberellin metabolism, and of day length responses (see Chapter 7) which is being exploited by physiologists and molecular biologists. The variety of developmental and morphological mutants that exist in peas (Marx, 1987; see Chapter 7) provide a wealth of variation which can be exploited by those seeking to unravel the processes underlying the development of complex structures in plants; the leaf and stipule mutants are a prime example of such variants. For all the above reasons peas will retain their attraction as a model plant system.

Peas will also have an important role in defining what it is possible to achieve with other legumes and in developing the methodology for these achievements. This will emerge, for example, from studies of its interaction with the nitrogen fixing symbiotic bacterium *Rhizobium* (Chapter 9). It will also serve the same role in providing a base line of information and DNA clones for the genetic mapping of other grain legumes. Together with soyabeans, peas are the best characterized grain legumes in terms of the provision and location of DNA markers for the chromosomes (Ellis *et al.*, 1992), and these two species will provide the body of information on which to examine the extent of synteny among grain legumes, the similarity of locations of genes of interest, the possible regions of conservation of gene order and of reorganization, and the identification of genes of agronomic interest which might be isolated and exchanged through transformation procedures.

What, then, of its future role as a crop plant? This is an era where the practices of sustainable agriculture are being increasingly demanded and the grain legumes, with their ability to yield well in the absence of nitrogenous fertilizers, are likely to be one of the crops that will be a beneficiary of this trend. Currently the pea is one of the world's most important grain legumes, serving a variety of roles as a source of both human food and animal feed. Its seed is harvested either as the dry mature form or in an immature state – the precise stage of maturity varying according to the end use. Within each of these two categories there are a number of specialized uses and markets. The dry mature seed is used extensively as animal feed, directly as human food and as an ingredient of a variety of prepared foods. Those harvested in an immature form are used directly as a fresh vegetable or are canned, dehydrated or frozen; within each of these four categories there are a number of classes which have been developed to satisfy different local or national requirements or niche markets. They include varieties with different colour, texture, sugar and starch contents, and size of seed – all of which have been selected for particular markets. Seed size can vary from 30 mg to more than 480 mg dry weight – although most cultivars are within the range of 180–300 mg. The sugar content can vary from 6% to 10% and starch from 30% to 50%. A small market exists for immature whole pods to be used as a vegetable and which involves harvesting them at an early stage of their development.

The ways in which the crop is handled reflect both the diversity of the farming systems under which the crop is grown and its end use. Peas are grown as an important source of animal feed and human food in many regions of the developed world where farming practices are among the most advanced. There they are harvested by a combine harvester if the dry mature crop is required; the crop usually has the added benefit in these circumstances of being a valuable break crop in a rotation that tends to be dominated by cereals. In similar regions, a sophisticated industry supports the production of the crop for canning, freezing and dehydrating or freeze-drying, with the first two now being the dominant methods. To ensure the necessary quality of the product, the crop has to be harvested at a precisely defined stage of development, treated within a few hours to prevent enzymic degradation, and canned, frozen or dried within a short interval of harvesting. The production and harvesting are therefore precisely controlled and the varieties grown strictly defined (Arthey, 1985).

The crop is also an important element of the agriculture and horticulture of developing countries, where it may be hand or mechanically cultivated and harvested. In these instances again it is either harvested as a mature dry seed or grown as a vegetable crop for harvesting at an immature stage.

The types of peas grown reflect the diversity of farming practices. The land races and primitive forms as well as the traditional garden peas are plants which can grow to a height of 1.5 m or more and as such have to be supported. They are also rather indeterminate in their pattern of flowering, which means that harvesting fresh pods is possible over an extended period of time. This is an advantageous feature, not only for the amateur gardener, for whom the pea is an important crop, but also in many areas of the developing world. Modern agricultural practices, particularly in areas of higher rainfall, have, however, demanded forms which are shorter and thus less prone to collapse, in order to facilitate mechanical harvesting. The latter in turn demands that all the pods on the plant mature much more simultaneously than in the primitive or traditional garden pea and a plant which has a more determinate habit. Genetic variation for internode length and thus for total height, as well as for degrees of determinacy, has been readily found and has allowed plant breeders to produce the range of modern cultivars that are currently available. These are often less than 70 cm tall, and produce a limited number of pods. Indeed, the pea provides a prime illustration of the rapid time-scale in which substantially changed forms more suited to modern agricultural practice can be produced.

The total world production of dry peas is currently estimated to be 16.3 million metric tons (MMT) making it the fourth most important grain legume worldwide. This value compares with that of 103.1 MMT for soyabean (*Glycine max*), 23.4 MMT for ground nuts (*Arachis hypogea*) and of 17.5 MMT for dry bean (*Phaseolus* spp) (FAO, 1991). The primary areas of production are the former USSR (an estimated 7.3 MMT), Europe (4.86 MMT, of which France produces 3.19 MMT), China (an estimated 2.0 MMT), India (0.6 MMT) and North America (0.57 MMT). While these represent the temperate regions, in which peas are mainly grown, it is worth noting that they are also grown, albeit to a much lesser extent, in warmer regions at higher altitudes or as a cool-season crop. The trends in production over the last 20 years indicate a fairly static output of dry peas from 1969 to 1980, with total world values being between 7.5 and 9.5 MMT, but then increasing steadily to values between 16.0 and 17.3 MMT for 1989 to 1991. Much of this increase is due to enhanced production in Europe and the former USSR; that in Europe can be attributed to a large extent to an increase in the area of the crop in France.

In comparison with the areas and yields of dry peas, the total values for fresh peas are very much less. The total world yield is estimated to be 4.8 MMT, with North America and Europe contributing 1.4 and 1.8 MMT respectively; these values represent peas with a very much higher water content than that of dry peas and cannot therefore be compared directly with the earlier values for the dry pea crop. Over the last 20 years there has been little change in the total production of green peas but any figures given for this crop are probably gross underestimates of the total production worldwide. This is because much of the crop is grown for home or local consumption and as such will not be recorded.

The extent to which the area of peas grown and the total output of the crop will change in the future is difficult to predict. In some regions this is a function of political forces which establish the extent to which the crop is subsidized. In all areas it is related to the relative merits of peas as opposed to any other competitor crop, legume or non-legume. In this context merit is judged in a variety of ways; it can be in terms of short term economic advantage, of short or long term agronomic benefit (for example as a break crop or as a provider of nitrogen to the succeeding crop) or of its particular value (for example, as a high protein crop, as a vegetable, or for any other specialized end use). The other criterion which will establish the extent to which it is grown in the future relates to the ability of breeders to improve the crop and remove the constraints that currently limit its yield or its exploitation for any given purpose. The recent history of the crop, the extent of the natural genetic variation available and the opportunities emerging from biotechnology all provide an encouraging picture of future opportunity.

The evolutionary ancestors of the crop are not known beyond its relatively recent domestication from primitive but closely related wild forms. The archaeological evidence regarding the occurrence of peas is substantial and interesting (Zohary and Hopf, 1988) but unfortunately does not assist us greatly in our understanding of the evolution of the crop. It is generally accepted that at most there are only two species within the genus – *Pisum sativum* and *P. fulvum* (Davis, 1970) – with their closest relatives probably being found in the genera *Vicia* and *Lathyrus*. *P. sativum* and *P. fulvum* can be crossed and, although chromosomal rearrangements have occurred within each (see Chapter 2), gene exchange can occur; thus it has been argued that the genus *Pisum* should be considered monospecific (Lamprecht, 1966). Primitive and wild forms of peas occur in the Near East, Ethiopia and Central Asia; introgression from these forms into cultivated peas has undoubtedly occurred (Ben Ze'ev and Zohary, 1973) and has helped to generate the variation seen in the species today. Some, although not all, of the primitive and wild forms, many of which were earlier given species status, are distinguished from modern varieties by their taller and more rambling habit and by their tough seed coat; they have dehiscing pods and generally have smaller seeds than cultivars, with often a bitter component in the seeds. The transition from primitive to advanced cultivar has been a progressive one, with the distinguishing characteristics being regulated mainly by simply controlled genetic differences (Marx, 1985).

As peas are diploid and self-fertilizing, the breeding strategies adopted to improve the crop have been those conventionally adopted for such species. The extent of natural outcrossing has been estimated to be less than 1% (Gritton, 1980). Hybridization among cultivars or between cultivars, land races and primitive forms, followed by pedigree, bulk or backcross methods of selection, has been traditionally used; more recently, single-seed descent methods have also been evaluated (McBride *et al.*, 1992) as three generations can be generated each year (Snoad, 1980). No method of producing haploids in peas has been devised and so the dihaploid route to the rapid production of homozygotes from heterozygotes is not available.

The ease with which all forms of peas, with the exception of *P. fulvum*, can be intercrossed allows plant breeders ready access to the large range of variation that exists in the wild, primitive and cultivated forms. While many of the wild forms as well as cultivars show chromosomal changes (see Chapter 2), which may cause partial sterility of hybrids, gene exchange is invariably possible. Intergeneric hybrids involving peas have never been produced; fertilization occurs when *Vicia faba* and peas are intercrossed but only a few cell divisions occur and embryo development ceases at a very early stage (Gritton and Wierzbicka, 1975). There has been no report of the successful embryo rescue of such hybrids. Substantial collections of peas are held at many centres, with the main ones being in Italy, ICARDA, Syria, Poland, UK, Sweden, USA, Germany and India (IBPGR, 1989), all of which have totals of between 1500 and 4000 accessions. In the period between 1950 and 1970 Swedish geneticists in particular devoted a great deal of effort to inducing mutations in peas, and these, together with the spontaneous forms, have constituted a resource in which over 2000 mutations are recognized (Blixt, 1974).

Even though the genus *Pisum* is monospecific, or at most bispecific, and given the inability to introduce genetic information from other genera by sexual crosses, there is no real evidence that the genetic base of *Pisum* is so narrow as to have been a constraint to its improvement and development as a crop. The transition from a wild to a domesticated crop has been referred to earlier and has involved changes in a number of attributes, most of which are under simple genetic control. A reduction in seed coat thickness is achieved by exploiting a recessive allele at the *gty* locus, plant stature is markedly influenced by the *le* locus and pod shattering can be prevented by the presence of a recessive allele at the *dpo* locus. The recessive allele at the *a* locus not only results in the production of white flowers but is also associated with the reduction of a bitter component in the mature seed.

Among the more recent examples of the incorporation of major gene mutations are those which result in higher sugar content of seed (*r* and *rb* loci) and which now form the basis of varieties used in the fresh food, freezing and much of the canning industries (see Chapter 5). A reduction in foliage, with a consequent improvement in standing ability, has been achieved by introducing a mutation at the *af* locus (Hedley and Ambrose, 1981), to generate the semi-leafless varieties which now represent a substantial proportion of the varieties grown in northern Europe (Fig. 1.1). Mutations at the *p* and *v* loci have been incorporated in order to generate pods lacking a parchment layer, thereby reducing pod shattering and seed loss at harvesting (Snoad, 1980); varieties with a thickened pod wall due to mutation at the *n* locus and pods without suture strings due to mutation at the *sin* locus are used as vegetables. Resistance to pea seed-borne mosaic virus (the *sbm* locus, Hagedorn and Gritton, 1973), numerous loci coding for resistance to fungal pathogens (Lewis and Matthews, 1985; Bernier *et al.*, 1988) and loci coding for flowering time and development (Murfet and Reid, 1985; see Chapter 7) have been exploited in cultivars.

There are, however, many constraints to increasing the utilization, acreage and profitability of the pea crop in the future. Again these differ according to the geographical location and end use. There is an unquestioned need to

Fig. 1.1. A field of semi-leafless peas.

improve the level of resistance to the main fungal, bacterial and virus diseases (Hagedorn, 1989). The root rot complex poses problems because of the number of organisms involved, although tolerance to at least some of the component pathogens can be found (Davis and Shehata, 1985). The pathogens in the complex can include *Fusarium solani* f. *pisi*, *Aphanomyces euteiches*, *Thielaviopsis basuola* and *Pythium* spp (Hagedorn, 1989). Leaf and pod spot caused by *Ascochyta pisi* is prevalent in many of the important pea-growing areas in the world, as are downy mildew (*Peronospora pisi*), grey mould (*Botrytis cinerea*), powdery mildew *(Erysyphe pisi)* and *Fusarium* wilt (*Fusarium oxysporum* f. sp. *pisi*). Viral pathogens, including pea seed-borne mosaic, pea enation mosaic, pea early browning, pea top yellows, pea mosaic and pea streak, are of concern to greater or lesser extents (Hagedorn, 1985). Bacterial blight (*Pseudomonas syringae*) is also widespread (Hagedorn, 1985). Genes for resistance to particular races or strains of many of these pathogens have been identified (Hagedorn, 1985; Lewis and Matthews, 1985; Bernier *et al.*, 1988) but only some are incorporated into cultivars. The main insect pests of peas are listed by van Emden *et al.* (1988) but sources of resistance have not been found as yet.

Yield *per se* remains a challenge and the increases achieved by cereal breeders have not been matched in peas. For example, in the United Kingdom over the period 1940 to 1980, the slope of the average increase in the yield of winter wheat was 0.082 tonnes per hectare per year, whereas that for peas was 0.044 tonnes per hectare per year (Heath and Hebblethwaite, 1985). In part this reflects the disparity of effort devoted to the crops but it also reflects the differing nature of the plants. The average yield of dry peas can be over 5 tonnes per hectare in some Western European countries but is much less in most, with the world average being less than 2 tonnes per hectare (FAO, 1991). The harvest index (i.e. the proportion of the biological yield that is economically valuable) is comparable in peas and modern varieties of wheat, and it therefore seems unlikely that there is much scope for improvement in this respect. There is one obvious biological constraint in that peas have a higher protein content than wheat, and the energy requirement for a unit gain in weight of protein is substantially higher than that for the equivalent gain of starch. There is a need to reduce the very high harvest losses that can sometimes occur in areas of high rainfall (Davies, 1977). In low rainfall areas, taller varieties of peas, which collapse as they mature, can be tolerated as they can still be harvested by combine harvesters under these climatic conditions. However, under higher rainfall conditions, there is a need to find varieties that will remain upright until they are harvested. The semi-leafless peas show an improvement in this respect (Snoad, 1985) but further progress demands a stiffer strawed form – an attribute which has not yet been identified in germplasm collections and remains a challenge for the future. Another approach to improving straw stiffness has been the exploitation of the rogue phenomenon (Bateson and Pellew, 1915). The genetic basis of this condition remains to be elucidated (Matthews, 1973) but its phenotypic consequences include a reduced density of foliage and a more wiry habit. A rogue mutant of the cultivar Maro was released as the variety Progreta in the UK, and was grown extensively for a time (Gent and Knight, 1978). Improved standing ability can be achieved by incorporating recessive alleles at both the *af* and *st* loci; the latter reduces the stipule area. Such plants have been termed leafless, and, although a cultivar, Filby (Fig. 1.2), of this kind was released in the UK, it had some disadvantages which mitigated against its eventual success (Hedley and Ambrose, 1981). It was found that leafless forms grown at low densities could not show compensatory growth for the decreased number of plants per unit area, probably due to their low leaf area index. It was therefore necessary to grow them at high planting densities, which in turn demanded plants with seed characteristics suited to such competitive environments. These studies highlighted the potential value of leafless forms and the need to ally breeding programmes more closely with detailed studies of crop physiology (Hedley and Ambrose, 1981). Winter hardiness has been sought by North American and French scientists (Auld *et al.*, 1983; Eteve, 1985) and although some level of tolerance to cold has been obtained it remains a challenge (Murray *et al.*, 1988; Murray and Swensen, 1991). The level of winter hardiness does not match that of the winter cereals.

Those constraints to the utilization of peas imposed by the composition of the seed have been in the past primarily related to their use as animal feed.

Fig. 1.2. A leafless pea, cultivar Filby.

In this context it was the storage protein content and composition that was felt to be the primary limiting factor. In spite of the fact that a substantial body of information has been generated regarding the nature and synthesis of the seed proteins (see Chapter 6), no practical improvement in either the amount or the quality of the protein synthezised has resulted from these studies. There has been no determined attempt to undertake a recurrent selection programme for protein content in pea seeds over an extended period of time. There have been numerous attempts to analyse the genetic control of protein content, but most have conceded defeat when faced with the significant effect of environmental variables on this character (Matthews and Arthur, 1985). At present, feed compounders limit the amount of peas (and indeed of other legumes) that is incorporated into the feed of young animals; this is related to the fact that digestive upsets and lower weight gain ensue if excessive amounts are incorporated (Marquardt and Bell, 1988). The basis of these problems may lie in the antigenic properties of the storage or other proteins in the seed or they may be due to the presence of metabolic inhibitors. Traditionally the latter have been assumed to be present at such low levels in peas, particularly in comparison with other legumes, that they have been ignored. However, critical feeding tests to establish which of the inhibitors present are of significance have not been done and will have to involve exploiting natural or induced variants of peas that have different levels of particular metabolic inhibitors (see Chapter 6).

The scope for improvement is therefore substantial and the challenges, although formidable in some instances, are certainly not insuperable given the advances in our knowledge and technologies that are now beginning to emerge. The succeeding chapters illustrate some of the information that is relevant in this context and which provides the basis for this optimism.

References

Arthey, D. (1985) Vining peas – processing and marketing. In: Hebblethwaite, P.D., Heath, M.C. and Dawkins, T.C.K. (eds), *The Pea Crop*. Butterworths, London, pp. 433–440.

Auld, D.L., Ditterline, R.L., Murray, G.A. and Swenson, J.B. (1983) Screening peas for winterhardiness under field and laboratory conditions. *Crop Science* 23, 85–88.

Bateson, W. and Pellew, C. (1915) On the genetics of 'rogues' among culinary peas (*Pisum sativum*). *Journal of Genetics* 5, 13–36.

Ben Ze'ev, N. and Zohary, D. (1973) Species relationships in the genus *Pisum*. *Israel Journal of Botany* 22, 73–91.

Bernier, C.C., Bijiga, G., Nene, Y.L. and Cousin, R. (1988) Breeding for disease resistance in pulse crops. In: Summerfield, R. (ed.), *World Crops: Cool Season Food Legumes*. Kluwer Academic Publishers, Dordrecht pp. 97–106.

Blixt, S. (1974) The pea. In: King, R.C. (ed.), *Handbook of Genetics*, vol. 2. Plenum Press, New York, pp. 181–222.

Davies, D.R. (1977) Restructuring the pea plant. *Science Progress* 64, 201–214.

Davis, D.W. and Shehata, M.A. (1985) Breeding for resistance to root-rot pathogens of peas. In: Hebblethwaite, P.D., Heath, M.C. and Dawkins, T.C.K. (eds), *The Pea Crop*. Butterworths, London, pp. 237–246.

Davis, P.H. (1970) *Flora of Turkey*, Vol. 3. pp. 370–373, Edinburgh University Press, Edinburgh.

Ellis, T.H.N., Turner, L., Hellens, R.P., Lee, D., Harker, C.L., Enard, C., Domoney, C. and Davies, D.R. (1992) Linkage maps in pea. *Genetics* 130, 649–663.

Eteve, G. (1985) Breeding for cold tolerance and winter hardiness in pea. In: Hebblethwaite, P.D., Heath, M.C. and Dawkins, T.C.K. (eds), *The Pea Crop*. Butterworths, London. pp. 131–136.

FAO (1991) *Production Yearbook*, vol. 5. FAO, Rome.

Gent, G.P. and Knight, R.F. (1978) Development of the 'rabbit-eared' rogue dried pea variety 'Progreta'. *Pisum Newsletter* 10, 105–106.

Gritton, E.T. (1980) Field pea. In: Fehr, W.R. and Hadley, H.H. (eds), *Hybridization of Crop Plants*. American Society of Agronomy, Madison, pp. 347–356.

Gritton, E.T. and Wierzbicka, B. (1975) An embryological study of a *Pisum sativum* x *Vicia faba* cross. *Euphytica* 24, 277–284.

Hagedorn, D.J. (1985) Diseases of peas: their importance and opportunities for breeding for disease resistance. In: Hebblethwaite, P.D., Heath, M.C. and Dawkins, T.C.K. (eds), *The Pea Crop*. Butterworths, London, pp. 205–214.

Hagedorn, D.J. (1989) *Compendium of Pea Diseases*. American Phytopathological Society, St Paul, Minnesota.

Hagedorn, D.J. and Gritton, E.T. (1973) Inheritance of resistance to the pea seed-borne mosaic virus. *Phytopathology* 63, 1130–1133.

Heath, M.C. and Hebblethwaite, P.D. (1985) Agronomic problems associated with the pea crop. In: Hebblethwaite, P.D., Heath, M.C. and Dawkins, T.C.K. (eds), *The Pea Crop*. Butterworths, London, pp. 19–30.

Hedley, C.L. and Ambrose, M.J. (1981) Designing 'leafless' plants for improving yields of dried pea crops. *Advances in Agronomy* 34, 225–277.

IBPGR (1989) Bettencourt, E., Konopka, J., Damania, A.B. (eds), *Directory of Germ Plasm Collections. II. Food Legumes*. IBPGR, Rome.

Lamprecht, H. (1966) *Die Entstelung der Arten und hoheren Kategorien*. Springer-Verlag, Vienna.

Lewis, B.G. and Matthews, P. (1985) The world germplasm of *Pisum sativum*: could it be used more effectively to produce healthy crops? In: Hebblethwaite, P.D., Heath, M.C. and Dawkins, T.C.K. (eds), *The Pea Crop*. Butterworths, London, pp. 215–230.

Marquardt, R.R. and Bell, J.M. (1988) Future potential of pulses for use in animal feeds. In: Summerfield, R.J. (ed.), *World Crops: Cool Season Food Legumes*. Kluwer Academic Publishers, Dordrecht, pp. 421–444.

Marx, G.A. (1985) The pea genome: a source of immense variation. In: Hebblethwaite, P.D., Heath, M.C. and Dawkins, T.C.K. (eds), *The Pea Crop*. Butterworths, London, pp. 45–54.

Marx, G.A. (1987) A suite of mutants that modify pattern formation in pea leaves. *Plant Molecular Biology Reporter* 5, 311–335.

Matthews, P. (1973) Genetic studies on spontaneous and induced rogues in *Pisum sativum*. PhD thesis, University of East Anglia.

Matthews, P. and Arthur, A.E. (1985) Genetic and environmental components of variation in protein content of peas. In: Hebblethwaite, P.D., Heath, M.C. and Dawkins, T.C.K. (eds), *The Pea Crop*. Butterworths, London, pp. 369–382.

McBride, J., Lacey, C. and Laws, R. (1992) A direct comparison of single seed descent and pedigree breeding methods in a range of pea (*Pisum sativum* L.) families. *1re Conference Européenne sur les Protéagineux*, Angers, pp. 55–56.

Murfet, I.C. and Reid, J.B. (1985) The control of flowering and internode length in *Pisum*. In: Hebblethwaite, P.D., Heath, M.C. and Dawkins, T.C.K. (eds), *The Pea Crop*. Butterworths, London, pp. 67–80.

Murray, G.A. and Swensen, J.B. (1991) Winterhardiness of pea. *Pisum Genetics* 23, 4–13.

Murray, G.A., Eser, D., Gusta, L.V. and Eteve, G. (1988) Winterhardiness in pea, lentil, faba bean and chickpea. In: Summerfield, R.J. (ed.), *World Crops: Cool Season Food Legumes*, Kluwer Academic Publishers, Dordrecht, pp. 831–843.

Snoad, B. (1980) The origin, performance and breeding of leafless peas. *ADAS Quarterly Review* 37, 69–86.

Snoad, B. (1985) The need for improved pea-crop plant ideotypes. In: Hebblethwaite, P.D., Heath, M.C. and Dawkins, T.C.K. (eds), *The Pea Crop*. Butterworths, London, pp. 31–41.

Sutcliffe, J.F. and Pate, J.S. (eds) (1977) *The Physiology of the Garden Pea*. Academic Press, London.

van Emden, H.F., Ball, S.L. and Rao, M.R. (1988) Pest, disease and weed problems in pea, lentil, faba bean and chickpea. In: Summerfield, R.J. (ed.), *World Crops: Cool Season Food Legumes*. Kluwer Academic Publishers, Dordrecht, pp. 519–534.

Zohary, D. and Hopf, M. (1988) *Domestication of Plants in the Old World*. Clarendon Press, Oxford, pp. 92–98.

The Nuclear Genome   2

T.H.N. ELLIS

John Innes Institute, Colney Lane, Norwich, NR4 7UH, UK

'Peas have great individuality'

George Orwell - *The Road to Wigan Pier*

Introduction

General comments

A genome is often regarded as a structure – the physical manifestation of the genetic composition of an organism. For the most part this is an acceptable view, but a close inspection of the pea genome suggests that a unitary structure is probably impossible to define. In some ways it seems more appropriate to think of the genome as a mechanism which transmits genetic information from one generation to the next. It is in terms of the process of transmission of genetic information that the organization of the genome needs to be understood. The coding function of the genome is separable from its role in transmission; this is most extravagantly demonstrated by the ciliated protozoans (reviewed by Gorovsky, 1980; Yao *et al.*, 1984; Cherry and Blackburn, 1985) where the genome has two distinct structures for these two different functions.

Perhaps this is pedantry, but at least a view of the genome as a mechanism allows a subtly different view of those classes of sequence that are not transcribed. Furthermore, the processes of replication and segregation are clearly important in attempts to understand the significance of a given sequence organization. Events to be understood in such terms are never witnessed, they are inferred from their consequences; it is in this process of inference that the idea of a function in relation to the genome can be helpful.

Scope

This chapter attempts to describe some of what is known about the pea nuclear genome. An understanding of the relationship between the structure and function of this genome is in its infancy; nevertheless we can expect a rapid increase in such information in the near future. So it is timely to take stock of what we know as well as the problems and issues before us.

Cytogenetics

The pea karyotype

Translocations

Peas are diploid and have seven chromosome pairs (Cannon, 1903; Fig. 2.1), tetraploids can be produced (Therman and Murashige, 1984; Kuma and Mercykutty, 1985; Mercykutty *et al.*, 1990), and mutants affecting meiosis have been described (reviewed by Kaul and Murthy, 1985). Non-disjunction products of such mutants can be trisomics, which can be useful in genetic mapping (Young *et al.*, 1987; Ellis and Cleary, 1988). Individual pea chromosomes are not easily recognized by banding patterns (Lamm, 1981; Monti *et al.*, 1985) although arm-length ratios and the presence of satellites have been used to distinguish pea chromosomes (Blixt, 1958, 1959, 1972).

Rearrangements in the pea karyotype were first identified by Pellew (1940); the availability of pea lines carrying translocations provided the raw material for the classical cytogenetic work, largely of Lamm and his colleagues, which led to the construction of a translocation tester-set for pea (Lamm and Miravalle, 1959; Snoad, 1966; Lamm, 1977). In principle such lines allow the assignment of linkage groups to chromosomes and the identification of chromosome arms. However, there appears to be some question about the structure of the karyotypic rearrangements in some of these tester stocks (Snoad, 1966; Lamm, 1977; Folkeson, 1984b, 1990c); apparently some rearrangements are not as simple as first thought.

Taxonomic relationships

Phylogenetic analysis of the genus *Pisum* in relation to karyotype has been undertaken by several laboratories (Ben-Ze'ev and Zohary, 1973; Errico *et al.*, 1991). The two species in the genus (*P. sativum* and *P. fulvum*) are distinguished by several translocations; these involve at least chromosomes 1, 3, 5 and 7, but there are other differences, for example in the number of satellite chromosomes (Errico *et al.*, 1991) and relative arm lengths.

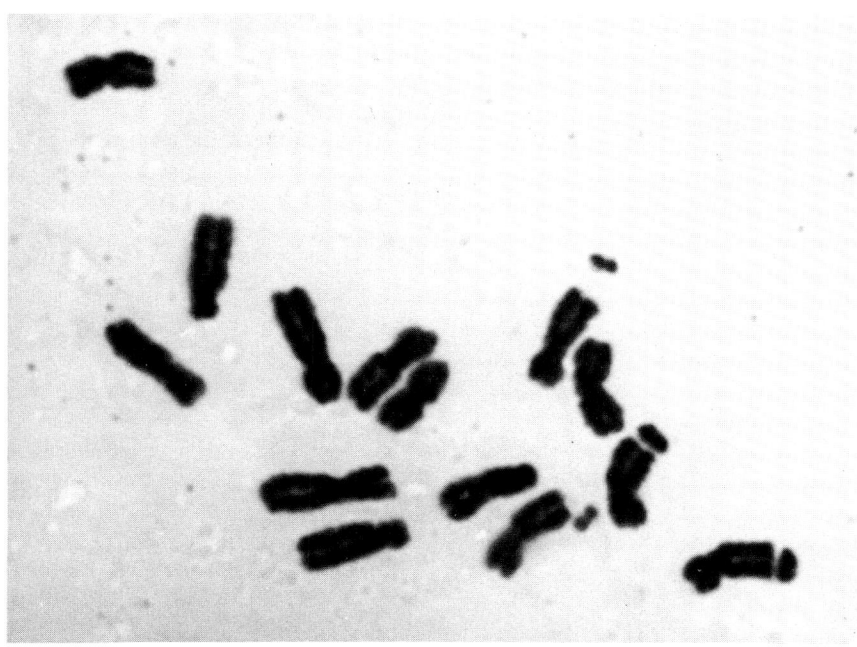

Fig. 2.1. Metaphase chromosomes of pea (×3000). (Photograph kindly provided by K. Hall and J.S. Parker of the University of Reading.)

It is difficult to argue that the two species are different in karyotype, when the karyotype of *P. sativum* has several forms (to some extent distinguishing the several subspecies; Ben-Ze'ev and Zohary, 1973), but there has been extensive karyotype evolution distinguishing *P. sativum* from *P. fulvum*. Ben-Ze'ev and Zohary (1973) showed that, although translocations could be found to distinguish some accessions of the subspecies *elatius*, *humile* and *sativum*, all were distinguished from *P.fulvum*.

Cytogenetic markers

Despite much cytological analysis, the pea karyotype remains poorly marked because there are no obvious chromosome banding patterns. This has led to an interest in both the analysis of polytene chromosomes (Davies and Cullis, 1982; Therman and Murashige, 1984) and *in situ* hybridization techniques (Simpson *et al.*, 1988, 1990). The potential of the polytene chromosomes as a source of chromosome-specific markers was investigated without success in this laboratory; the polytene chromosomes also failed to give a great improvement in the *in situ* hybridization signal. Taken together, these suggest that the structure of these polytene chromosomes may pack the DNA in a relatively inaccessible form.

In situ hybridization to metaphase spreads of pea chromosomes can be achieved with great sensitivity (Simpson et al., 1988), but the greatest practical benefit of this technique is its use with tandemly repeated DNA sequences. These give strong, reproducible and easily identified signals that have allowed all but one (chromosome 1) of the chromosomes in at least one karyotype to be marked by hybridization with the 5S rRNA genes, the large rRNA genes and a satellite-like sequence (Simpson et al., 1990). The Arabidopsis thaliana telomere sequence can also be used successfully as an in situ hybridization probe to mark chromosome ends (Simpson et al.,1990; Rawlins and Shaw, 1991).

Genetics

General introduction

The origin of genetics is generally traced back to Mendel's (1866) study on factorial inheritance in pea (see Mendel, 1965, for a translation). In his analysis Mendel used seven markers, now designated a, fa, gp, i, le, r and v (Blixt, 1974). All of these, with the exception of fa (which is expected to be linked to Rrn1; see Polans et al., 1986; Folkeson, 1990b,c), are placed on the accompanying skeleton linkage map of Ellis et al. (1992) (Fig. 2.2). Three of these markers are well characterized at a biochemical level. The mutation at the r locus has been shown to be due to the insertion of a Ds-like element into the coding region of the gene encoding isoform I of the starch branching enzyme, which is expressed early in seed development (Bhattacharyya et al., 1990; see Chapters 5 and 6). The a locus regulates the abundance of one class of chalcone synthase transcript, but is not a chalcone synthase structural gene (Harker et al., 1990). The recessive allele at the le locus blocks the 3β-hydroxylation of GA_{20} which converts it to GA_1 (Potts et al., 1982; Ingram et al., 1983; Potts and Reid, 1983); but it is not clear whether this is a structural or regulatory gene mutation.

Pea is amenable to genetic analysis for several reasons:

1. Inbred stocks are readily available.
2. Flowers are large and easily emasculated.
3. There are no barriers to cross pollination.
4. There is an enormous wealth of genetic variability for clear and easily scored morphological characters (Blixt, 1972; see Chapter 7).

Despite these advantages, genetic analysis in pea is at a disadvantage with respect to maize or tomato for example, because each cross pollination generates only a few seeds, and the number of seeds per individual F_1 plant is fairly small and quite variable according to genotype.

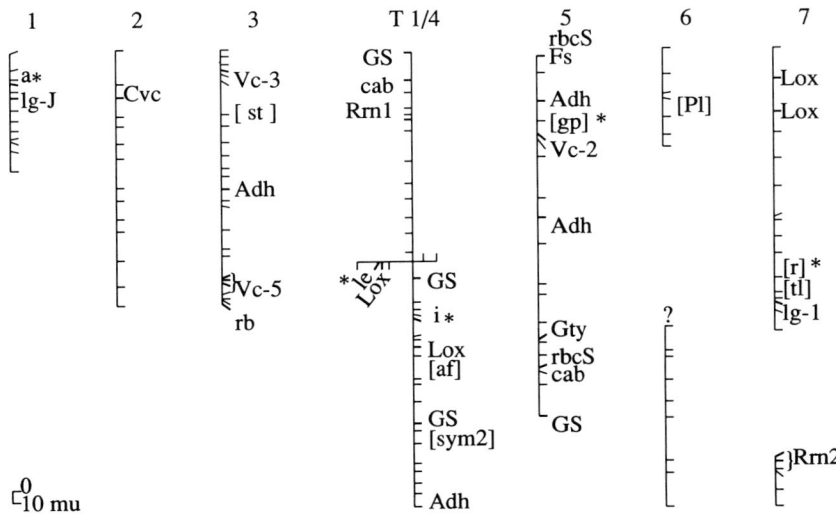

Fig. 2.2. A pea linkage map derived from the cross JI281 × JI399. This map is essentially as described in Ellis *et al.* (1922), but with some new markers. Markers that can be used to relate this and other linkage maps are named. Those in square brackets did not segregate in the cross JI281 × JI399, but show tight association with one or other markers in another cross. The markers indicated * were those used in Mendel's experiments.

The cross-shaped map designated T 1/4 appears to be a 1/4 reciprocal translocation. The segment marked '?' has some features of a duplicated segment, but it is not clear with which of the other linkage groups it is associated. The symbols are explained below, but a fuller description of the classical markers/genes is given by Blixt (1972, 1974); see also Chapter 7.

Classical markers		Molecular markers	
a	acyanic	Adh	alcohol dehydrogenase
af	afila, leaflets modified	cab	chlorophyll a/b binding protein
Fs	violet spots on testa	Cvc	convicilin
gp	yellow pods	GS	glutamine synthase
Gty	gritty testa	lg-1	legumin
i	green cotyledons	lg-J	legumin
le	short internodes	Lox	lipoxygenase
Pl	black hilum	rbcS	rubp carboxylase
r	wrinkled seed	Rrn1	rRNA
rb	wrinkled seed	Rrn2	rRNA
st	small stipules	Vc-2	vicilin
sym2	symbiotic association	Vc-3	vicilin
tl	tendril-less	Vc-5	vicilin

Linkage maps

Although the markers *le* and *v* are linked, it appears that Mendel did not study their joint segregation. Mendel could have discovered linkage, but it was not until de Vilmorin and Bateson (1912) described 'gametic coupling' of *r* and *tl* that the first case of linkage in pea was established. Wellensiek (1925) presented a map of six linkage groups, Winge (1936) later presented one of eight linkage groups, and this together with the work of many others was compiled into a map of seven linkage groups by Lamprecht (1948). Lamprecht's map provided the basis for Blixt's (1974) extensive map which in turn has provided the basis of the pea linkage maps elaborated by Weeden and Marx (1987) and Weeden and Wolko (1990).

The development of these views of linkage associations in pea, together with a detailed discussion of contradictory information, has been reviewed by Folkeson (1990c). All of these linkage maps are compiled from the analysis of several different crosses; it is only with the advent of RFLP analysis that it has been possible to generate extensive linkage maps from the analysis of a single cross. This sort of linkage map for pea has been generated by Dirlewanger (1991) and Ellis *et al.* (1992). The relationship between the classical and RFLP linkage maps will be discussed below, and with reference to the revisions suggested by Folkeson (1984a,b; 1990a,b,c) on the basis of genetic and cytogenetic data.

The two RFLP maps (Dirlewanger, 1991; Ellis *et al.*, 1992) are difficult to relate although several of the RFLP markers are in common. The only common linkage confirmed between these two maps is between the classical markers *i* and *af*. The two maps are discussed in relation to repeated sequences below.

A comparison between the classical linkage map (taken from Weeden and Wolko, 1990) and the RFLP map of Ellis *et al.* (1992) is shown in Fig. 2.3. The map of Weeden and Wolko (1990) is derived from the analysis of several different crosses and is based on the map of Blixt (1974), while the skeletal map presented in Figs 2.2 and 2.3 is derived from a single cross (JI281 × JI399; Ellis *et al.*, 1992). There are two simple conclusions to draw from this comparison; first, there is extensive rearrangement between the two maps and, secondly, the RFLP map is longer than the classical map. These two differences are easily explained by assuming that karyotypic rearrangements are very common in pea (see Blixt, 1972).

If a map is compiled from linkage analysis in several crosses (as classical maps need to be), then the chance that some karyotypic rearrangement lies between chosen markers in at least one of these crosses is fairly high. The appearance of recombinant genotypes for these markers may be suppressed, for example if there is an inversion between them in one line with respect to another. This will condense a linkage map compiled from several crosses. If translocations distinguish the lines used in the crosses, then association of

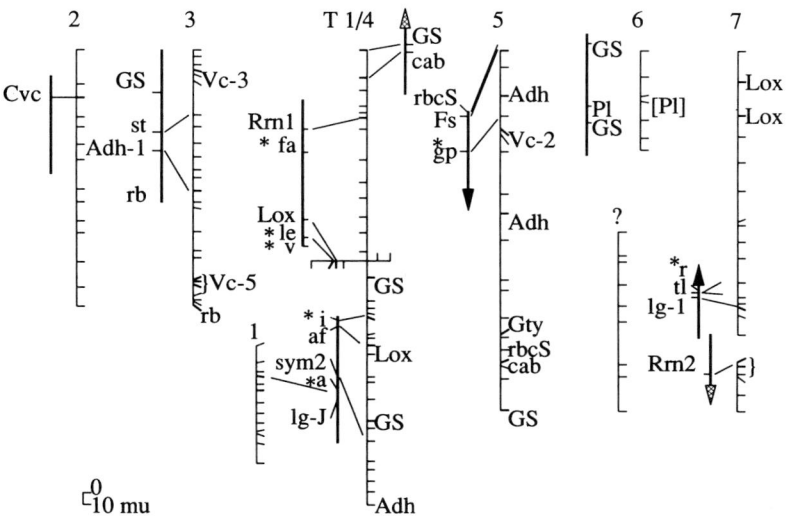

Fig. 2.3. Comparison of the map shown in Fig. 2.2 with that of Weeden and Wolko (1990). Gene names and symbols are as for Fig. 2.2, but duplication has been avoided where possible. The marker *fa* is responsible for a fasciated phenotype. The map of Weeden and Wolko has been broken in two places to allow this alignment, and the points of breakage are marked with arrowheads. There are several other inconsistencies, notably in the placement of the various GS markers. The breaking of Weeden and Wolko's map is for convenience of drawing and not intended to imply a revision.

markers across a translocation point may suggest tight linkage between markers which are often on different linkage groups. Detailed karyotypic analysis of several multiply-marked genetic stocks is required to assess the extent to which these concerns need to be accommodated in linkage maps of pea.

The relationship between the different linkage maps proposed for pea has been confused by the presence of translocated segments. These have been turned to advantage by several groups attempting to exploit chromosomal rearrangements to associate genetic and cytogenetic maps (reviewed by Monti *et al.*, 1985). By convention roman numerals usually designate linkage groups, and arabic numerals chromosomes; the hope is that the numbers are the same, but it is clear which is being discussed. In pea the lack of correspondence between linkage group and chromosome number designation causes much difficulty; in this chapter classical linkage groups (Blixt, 1974) are referred to using roman numerals while chromosomes or other linkage groups use arabic numerals. Folkeson's (1990c) re-examination of the behaviour of translocation stocks suggests that the relationship between chromosomes and linkage groups is quite complex, and that nomenclature can be quite misleading, especially in relation to groups III and V and chromosomes 3 and 5.

A detailed comparison of the relationship between the linkage groups and chromosomes 3, 5 and 7 is illustrated in Fig. 2.4. This comparison is intended to illustrate the detailed analysis that is needed to resolve the relationships between different linkage maps in pea. It seems fairly clear that, in general, no one map will suffice for the description of linkage relationships in pea, although it is possible that large segments of the genome share the same structure in most lines. This needs to be investigated both within *Pisum* and in relation to the genetic maps of other legumes.

As far as pea is concerned, we can generate a reference map from these recombinant inbred stocks which can be used to integrate maps using different molecular markers, because they correspond to a set of inbred lines that have defined interrelationships. In principle, the segregation of many markers can be scored among these lines, and they can therefore combine the scores for molecular markers exploited in different laboratories.

PROCEDURES IN RFLP ANALYSIS

Many different population structures can be used in genetic analysis and in pea, which naturally selfs, two types of population have been used in segregation analysis; these are F_2 and recombinant inbred lines. Near isogenic lines have also been constructed for the analysis of traits such as seed development (Hedley *et al.*, 1986), leaf morphology (Marx, 1974; Meicenheimer *et al.*, 1983) and nodulation specificity (T. Bisseling, personal communication).

The first use of F_2 populations was, of course, by Mendel (1866), and this population structure has since been widely used in genetic analysis of pea. Recombinant inbred lines of pea have been developed mainly in this laboratory, but are easily derived by the systematic selfing of an F_2 population. Techniques for genetic analysis in pea have been described in detail elsewhere (Ellis *et al.*, 1992; Ellis, 1993), and will not be reiterated here. The

Fig. 2.4. The relationship between Folkeson's (1990a, c) map of chromosome 3 and various linkage maps is illustrated. The marker names are as for Fig. 2.2, with the addition of *bt* (blunt pod apex), *ce* (cerise flower colour), and *coch* (stipules leaf-like). Folkeson (1990a, c) identified three translocation points on chromosome 3 with respect to several other karyotypes; these are marked 'T'. Folkeson's revised linkage group V corresponds to his chromosome 3. Weeden and Wolko's map splits Folkeson's linkage group V into two parts, and these segments appear to be colinear with the ends of groups 5 and 7 of the cross JI281 × JI399 described by Ellis *et al.* (1992), who unite these two segments in a single linkage group derived from the cross JI15 × JI1194. Thus the segment *Vc-2* to *r* of the cross JI15 × JI1194 appears to correspond to Folkeson's linkage group V and chromosome 3. Ellis *et al.* (1992) described a 3/5 translocation in one cross; the group 3 part of this segment corresponds to one end of linkage group 3 from JI281 × JI399 (Fig. 2.2). This whole group appears to be in good agreement with Folkeson and Weeden and Wolko's map of linkage group III.

The inclusion of *Gty* on linkage group III by Folkeson is at variance with Weeden and Wolko (who assign it to group 6) and Ellis *et al.* (1992) (who assign it to group 5). Wolko and Weeden (1990) place *Gty* next to Prx-3, which is in turn linked to a GS marker and *sbm-1* according to Timmermann *et al.* (1993); this is of note because *Gty* shows linkage to a GS marker on the group 5 map of Ellis *et al.* (1992) (see Fig. 2.2), suggesting that group 5 of Ellis *et al.* (1992) may be included as part of linkage group VI in some maps.

The Nuclear Genome

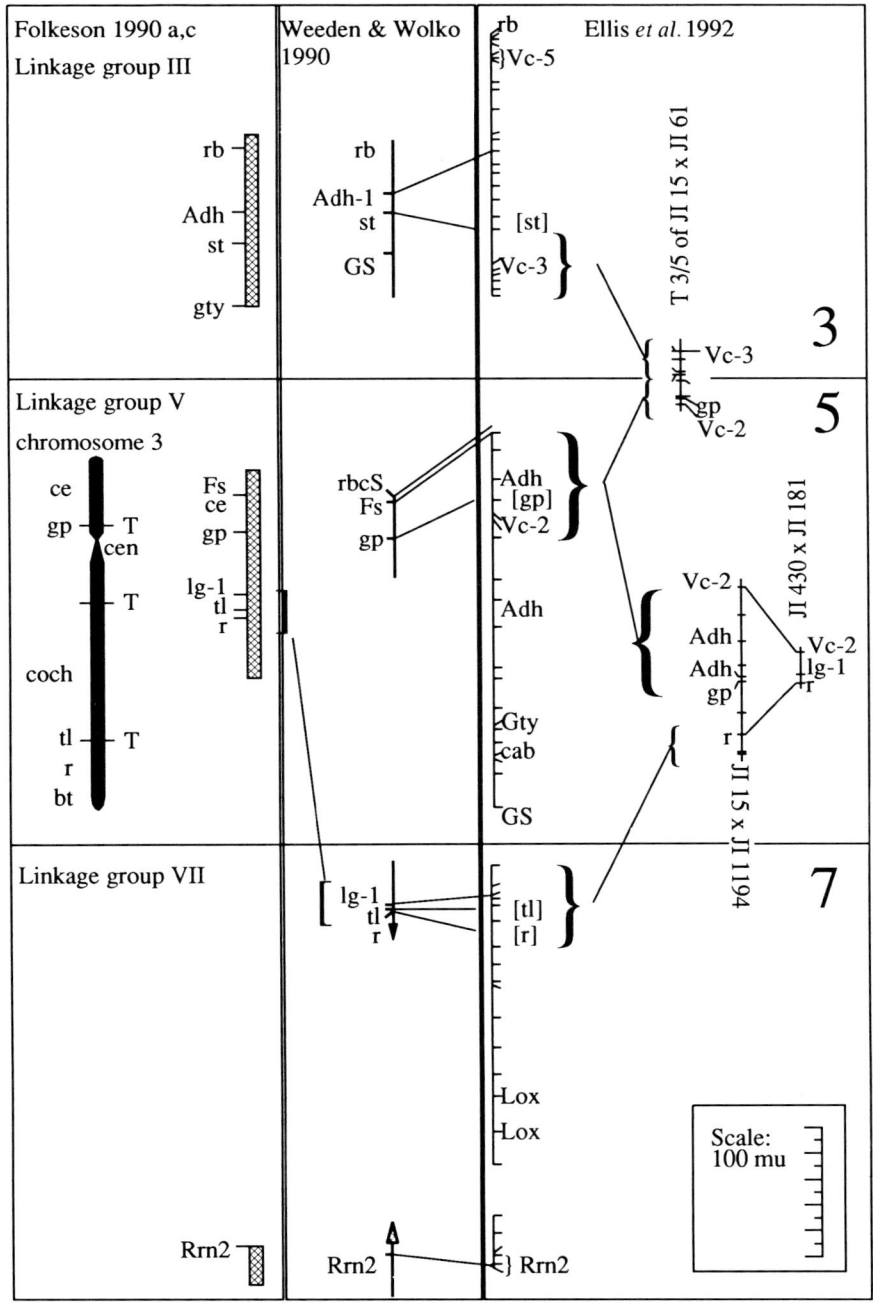

advantages of recombinant inbred lines in plant genetics have been reviewed by Burr and Burr (1991); these populations can be used for the analysis of any type of molecular marker in current use as well as having the potential of allowing the construction of replicated trials.

RFLP markers used in pea genetic analysis include isozymes, cDNAs, genomic clones, oligonucleotide hybridization probes, PCR markers and RAPDS (Lee et al., 1990; Weeden and Wolko 1990; Dirlewanger, 1991; Ellis et al.,1992). All appear to detect polymorphism fairly easily, so it is largely a matter of convenience which is to be used. The distribution of molecular markers between laboratories is complicated by proprietary considerations, so sequence-defined markers have the greatest utility; microsatellites, defined gene sequence probes, oligonucleotide probes and sequence-defined PCR techniques are readily transferred between investigators.

PATTERNS OF DIVERSITY

The diversity within the species *Pisum sativum* is widely recognized; this is reflected by the abundance of genetic markers (Chapter 7; Blixt, 1972), and has allowed the characterization of subspecies. Structural variability in the pea karyotype was first documented by Pellew (1940) and many variants have been found (see Blixt, 1972; Monti et al., 1985); quantitative statements on the frequency of karyotypic variants cannot be made however, because of the lack of a systematic survey. The subspecies structure of *Pisum* could be considered as a reflection of the underlying pattern of karyotypic diversity, but this is probably an over-simplification; the different subspecies are likely to be karyotypically diverse (see Ben-Ze'ev and Zohary, 1973). Phylogenetic relationships among legumes deduced from variation at the DNA level have been discussed by Doyle (1987), and within *Pisum*, on the basis of plastid DNA polymorphism, by Palmer et al. (1985) (see Chapter 3).

Molecular markers can give a guide to phylogenetic relationships. The distribution of allelic types, as judged from the RFLPs used to construct a linkage map by Ellis et al. (1992), provides the information needed to construct a matrix of genetic distances between a set of diverse accessions of *Pisum sativum*. This distance matrix is a record of the number of allelic differences between all pairwise combinations of these lines, and is represented graphically in Fig. 2.5, where the branching pattern is constructed according to the neighbour-joining method of Saitou and Nei (1987). Similar attempts to analyse the patterns of diversity within the genus as a whole, combining information from molecular markers, including RAPDs, isozymes and classical markers, have been undertaken by N.O. Polans (personal communication).

SEGREGATION

In general, markers are expected to segregate in a simple Mendelian fashion; however there are some exceptions. The first major class of non-Mendelian markers are cytoplasmic markers, of which there are some examples in pea (de Haan, 1930, discussed by Corriveau et al., 1989; see Chapter 3). A classical

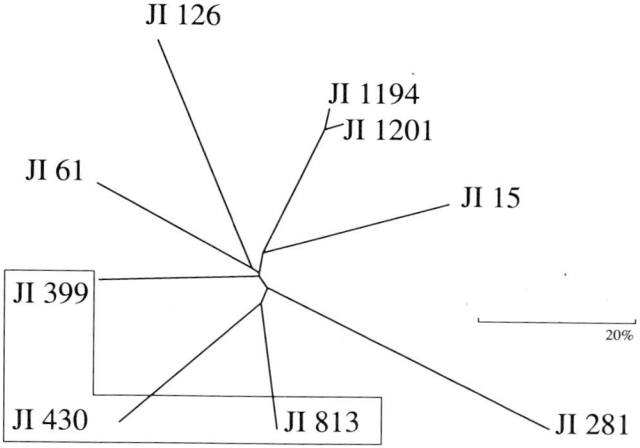

Fig. 2.5. The relatedness of the various pea lines used by Ellis *et al.* (1992) is illustrated. The branching pattern was deduced by the neighbour-joining method of Saitou and Nei (1987) from a data set of 72 RFLP markers. Lines were judged as different if the RFLP pattern was different between lines. A scale is shown corresponding to a line length which would be produced if a pair of lines differed for 20% of markers tested. Note that some of these RFLP probes identify more than one genetic locus, so that the percentage difference is not per locus.

gene that has defied Mendelian analysis is the rogue mutant of pea (Bateson and Pellew, 1915; Blixt, 1972; Matthews, 1973); the nature of this mutation remains a mystery.

The discussions above have highlighted the problems associated with karyotypic variability in pea. Such variation can have an influence on patterns of segregation; for example, a distortion in segregation ratios in one cross was attributed to a duplicated segment by Ellis *et al.* (1992). Reciprocal translocations lead to complex meiotic associations of chromosomes, and rings of as many as six chromosomes have been reported in pea (Sansome, 1932, 1933). The segregation of such rings can lead to the formation of gametes with duplications or deficiencies. However, the co-segregation of alternate centromeres preserves balanced gametes. In such situations the co-segregation of markers on different chromosomes will be observed. This is 'gametic coupling', to borrow Bateson's term, rather than linkage. It seems to be important in the genetic analysis of pea to realize that the co-segregation of genes in a given cross can be the consequence of a variety of genomic mechanisms, of which linkage, in the usual sense, is only one.

For a recombinant inbred population, from which a linkage map has been derived, it is possible to display the genotype of each line for any given linkage group (or groups). A display of this type shows the pattern of segregation of markers along a proposed linkage group. An example of this is illustrated (Fig. 2.6) by showing the genotype of a selected set of recombinant inbred

Cross JI281 x JI399 linkage group 3
Ellis et al. 1992

```
Marker:                       Number of              RI
          111111111122     recombination events     line
123456789012345678901                              number
+++++++++++++++++-----          1                    14
+++++++++++++++++---+++         2                    78
+++++++++++++++++------         1                    17
+++++++++++++++++------         1                    61
++++++++++++++-++++++++           2                  81
++++++++++++--+-----+---                5             6
++++++++++--++++++++++---             3              44
+++++++++-----++++++++++          2                  71
++++++-+++++++++++++--              3                 1 *
++++++-+++++++++++-++---                5             2
++++++------+---------                3              76
++++++---------+--+++++               4              39
++++++----------------          1                    11
+++++-+-------------++             4                 55
+++++---------+----++---                5            32
++++-+---------++++++++            4                 74
++++-----+------++-++-                    7          27
+++-+----+------------                  5            13
++------+++++++++++++++           2                  48
+-++++-----------------             3                 4
+-------++++-----+----                  5            10
+--------++++++-+++++                 4              45
-++++++-++++++++------                4               9
-----+++++--++--------                4              40
----------+++++++++++-+             3                35
----------+---+-++++---                   6          21
-------------+++++++---           2                   8
--------------++++++--            2                  28
-------------------++           1                    63
----------------------          0                    18
                              mean         3.13
                              variance     2.84
```

Marker number   Marker name
       1   cDNA 34
       2   cDNA 125/1
       3   cDNA 23
       4   Vc-3   vicilin
       5   5S/2 5S   rRNA gene array
       6   cDNA 40/2
       7   cDNA 137
       8   cDNA 67a
       9   cDNA 40/7
      10   CHS 2/1   chalcone synthase
      11   DR 14
      12   cDNA 194/1
      13   pPSR 546/1   alcohol dehydrogenase
      14   gr-16/7
      15   gr-16/2
      16   cDNA 331
      17   DR 10
      18   DR 4
      19   Vc-5   vicilin
      20   cDNA 53
      21   rb

Fig. 2.6.(a) The allelic type of a selection of recombinant inbred lines from which the linkage map shown in Fig. 2.2 was derived is illustrated for group 3 markers.

lines for linkage group 3 of Ellis *et al.* (1992). This pattern is, in essence, an attempt to display the way in which individual recombination events have segregated among the progeny; the recombination events are inferred to be single events located between the markers. The relative frequencies of recombination events are plotted in Fig. 2.6b.

The Organization of DNA Sequences

The scale of the genome

The pea nuclear genome is about the same size as the human genome [1c = 4–5 pg (Bennett, 1976; Bennett and Smith, 1976; Murray *et al.*, 1978; Arumuganathan and Earle, 1991; Michaelson *et al.*, 1991); this corresponds to about 3.8 to 4.8×10^9 base pairs per haploid genome; see Fig. 2.7]. Pea has a haploid complement of seven chromosomes; about one third of the number in the human genome. Most of the pea genome (*c.* 90%) is comprised of repeated DNA sequences (Murray *et al.*, 1980; Thompson and Murray, 1980); and these repeated sequences are present in tens of thousands of copies, represented by discrete sequence classes, as judged from their thermal

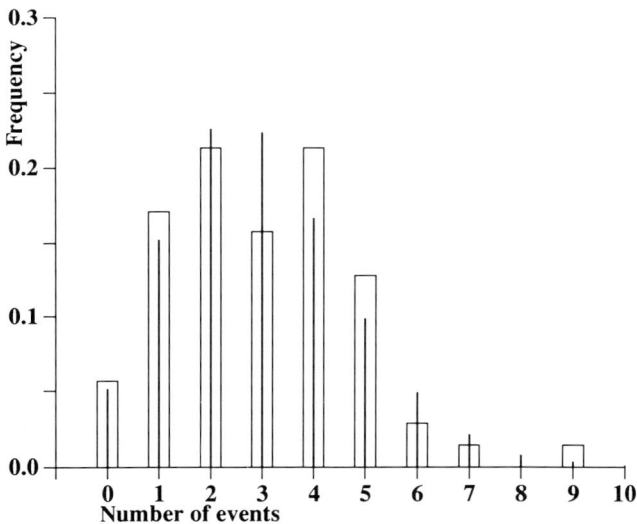

Fig. 2.6(b) The frequency distribution of the number of exchanges in this group for the whole recombinant inbred population is plotted (open boxes). This frequency distribution has a mean of 2.97 and variance of 3.18. The expected Poisson frequency distribution of the same mean is plotted (vertical lines). The frequency, derived from a recombinant inbred population, is not the same as the expected frequency of exchanges in a single meiosis (see Haldane and Waddington, 1931).

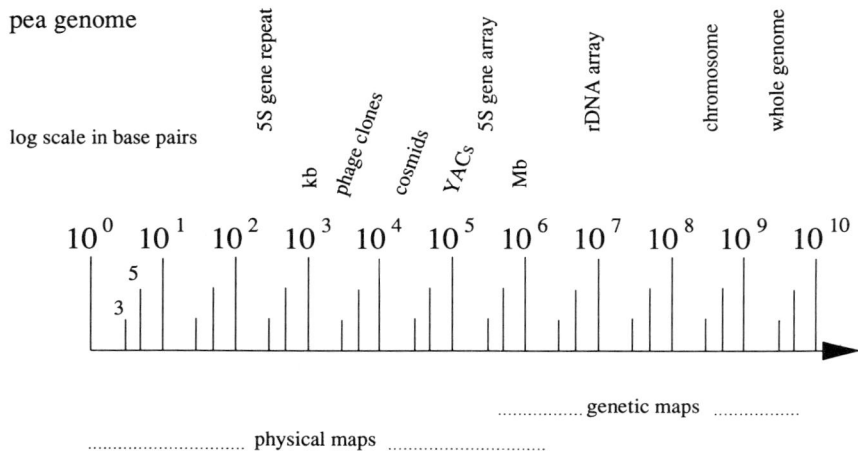

Fig. 2.7. The scale of the pea genome is illustrated and the range of resolution of different techniques marked, as is the size of several structures within the genome. The size of DNA fragment that can be isolated in some vector types is also shown.

denaturation profile (Thompson and Murray, 1980). The interspersion pattern of repeated and single copy sequences is of the 'short period' type; most regions of single copy DNA are very short (Thompson and Murray, 1980). In practice this means that in the construction of genomic libraries (for example, in phage λ), most recombinant DNA clones contain high-copy sequences. This is a convenient attribute when attempting to assess the success or otherwise of the construction of such libraries, but leads to many technical difficulties.

Bulk properties

Base composition

Recent analyses of the density distribution of higher plant DNAs (Montero *et al.*, 1990) has led to the suggestion that many plant genomes, and demonstrably pea, have an isochore organization. Isochores are long tracts of DNA of relatively homogeneous base composition as judged by their behaviour in isopycnic $CsSO_4$ density gradients containing 3,6 bis (acetato-mercuri-methyl) dioxane (Salinas *et al.*, 1986, 1988; Bernardi, 1989). The most obvious structure of homogeneous and unusual base composition is the rDNA, which is very G/C-rich and can be isolated and extensively purified from the rest of the pea genome on density gradients; it comprises about 0.1–0.2% of the genome, representing *c.* 2000 copies (Cullis and Davies, 1975). The two tracts of this rDNA are visible as secondary constrictions on the chromosomes.

The relationship between isochores in pea and the observed major classes of repeated DNA sequence, of discrete thermal denaturation properties and therefore of base composition (Thompson and Murray, 1980) is not known, but it is tempting to speculate that it reflects some underlying order within the genome. Individual genes are located within different isochores (Montero et al., 1990), but little functional relationship seems to be reflected in this pattern; for example, the genes encoding the large subunit of ribulose-bisphosphate carboxylase and the chlorophyll a/b binding protein are found at opposite ends of the gradient (one pair of these two genes lies close together on the genetic map; see Fig. 2.2). Genes present in multiple copies may have their various members distributed among different density classes, but the relationship between this distribution and the distribution of genes between loci has not been studied.

These bulk properties of the pea genome lead directly to genetic and cytogenetic questions; it is only in this context that their significance can be understood. The pattern of distribution of various classes of repeated sequences has been argued to represent the molecular basis of the cytogenetic banding pattern of human chromosomes (Kornberg and Rykowski, 1988); how the distribution of repeated sequences in pea relates to the lack of a defined chromosome banding pattern remains to be established.

DNA methylation

In many plant genomes about 20% of cytosine (C) residues are methylated; the pea genome is no exception. In higher plants methylated C residues (^{5mc}C) are found in the symmetric sequences CpG and CpXpG (Gruenbaum et al., 1981; Cooper, 1983). The incorporation of ^{5mc}C into DNA in vivo is a post-replicative base modification. Interest in DNA methylation arises from the suggestion by Holliday and Pugh (1975) that DNA methylation may play an important role in the developmental regulation of gene expression. The sequence specificity of the distribution of ^{5mc}C residues presumably reflects the substrate specificity of the maintenance methylase. The distribution of ^{5mc}C residues in the pea genome and their influence on the binding of transcription factors are therefore of interest. Klaas and Amasino (1989) have shown that DNaseI-hypersensitive regions are preferentially undermethylated, and Inamdar et al. (1991) demonstrated that CpG methylation inhibits the DNA binding of a pea nuclear factor related to the cAMP response element binding protein. These two observations are consistent with the view that methylated DNA sequences tend to be transcriptionally inactive and suggest an obvious mechanism; however, the causal connection between DNA methylation and inhibition of transcription is notoriously difficult to establish (Macleod and Bird, 1983).

The coexistence of two classes of methylated DNA sequences in plants makes it difficult to interpret the patterns of methylation. In some animals, especially mammals, the high frequency of methylated CpG dinucleotides is correlated with a lower than expected frequency of CpG. This deficiency, called CpG suppression, is associated with a concomitant elevation in the frequency of TpG and CpA dinucleotides, as expected if deamination of ^{5mc}C is a common pathway of mutation (Bird, 1980). The two patterns of DNA methylation in higher plants also complicate any investigation of the phenomenon of CpG suppression.

The activity and properties of the pea DNA methylase have been investigated in some detail (Yesufu et al., 1991). There appears to be only one enzyme which methylates C residues, and it incorporates ^{5mc}C at all possible dinucleotide sequences with a 5' C. This poses a problem in attempting to account for the methylation of CpG dinucleotides not present in CpGpG (also CpCpG) trinucleotides; Yesufu et al. (1991) propose that the methylase they have characterized may methylate CpG dinucleotides only in the CpGpG trinucleotide. They also comment that pea DNA is the best natural substrate for the methylase, while, for the mouse DNA methylase, homologous DNA (i.e. mouse DNA) is a poor substrate. This suggests that pea DNA, unlike that of mammals, is probably rich in hemi-methylated sites (i.e. the methylation is on only one strand). As they showed that pea DNA is a poor substrate for the mouse methylase, these hemi-methylated sites should be CpApG or CpTpG trinucleotides.

The most convenient assays for the methylation status of given DNA sequences exploit the behaviour of certain restriction endonucleases (Bird and Southern, 1978). These recognize palindromic sequences, so the possible influence of rare ^{5mc}C residues outside these sequences is often ignored, as indeed is the possible influence of other base modifications. The observation that pea DNA contains a large amount of hemi-methylated DNA could be explained if some of the ^{5mc}C residues occur outside CpG or CpXpG sequences. The methylation status of pea DNA is thus not entirely clear, but it is certainly the case that CpXpG sequences can be methylated, and many of them are hemi-methylated. It is not clear to what extent CpG sequences not in the CpGpG trinucleotides are methylated, nor is it clear to what extent (if any) C residues which do not occur in these symmetrical di- or trinucleotides are methylated.

The correlation between DNaseI-sensitive regions and lack of DNA methylation in pea (Klaas and Amasino, 1989) implies that there is an association between CpG and CpXpG methylated sequences. In pea rDNA both types of methylation appear to be distributed at random within a nucleolus organizer (Ellis et al., 1989), but there can be a considerable difference in the extent of methylation between the two ribosomal gene arrays (Watson et al., 1987). The coincidence between *Bgl*II and *Hpa*II sensitive sites suggests that, at least in the rDNA, the CpG and CpXpG methylated tracts overlap each other. The difference between loci includes a differential susceptibility to an *Hpa*II site that maps near to the presumed point of transcript initiation (Kato *et al.*, 1990), and is correlated with differential developmental regulation of the two rRNA gene arrays (Watson *et al.*, 1987).

Repeated sequences

Most of the pea genome is composed of repeated sequences (Murray *et al.*, 1978, 1980; Thompson and Murray, 1980) and most of these are high-copy dispersed repeated sequences (Murray and Thompson, 1982). Only four classes of tandemly repeated sequences have been characterized: the 5S and large rRNA genes (see later), a single array of a 1-kb repeat unit (Ellis *et al.*, 1989; Simpson *et al.*, 1990), and the telomeric sequences which are very similar in structure to those of *Arabidopsis* (Richards and Ausubel, 1988; D. Lee, personal communication). Simple sequence DNAs appear to be abundant in the pea genome (unpublished observations) and have been successfully employed in genetic mapping experiments (Dirlewanger, 1991).

There are two major areas of interest in characterizing dispersed repeated sequences. One relates to the fact that a class of these sequences is represented by transposable elements (discussed below), and the other is their potential for use in PCR-based genetic mapping as discussed elsewhere (Lee *et al.*, 1990).

Transposable elements and unstable genes

Several genetically unstable mutations have been characterized in pea (see Blixt, 1972). Perhaps the clearest available example is *purple patched*, an allele of *pur*, which exhibits a variegation of purple pigment on the pod wall and occasionally throws green-podded progeny. The *arg* mutation is responsible for lifting the epidermis from the leaf lamina, creating an air space and giving the leaf a silvery appearance. Occasional *arg/arg* plants show sectors of normal appearance, and lines of wholly normal appearance can arise. The genetics of this seems less clear than that of *purple patched* but may be a consequence of the more complex interaction between cell layers that is responsible for the phenotype.

Lines that exhibit chlorophyll variegation have also been isolated (de Haan, 1930; D.R. Davies, personal communication), but the example of an unstable gene which has been most clearly characterized genetically is an allele of the *a* locus (also described by de Haan, 1930). Regrettably this stock has been lost, but there are several interesting features of the mutation that bear repeating.

The unstable flower colour mutation characterized by de Haan is clearly an allele of *a* (de Haan, 1930). The *a* mutation regulates the level of transcript of the pea chalcone synthase gene responsible for the production of floral pigmentation (Harker *et al.*, 1990). The prospects for the isolation of this gene

Table 2.1. Transposable element-related sequences identified in pea.

Class	Name	Reference
Ac/Ds	Ips-r	Bhattacharyya et al. (1990)
En/Spm	Pis1	Shirsat (1988)
Mu1	unnamed	Bown (1992)
Ty1-copia	PDR1	Lee et al. (1990)
	PIS179, PIS182	Flavell et al. (1992)

on the basis of DNA sequence-relatedness to chalcone synthase transcriptional regulators in maize and *Antirrhinum* thus appear to be quite good. The pattern of sectors observed by de Haan is interesting on at least two counts. First the pigmented sectors can be as small as single cells, indicating that the wild type allele of *a* is expressed late in the development of flowers. Second, moderately large coloured sectors on a white background, of the order of 100 cells, contain internal white patches. This latter behaviour suggests that the condition of the *a* gene is cycled between two states and is perhaps responding to the state of a second (possibly unlinked) factor in the genome, as is the case for a similar mutation in *Lathyrus odaratus* (Punnett, 1936).

None of the genetically unstable mutations in pea have been characterized at a molecular level. This is a major block to the development of a usable endogenous transposon-tagging system in pea. However, several classes of DNA sequence related to known transposable elements have been characterized at the molecular level, and these are summarized in Table 2.1.

The list of identified DNA sequences related to transposable elements is undoubtedly incomplete, and some uncharacterized high-copy number sequences have been shown to disrupt *PDR1* elements, suggesting that this uncharacterized sequence class, abundant in the pea genome, may contain many active transposons (Lee *et al.*, 1990).

The sequence identified as *Mu1*-like is so characterized only on the basis of the similarity of the ends of the maize *Mu1* element (Barker *et al.*, 1984) and of the DNA sequence very close to the point of disruption of the vicilin gene from which this sequence was isolated. The sequence similarity is 5'-GAGATAATTG-3' (*Mu1*) vs. 5'-GAGATcAATTG-3', where the 5' GA overlaps the in-frame stop codon (TGA) of the vicilin pseudogene. This sequence comparison is not made in Bown (1992).

Gene clusters

GENETICALLY DEFINED GENE CLUSTERS

Genes involved in recognition generally seem to be represented by complex allelic series that map to one or a few loci, as exemplified by self-incompatibility loci (reviewed by Thompson and Kirch, 1992). Two classes of loci in the pea genome obey this general rule. One of these is the genes involved in recognition of strain specificity of *Rhizobium*-host interaction, notably the *sym2* cluster on linkage group I (Young, 1985; Weeden *et al.*, 1990). The second group of genes are those involved in response to different strains of pea seed-borne mosaic virus, especially the *sbm* gene cluster that maps to linkage group VI (Provvidenti and Alconero, 1988a,b; Provvidenti, 1990; Timmerman *et al.*, 1993).

MOLECULARLY DEFINED GENE CLUSTERS

Many cDNAs that have been placed on the pea RFLP map detect sequences at several loci, which may reflect ancient duplications in the legume genome. Approximately 35% (20/57) of the cDNAs mapped by Ellis *et al.* (1992) exhibit this behaviour, but most detect more than one gene copy. Because these cDNAs detect more genes than loci, even when multiple loci are detected, it follows that most cDNAs correspond to groups of genes that are very close to one another in the genome, although for a single sequence class there may be more than one cluster. This situation has been investigated most extensively for storage-protein genes (Casey *et al.*, 1986; see below and Chapter 6).

One of the most intriguing examples is that of the genes encoding chalcone synthase in pea; these are present at two genomic locations. A single locus (which contains two genes; R. Hellens, personal communication), which maps to linkage group 3 (Harker *et al.*, 1990; Ellis *et al.*, 1992), is regulated by the allelic state of the *a* or *a2* locus and is thus involved in pigment biosynthesis. All the rest of these genes map close together on linkage group 1. Thus there appears to be functional segregation of the genes encoding the same enzyme activity, which is correlated with their pattern of regulation.

rRNA GENES

The genes encoding the large rRNAs and the 5S rRNA have received attention in many organisms because they are amenable to many different sorts of analysis. In pea these sequences have been studied in terms of their organization and genetics, as well as in relation to replication, recombination and their three-dimensional organization in interphase nuclei. A brief discussion of the organization of these sequences follows.

45S RNA TRANSCRIPTION UNIT

In *Pisum sativum* there are two nucleolus-organizing chromosomes corresponding to the location of the genetic loci encoding the PolI-transcribed structural ribosomal RNAs (Ellis *et al.*, 1984, 1992; Polans *et al.*, 1986; Simpson *et al.*, 1990). In *Pisum fulvum*, there appear to be three such chromosomes (Errico *et al.*, 1991; see also the discussion of linkage maps given above); interestingly, the number of length variants of the rRNA gene repeat in *P. fulvum* is generally greater than in *P. sativum* (unpublished observations), but the relationship between these two observations is not clear.

For *P. sativum*, the tandemly repeated rRNA genes (rDNA) are contained within two arrays each of approximately 1000–3000 (Cullis and Davies, 1975) 9–10-kb repeat units. It is usually possible to recognize only one type of repeat unit in any tandem array (nucleolus organizer, locus; Ellis *et al.*, 1984). Some pea lines, however, appear to contain more than one type of repeat per array (this is especially common in *P. fulvum*). The arrangement of length variants within an array has been deduced from the examination of partial-digest patterns (Ellis *et al.*, 1984); these variants appear to be arranged in long homogeneous stretches (Fig. 2.8). This 'segmental' organization is in marked contrast to many other organisms, notably *Vicia faba* (Rogers and Bendich, 1987). The structure and organization of these repeat units has been described by Kato *et al.* (1990) and Piller *et al.* (1990).

5S RNA TRANSCRIPTION UNIT

The 5S rRNA genes are arranged as a tandem repeat of a *c*. 300-bp sequence, and these are found at three major chromosomal sites; usually these sites are not found on the same chromosome (Simpson *et al.*, 1990). Each locus corresponding to one of these sites represents an array of about 2000–3000 tandem repeats. Variants of the 5S rRNA gene repeats are of two types: length variants and sequence variants; unlike the rDNA these variants are interspersed with one another. This is most clearly seen for sequence variants at restriction sites. Variant repeats that carry *Eco*RI sites are very rare, and may be as much as several hundred kilobases apart (Ellis *et al.*, 1988).

Recombination

Background information

A linkage map is generally a representation of the inferred distribution of recombination events in an F_1 hybrid. The linkage map presented in Fig. 2.2 for pea also represents the pattern of recombination events within lineages leading to the recombinant inbred lines. The general structure of this map has been discussed earlier (and see Ellis *et al.*, 1992), but at this point we shall consider whether there are structural features of the genome that determine whether or where recombination events occur.

The Nuclear Genome

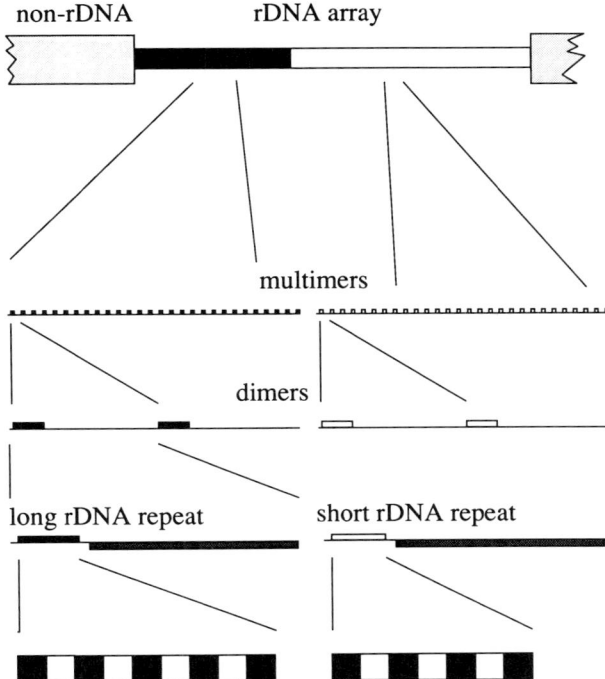

Fig. 2.8. The arrangement of DNA sequences within an rDNA array is illustrated, showing the lack of interspersion of long vs short rDNA repeats. The length difference between individual repeat types is often the consequence of variation in the number of subrepeats in the spacer (boxes).

The chemical mechanism of recombination and the nature of recombination intermediates have been extensively studied in other organisms (reviewed by Stahl, 1979). Until evidence is available to the contrary we can only assume that recombination in pea obeys these general rules. There is a considerable body of data to suggest that recombination does not occur at random with respect to DNA sequences; rather, the frequency of recombination events reflects features of chromosomal architecture. For example, in the human genome, sequences that are near to the synaptonemal complex are rich in recombinogenic simple sequences (Pearlman *et al.*, 1992). The $(GT)_n$ dinucleotide repeats are thought to promote recombination in several organisms (Murphy and Stringer, 1985; Treco and Arnheim, 1986; Pardue *et al.*, 1987; Stallings *et al.*, 1991;) and short $(GT)_n$ tracts can be found in several pea DNA sequences in the EMBL data base. Chromosome structure at a higher order also has a profound effect on the distribution of recombination events, as can be seen, for example, by comparing cytogenetic and linkage maps of cereal chromosomes (Heslop-Harrison, 1991).

At one level of chromosome organization, chromatin is thought to be organized in loops (Callan and Lloyd, 1960; Cook and Brazell, 1975; Small et al., 1982), which are considered to represent independent elements of chromosome function (Pardue et al., 1987). Transcriptionally active chromatin loops are more accessible to nucleic acid binding factors or processing enzymes (transcription factors, RNA polymerase), so it might be expected that such regions are also readily accessible to the enzymatic apparatus responsible for the generation of recombination events.

In pea the frequency of recombination events in a given interval is variable between crosses (Rasmusson, 1928; Lamprecht, 1948) and the segregation ratios for some markers are variable even within the progeny of the same set of F_1 seed (Winge, 1936). We can conclude that there is some control of the frequency and distribution of recombination, which is more than simply the length of DNA between the markers in question. Furthermore, the classical cytogenetic investigations of Sansome (1932, 1933) show that crossing-over in interstitial chromosome segments is common in pea (Darlington, 1965), implying that the distribution of recombination events along pea chromosomes is not as uneven as, for example, in cereals.

Recombination in a legumin gene cluster

Very few recombination events in pea have been studied in detail. The best characterized are events within the 5S rRNA genes and rDNA, and one event observed close to the 5' end of a legumin gene. The last event (Turner et al., 1993) is of interest because its location was at a position of a major sequence difference between the two chromosomes involved (there was effectively a deletion in one). Further information is needed to establish the relationship of the sequence difference and the location of the crossover.

Recombination in the rDNA

Genetical analyses have been undertaken of repeat length differences in arrays of rDNA (Ellis et al., 1984, 1992; Polans et al., 1986). In all of these studies the rDNA segregates as a simple Mendelian character, which is perhaps surprising as the total length of DNA involved is greater than that in any yeast (S. cerevisiae) chromosome. If recombination occurred within the rDNA array at an appreciable frequency then the length difference would not segregate in a simple fashion. Such recombination events would generate new arrays that appear like heterozygotes; this sort of event has been proposed to be the origin of the 'segmental' arrangement of some rDNA arrays (Ellis et al., 1984). Close examination of an rDNA array where there was no length difference between the homologues identified some minor sequence variants. These mapped to approximately the same location, but there were a few recombinant types

(Ellis *et al.*, 1992). Recombination can occur within an rDNA array, but it appears to be suppressed when the repeat length differs between homologues. rDNA repeat length has been associated with differential nucleolar activity (reviewed by Flavell, 1986), and the difference in the activity of different nucleolus organizer regions can be reflected in a difference in the degree of condensation of this chromosomal region. The differential activity of different nucleolus organizing regions is known as nucleolar dominance. The difference between the behaviour of the two rDNA loci in pea cannot be attributed to nucleolar dominance, that is to say differential decondensation of the two arrays, because the two loci behave similarly when length variants are present.

Recombination in the 5S rRNA genes

One of the 5S rRNA loci (5S/2) is flanked by genetic markers. The minor variants that map to this region should be able to be ordered with respect to these markers. However, examination of the patterns of segregation of these markers suggests that it is not possible to generate a simple map without an unexpectedly high frequency of double recombinants (D. Lee and N. Ellis, unpublished observations). It is possible to rationalize this observation by suggesting that exchanges between homologues may be unequal; therefore there is no unambiguous linear order of markers derived from the tandem array (Fig. 2.9). All exchanges between arrays of different length are necessarily unequal.

The recombinational behaviour of tandemly repeated sequences has been widely discussed (e.g. Tartof, 1974; Szostak and Wu, 1980; Ohta, 1981;

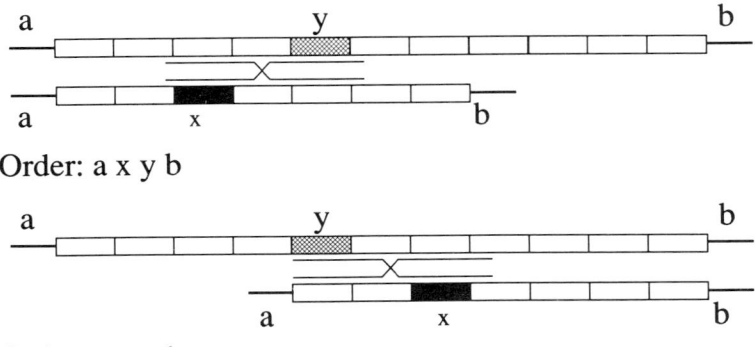

Fig. 2.9. Unequal exchange in a tandem repeat. If two tandem arrays contain different numbers of individual repeat units, then precise alignment is not possible (misalignment may also occur when arrays are of the same length). Alignment of the arrays in different ways leads to an apparent difference in the order of markers within the array.

Nagylaki, 1984) but usually in relation to sister chromatid exchange, and with reference to the influence of unequal sister chromatid exchange. The unequal exchanges proposed here are non-sister exchanges, and suggest that structural differences between alleles of the 5S rRNA genes do not necessarily inhibit recombination.

Summary of factors influencing the position of recombination events

Radman (1989) argued that sequence mismatch can abort strand exchange during recombination, which could explain the observations on the rDNA discussed above. However, given the information from the legumin gene cluster and the 5S rRNA genes, it is clear that structural differences do not necessarily inhibit recombination. It could be argued that transcriptional activity, or the potential for transcriptional activity, has a strong influence on recombination, but in turn this does not account for the behaviour of the rDNA when both loci exhibit repeat length differences. There are still insufficient data to draw any general conclusion, but it seems, on the basis of these preliminary observations, that we can expect the major influences on the frequency and positioning of recombination events to vary from one genomic location to another.

Replication

The replication of plant nuclear DNAs has been extensively studied in pea by Van't Hof and colleagues (Van't Hof *et al.*, 1987; Hernández *et al.*, 1988a,b; Van't Hof and Lamm, 1991, 1992), partly because it is possible to synchronize cell division in the primary roots of pea seedlings (Van't Hof, 1966). These studies have shown that a major mode of DNA replication in the pea rDNA uses a displacement loop. The site of initiation of rDNA replication has been located fairly precisely (Van't Hof and Lamm, 1992); it lies close to the end of the 180-bp subrepeats nearest to the point of initiation of transcription (Fig. 2.10). Replication proceeds from this point in opposite directions, on opposite strands, generating two displacement loops (Fig. 2.11). When two loops, initiated at different origins, converge, the supercoiled structure between them is resolved and replication proceeds, generating two fully replicated daughter strands. The discontinuities in the newly replicated DNA are sealed by ligation.

The factors governing the choice of which rDNA repeat is chosen to initiate replication are unknown, but it is clear that only a selection of origins is used. The joining of adjacent replicons involves the resolution of the intervening supercoiled DNA between the newly synthesized DNA strands. There appears to be some specificity in the terminus of DNA replication, and presumably

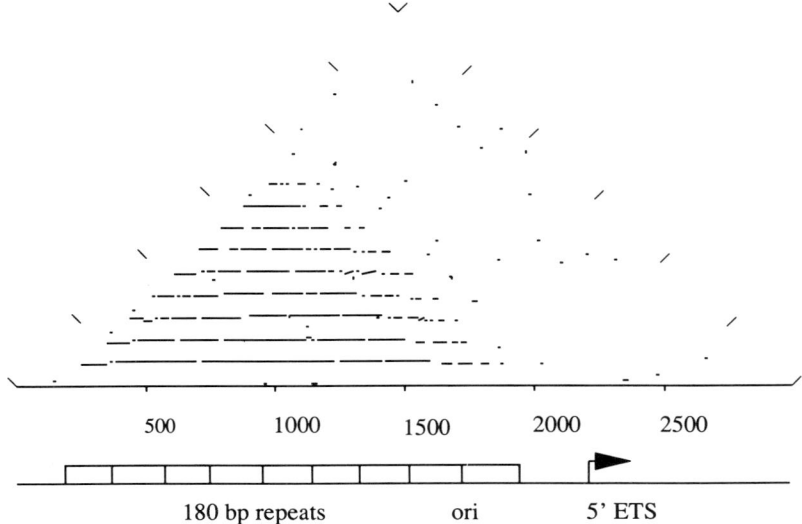

Fig. 2.10. The internal structure of the pea rDNA spacer (Kato *et al.*, 1990) is illustrated. Numbers are a scale in base pairs. Above the scale is a triangular region showing this sequence compared with itself using the program COMPARE and DOTPLOT of the University of Wisconsin package of sequencing programs (Devereux *et al.*, 1984). With the sequence written along the two sides of the triangle leading to the right-angled vertex, a dot is placed within the enclosed space if the sequence matches along these two axes. The stringency of the match was chosen as at least 14/21 base pairs. The consequence of this procedure is that the 180-bp repeating unit is overscored by a series of broken lines. The breaks in these lines reveal irregularities in the repeated sequence. Below the scale, the 180-bp repeats are illustrated as boxes, and the approximate positions of the replication origin (ori; Van't Hof and Lamm, 1992) and the initiation point for the 45S rRNA precursor (arrow) are marked.

this corresponds to a preferred meeting point for replication forks (Hernández *et al.*, 1988b). In the pea rDNA there appear to be two such places, one corresponding approximately to the beginning of the 18S rRNA sequence and the other mapping near to the origin. Thus the replication of the pea rDNA is a highly ordered process, but the relationship between DNA replication and other functions (transcription, recombination) is not known.

Three-dimensional Structure

Presumably chromosomal architecture has some degree of control over the pattern of gene expression, so the involvement of DNA sequences that modify the organization of chromatin loops (reviewed by van Driel *et al.*, 1991), for example by the formation of Z-DNA, will have an indirect influence on patterns of gene expression. Pardue *et al.* (1987) have argued that the distribution

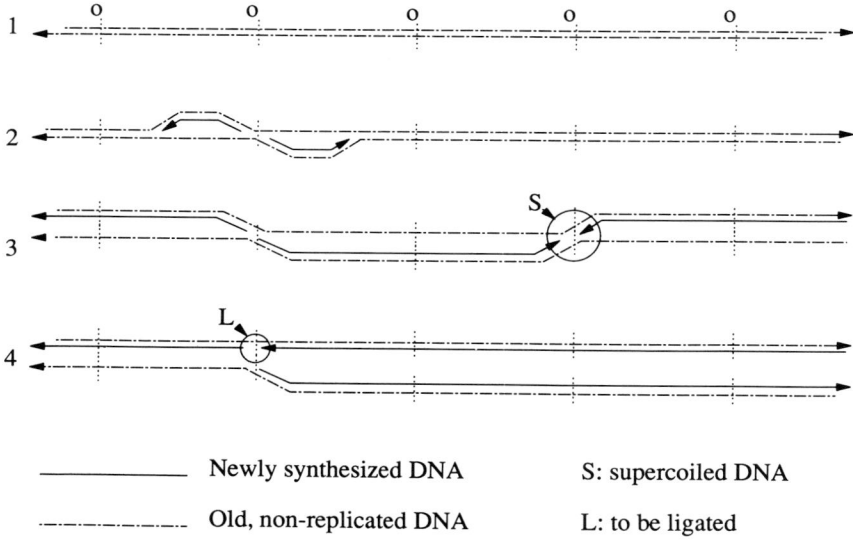

Fig. 2.11. Van't Hof and Lamm's (1991) scheme for the replication of pea rDNA is illustrated for a series of five rDNA repeats.

of $(GT)_n$ repeats in *Drosophila* is related to dosage compensation. The relationship between chromosome loops attached to the synaptonemal complex and domains of chromosome function in other cell types, or stages of the cell cycle, is likely to be complex; for example, Pearlman *et al.* (1992) have shown that sequences associated with the rat synaptonemal complex are different from scaffold- or matrix-associated DNA. An investigation of the nature of DNA associated with the pea nuclear scaffold has been undertaken by Slatter *et al.* (1991), who showed that an abundant A/T-rich repeated sequence is often associated with scaffold attachment regions.

There has been much discussion about the location of plant chromosomes with respect to each other in interphase and mitotic nuclei (see, for example, Bennett, 1982; Rawlins and Shaw, 1988; and references therein). There are some obvious constraints: the location of the telomeres at the nuclear membrane (Rawlins and Shaw, 1991), and the association of the rDNA with the nucleolus (Rawlins and Shaw, 1990).

The nucleolus itself provides an opportunity to study the three-dimensional organization of transcriptional activity and ribosomal assembly (Highett, 1993; Highett *et al.*, 1993); both classes of the ribosomal RNA genes are amenable to structural analysis by virtue of their large copy number. These two types of gene are functionally related and their gene products are assembled in the nucleolus, providing a unique opportunity for the investigation of structure–function relationships in the interphase nucleus (Rawlins and Shaw,

1990; Highett, 1993; Highett *et al.*, 1993). It is clear from these studies that there is a great deal of order in the three-dimensional organization of the rDNAs, and that the observed patterns can be maintained through serial cell divisions, as seen when files of cells contain identical or mirror-imaged organization of perinucleolar spots (Rawlins and Shaw, 1990).

Prospects

The pea is likely to continue to be a source of cloned genes, particularly as it is relatively amenable to biochemical analysis (Chapter 6). Furthermore, the large number of mutants available in pea (Blixt, 1972; Chapter 7) represent a resource of potentially clonable genes. There are, at present, three general procedures for cloning such genes: (i) chromosome walking; (ii) transposon tagging; and (iii) differential cDNA cloning. Of these procedures, only the last is practicable in pea. The possibility of developing a transposon-tagging system exists, either through the characterization of endogenous elements that can disrupt genes (such as *Ips-r* elements; Bhattacharyya *et al.*, 1990) or by the introduction of a characterized system in transgenic plants (Chapter 10).

Chromosome-walking experiments seem an unlikely method for gene cloning in pea, at least with our present state of knowledge of the organization of DNA sequences around genes. Nevertheless, such experiments will be informative about the organization of DNA sequences around genes and perhaps go some way towards telling us about the significance of isochores and the pattern of distribution of repeated sequences in the genome. It is only with the availability of detailed maps and sequence information from large segments of the genome that we will be able properly to address the questions raised about the control of the positioning of recombination events and the structure of replicons. The study of tandemly repeated sequences in connection with these two issues is simply a technical convenience; tandem repeats are by definition long tracts of sequence and are relatively easy to characterize (at least to a first approximation). We must wonder whether such homogeneous DNA sequences are typical of the behaviour of the rest of the genome.

The most peculiar feature of the pea genome is the apparent frequency of translocated segments. It seems an urgent matter to characterize the variability of the karyotype in detail, both in terms of structure, for example by the analysis of synaptonemal complexes, and in terms of the frequency distribution of different chromosome types. The relationship between the genetic and cytogenetic map of *Pisum* needs clarification. It is also of interest to determine whether the exchanges identified in pea are unique to this organism. Chromosome segments which have a conserved gene order may exist in a variety of legume species and these segments could exist in a variety of chromosomal distributions. This means that genetic and cytogenetic maps need to be derived from a range of legume species, employing the same genetic markers. The basic chromosome number in legumes is quite varied (1x = 5, 6, 7, 8, 10 or 11; Bennett and Smith, 1976), so comparative mapping experiments may reveal many conserved regions of synteny.

Conclusion

In this chapter the words 'structure', 'order', 'organization', 'pattern' and their allies have often been used. These words have powerful associations and, used in combination, create an air of purposefulness or economy of design. Their use is a symptom of an attempt to rationalize observations of the genome and to understand these observations in the context of its perceived function. These associations cannot be attributed to the genome itself, which is simply a survivor. It is not the purpose of the genome to survive, but, if we ascribe to it the function of transmission, then at least we can hope that the allure of such a trap is only a little misleading.

These are rough times for the pea genome, assailed as it is by bouts of activity of retroelements and massive karyotypic rearrangements. What we can hope to gain, by forcing observations into any framework, is an understanding of the historical events that have moulded the genome and recognition of the mechanisms that have extricated this species from such outrageous fortune.

References

Arumuganathan, K. and Earle, E.D. (1991) Estimation of nuclear DNA content of plants by flow cytometry. *Plant Molecular Biology Reporter* 9, 229-233.

Barker, R.F., Thompson, D.V., Talbot, D.B., Swanson, J. and Bennetzen, J.L. (1984) Nucleotide sequence of the maize transposable element *Mu1*. *Nucleic Acids Research* 12, 5955-5967.

Bateson, W. and Pellew, C. (1915) On the genetics of 'Rogues' among culinary peas. *Journal of Genetics* 5, 13-36.

Bennett, M.D. (1976) DNA amount, latitude, and crop plant distribution. *Environmental and Experimental Botany* 16, 93-108.

Bennett, M.D. (1982) Nucleotypic basis of the spatial ordering of chromosomes in eukaryotes and the implications of the order for genome evolution and phenotypic variation. In: Dover, G.A. and Flavell, R.B. (eds), *Genome Evolution*. Academic Press, London, New York, pp. 239-261.

Bennett, M.D. and Smith, J.B. (1976) Nuclear DNA amounts in angiosperms. *Philosophical Transactions of the Royal Society of London series B* 274, 227-274.

Ben-Ze'ev, N. and Zohary, D. (1973) Species relationships in the genus *Pisum* L. *Israel Journal of Botany* 22, 73-91.

Bernardi, G. (1989) The isochore organization of the human genome. *Annual Review of Genetics* 23, 637-661.

Bhattacharyya, M.K., Smith, A.M., Ellis, T.H.N., Hedley, C. and Martin, C. (1990)

The wrinkled-seed character of pea described by Mendel is caused by a transposon-like insertion in a gene encoding starch-branching enzyme. *Cell* 60, 115–122.
Bird, A.P. (1980) DNA methylation and the frequency of CpG in animal DNA. *Nucleic Acids Research* 8, 1449–1504.
Bird, A.P. and Southern, E.M. (1978) The use of restriction enzymes to study eukaryotic DNA methylation: I. The methylation patterns in ribosomal DNA from *Xenopus laevis*. *Journal of Molecular Biology* 118, 27–47.
Blixt, S. (1958) Cytology of *Pisum* II. The normal karyotype. *Agri Hortique Genetica* 16, 221–237.
Blixt, S. (1959) Cytology of *Pisum* III. Investigation of five interchange lines and coordination of linkage groups with chromosomes. *Agri Hortique Genetica* 17, 47–75.
Blixt, S. (1972) Mutation genetics in *Pisum*. *Agri Hortique Genetica* 30, 1–293.
Blixt, S. (1974) The pea. In: King, R.C. (ed.), *Handbook of Genetics*, vol. 2. Plenum Press, New York. pp. 181–221.
Bown, D. (1992) Characterisation of genes expressed in various tissues of pea (*Pisum sativum* L.): correlation of genotype and phenotype. PhD thesis, University of Durham.
Burr, B. and Burr, F.A. (1991) Recombinant inbreds for molecular mapping in maize: theoretical and practical considerations. *Trends in Genetics* 7, 55–60.
Callan, H.G. and Lloyd, L. (1960) Lampbrush chromosomes of the crested newt *Triturus cristatus*. *Philosophical Transactions of the Royal Society of London series B* 243, 135–219.
Cannon, W.A. (1903) Studies on plant hybrids: the spermatogenesis of hybrid peas. *Bulletin of the Torrey Botanical Club* 30, 519–543.
Casey, R., Domoney, C. and Ellis, N. (1986) Legume storage proteins and their genes. *Oxford Surveys of Plant Molecular and Cell Biology* 3, 1–95.
Cherry, J.M. and Blackburn, E.H. (1985) The internally located telomeric sequences in the germ-line chromosomes of *Tetrahymena* are at the ends of transposon-like elements. *Cell* 43, 747–758.
Cook, P.R. and Brazell, I.A. (1975) Supercoils in human DNA. *Journal of Cell Science* 19, 261–279.
Cooper, D.N. (1983) Eukaryotic DNA methylation. *Human Genetics* 64, 315–333.
Corriveau, J.L., Polans, N.O. and Coleman, A.W. (1989) Cultivar variability for the presence of plastid DNA in pollen of *Pisum sativum* L: implications for plastid transmission. *Current Genetics* 16, 47–51.
Cullis, C.A. and Davies, D.R. (1975) Ribosomal DNA amounts in *Pisum sativum*. *Genetics* 81, 485–492.
Darlington, C.D. (1965) *Cytology*. J. and A. Churchill, London.
Davies, D.R. and Cullis, C.A. (1982) A simple plant polytene chromosome system, and its use for *in situ* hybridization. *Plant Molecular Biology* 1, 301–304.
de Haan, H. (1930) Contributions to the genetics of *Pisum*. *Genetica* 12, 321–439.
Devereux, J., Haeberli, P. and Smithies, D. (1984) A comprehensive set of sequence analysis programs for the VAX. *Nucleic Acids Research* 12, 387–407.
de Vilmorin, P. and Bateson, W. (1912) A case of gametic coupling in *Pisum*. *Proceedings of the Royal Society series B* 84, 9–11.
Dirlewanger, E. (1991) Recherche de marqueurs moléculaires liés à des gènes de résistance à quatre maladies du pois (*Pisum sativum* L.): fusariose, oidium, anthracnose et mosaïque commune du pois. Doctorate thesis, Université de Paris-sud, Centre d'Orsay.

Doyle, J.J. (1987) Variation at the DNA level: uses and potential in legume systematics. In: Stirton, C.H. (ed.), *Advances in Legume Systematics*, Royal Botanic Gardens, Kew, pp. 1–30.

Ellis, T.H.N. (1993) Approaches to the genetic mapping of pea. In: Jackson, J.F. and Linskens, H.F. (eds), *Modern Methods of Plant Analysis: Vegetables and Vegetable Products*, Vol. 16. Springer-Verlag, Heidelberg.

Ellis, T.H.N. and Cleary, W. (1988) Trisomy: a useful adjunct to RFLP mapping in pea. *Chromosoma* 96, 91–94.

Ellis, T.H.N., Davies, D.R., Castleton, J.A. and Bedford, I.D. (1984) The organization and genetics of rDNA length variants in peas. *Chromosoma* 91, 74–81.

Ellis, T.H.N., Lee, D., Thomas, C.M., Simpson, P.R., Cleary, W.G., Newman, M.-A. and Burcham, K.W.G. (1988) 5S rRNA genes in *Pisum*: sequence, long range and chromosomal organization. *Molecular and General Genetics* 214, 333–342.

Ellis, T.H.N., Delseny, M., Lee, D. and Burcham, K.W.G. (1989) Methylated and undermethylated rDNA repeats are interspersed at random in two higher plant species. *Plant Molecular Biology* 14, 73–80.

Ellis, T.H.N., Turner, L., Hellens, R.P., Lee, D., Harker, C.L., Enard, C., Domoney, C. and Davies, D.R. (1992) Linkage maps in pea. *Genetics* 130, 649–663.

Errico, A., Conicella, C. and Venora, G. (1991) Karyotype studies on *Pisum fulvum* and *Pisum sativum*, using a chromosome image analysis system. *Genome* 34, 105–108.

Flavell, R.B. (1986) Ribosomal RNA genes and control of their expression. *Oxford Surveys of Plant Molecular and Cell Biology* 3, 251–276.

Flavell, A.J., Smith, D.B. and Kumar, A. (1992) Extreme heterogeneity of Ty1-copia group retrotransposons in plants. *Molecular and General Genetics* 231, 233–242.

Folkeson, D. (1984a) The use of BSG-staining in making a more detailed nomenclature possible for interchange systems in *Pisum sativum* L. *Hereditas* 101, 119–121.

Folkeson, D. (1984b) Free segregation between a (3S-7S) interchange and genes within linkage group VII in *Pisum sativum*. *Hereditas* 101, 227–233.

Folkeson, D. (1990a) Assignment of linkage segments to chromosomes 3 and 5 in *Pisum sativum*. *Hereditas* 111, 249–255.

Folkeson, D. (1990b) Assignment of linkage segments to the satellite chromosomes 4 and 7 in *Pisum sativum*. *Hereditas* 112, 257–263.

Folkeson, D. (1990c) A revised genetic map of *Pisum sativum*. PhD thesis, University of Lund.

Gorovsky, M.A. (1980) Genome organization and reorganization in *Tetrahymena*. *Annual Review of Genetics* 14, 203–239.

Gruenbaum, Y., Naveh-Many, T., Cedar, H. and Razin, A. (1981) Sequence specificity of methylation in higher plant DNA. *Nature* 292, 860–862.

Haldane, J.B.S. and Waddington, C.H. (1931) Inbreeding and linkage. *Genetics* 16, 357–374.

Harker, C.L., Ellis, T.H.N. and Coen, E.S. (1990) Identification and genetic regulation of the chalcone synthase multigene family in pea. *Plant Cell* 2, 185–194.

Hedley, C.L., Smith, C.M., Ambrose, M.J., Cook, S. and Wang, T.L. (1986) An analysis of seed development in *Pisum sativum* II. The effect of the *r*-locus on the growth and development of the seed. *Annals of Botany* 58, 371–379.

Hernández, P., Bjerknes, C.A., Lamm, S.S. and Van't Hof, J. (1988a) Proximity of an ARS consensus sequence to a replication origin of pea (*Pisum sativum*). *Plant Molecular Biology* 10, 413-422.

Hernández, P., Lamm, S.S., Bjerknes, C.A. and Van't Hof, J. (1988b) Replication termini in the rDNA of synchronized pea root cells (*Pisum sativum*). *EMBO Journal* 7, 303-308.

Heslop-Harrison, J.S. (1991) The molecular cytogenetics of plants. *Journal of Cell Science* 100, 15-21.

Highett, M.I. (1993) The organization of ribosomal DNA and transcripts in *Pisum sativum* nucleoli: a study using 3-D confocal microscopy. PhD thesis, University of East Anglia.

Highett, M.I., Rawlins, D.J. and Shaw, P.J. (1993) Different patterns of rDNA distribution in *Pisum sativum* nucleoli correlate with different levels of transcriptional activity. *Journal of Cell Science* 104, 843-852.

Holliday, R. and Pugh, J.E. (1975) DNA modification mechanisms and gene activity during development. *Science* 187, 226-232.

Inamdar, N.M., Ehrlich, K.C. and Ehrlich, M. (1991) CpG methylation inhibits binding of several sequence-specific DNA-binding proteins from pea, wheat, soybean and cauliflower. *Plant Molecular Biology* 17, 111-123.

Ingram, T.J., Reid, J.B., Potts, W.C. and Murfet, I.C. (1983) Internode length in *Pisum*. IV. The effect of the *Le* gene on giberellin metabolism. *Physiologia Plantarum* 59, 607-616.

Kato, A., Nakajima, T., Yamashita, J., Yakura, K. and Tanifuji, S. (1990) The structure of the large spacer region of the rDNA in *Vicia* and *Pisum sativum*. *Plant Molecular Biology* 14, 983-993.

Kaul, M.L.H. and Murthy, T.G.K. (1985) Mutant genes affecting higher plant meiosis. *Theoretical and Applied Genetics* 70, 449-466.

Klaas, M. and Amasino, R.M. (1989) DNA methylation is reduced in DNaseI-sensitive regions of plant chromatin. *Plant Physiology* 91, 451-454.

Kornberg, J.R. and Rykowski, M.C. (1988) Human genome organization: Alu, lines and the molecular structure of metaphase chromosome bands. *Cell* 53, 391-400.

Kuma, H., and Mercykutty, V.C. (1985) Hyper-tetraploids in pea. *Experientia* 41, 1595-1596.

Lamm, R. (1977) Transpositions in *Pisum*. *Pisum Newsletter* 9, 28-29.

Lamm, R. (1981) Giemsa C-banding and silver staining for cytological studies in *Pisum*. *Hereditas* 94, 45-52.

Lamm, R. and Miravalle, R.J. (1959) A translocation tester set in *Pisum*. *Hereditas* 45, 417-440.

Lamprecht, H. (1948) The variation of linkage and the course of crossing-over. *Agri Hortique Genetica* 6, 10-48.

Lee, D., Ellis, T.H.N., Turner, L., Hellens, R.P. and Cleary, W.G (1990) A *copia*-like element in *Pisum* demonstrates the uses of dispersed repeated sequences in genetic analysis. *Plant Molecular Biology* 15, 707-722.

Macleod, D. and Bird, A.P. (1983) Transcription in oocytes of highly methylated rDNA from *Xenopus laevis* sperm. *Nature* 306, 200-203.

Marx, G.A. (1974) A scheme for demonstrating some classical genetic principles in the classroom. *Journal of Heredity* 65, 252-254.

Matthews, P. (1973) Genetic studies on spontaneous and induced rogues in *Pisum sativum*. PhD thesis, University of East Anglia.

Meicenheimer, R.D., Muehlbauer, F.J., Hindman, J.L. and Gritton, E.T. (1983) Meristem characteristics of genetically modified pea (*Pisum sativum*) leaf primordia. *Canadian Journal of Botany* 61, 3430–3437.

Mendel, G. (1866) Versuche uber Pflanzen-hybriden. *Verhandlungen der Naturforschung Vereins in Brunn* 4, 3–47.

Mendel, G. (1965) Experiments in plant hybridisation. Bennet, J.H. (ed.), Oliver and Boyd, London.

Mercykutty, V.C. Kumar, H. and Srivastava, C.P. (1990) Chromosome manipulation for protein improvement in pea (*Pisum sativum* L.). *Journal of Genetics and Breeding* 44, 249–252.

Michaelson, M.J., Price, H.J., Ellison, J.R. and Johnston, J.S. (1991) Comparison of plant DNA contents determined by fulgen microspectrophotometry and laser flow cytometry. *American Journal of Botany* 78, 183–188.

Montero, L.M., Salinas, J., Matassi, G. and Bernardi, G. (1990) Gene distribution and isochore organization in the nuclear genome of plants. *Nucleic Acids Research* 18, 1859–1867.

Monti, L.M., Saccardo, F. and Rao, R. (1985) Chromosome variation in peas and its use in genetics and breeding. In: Hebblethwaite, P.D., Heath, M.C. and Dawkins, T.C.K. (eds), *The Pea Crop*. Butterworths, London.

Murphy, K.E. and Stringer, J.R. (1985) *RecA* independent recombination of poly(d(GT)-d(CA)) in pBR322. *Nucleic Acids Research* 14, 7325–7340.

Murray, M.G. and Thompson, W.F. (1982) Repeat sequence interspersion in coding DNA of peas does not reflect that in total pea DNA. *Plant Molecular Biology* 1, 143–153.

Murray, M.G., Cuellar, R.E. and Thompson, W.F. (1978) DNA sequence organization in the pea genome. *Biochemistry* 17, 5781–5790.

Murray, M.G., Peters, D.L. and Thompson, W.F. (1980) Ancient repeated sequences in the pea and mung bean genomes. *Carnegie Institute Washington Year Book* 79, 112–114.

Nagylaki, T. (1984) The evolution of multigene families under intrachromosomal gene conversion. *Genetics* 106, 529–548.

Ohta, T. (1981) Population genetics of selfish DNA. *Nature* 292, 648–649.

Palmer, J.D., Jorgensen, R.A. and Thompson, W.F. (1985) Chloroplast DNA variation and evolution in *Pisum*: patterns of change and phylogenetic analysis. *Genetics* 19, 325–354.

Pardue, M.L., Lowenhaupt, K., Rich, A. and Nordheim, A. (1987) $(dC-dA)_n$-$(dG-dT)_n$ sequences have evolutionarily conserved chromosomal locations in *Drosophila* with implications for roles in chromosome structure and function. *EMBO Journal* 6, 1781–1789.

Pearlman, R.E., Tsao, N. and Moens, P.B. (1992) Synaptonemal complexes from DNase treated rat pachytene chromosomes contain $(GT)_n$ and LINE/SINE sequences. *Genetics* 130, 865–872.

Pellew, C. (1940) Genetical studies on the first reciprocal translocation found in *Pisum sativum*. *Journal of Genetics* 39, 125–131.

Piller, K.J., Baerson, S.R., Polans, N.O. and Kaufman, L.S. (1990) Structural analysis of the short ribosomal DNA variant fron *Pisum sativum* L. cv. Alaska. *Nucleic Acid Research* 18, 3135–3145.

Polans, N.O., Weeden, N.F. and Thompson, W.F. (1986) Distribution, inheritance and linkage relationships of ribosomal DNA spacer length variants in pea. *Theoretical and Applied Genetics* 72, 289-295.

Potts, W.C. and Reid, J.B. (1983) Internode length in *Pisum* III. The effect and interaction of the *Na/na and Le/le* gene differences on endogeneous giberellin-like substances. *Physiologia Plantarum* 57, 448-454.

Potts, W.C., Reid, J.B. and Murfet, I.C. (1982) Internode length in *Pisum* I. The effect of the *Le/le* gene difference on endogeneous giberellin-like substances. *Physiologia Plantarum* 55, 323-328.

Provvidenti, R. (1990) Inheritance of resistance to pea mosaic virus in *Pisum sativum*. *Journal of Heredity* 81, 143-145.

Provvidenti, R. and Alconero, R. (1988a) Inheritance of resistance to a lentil strain of pea seed-borne mosaic virus in *Pisum sativum*. *Journal of Heredity* 79, 45-47.

Provvidenti, R. and Alconero, R. (1988b) Inheritance of resistance to a third pathotype of pea seed-borne mosaic virus in *Pisum sativum*. *Journal of Heredity* 79, 76-77.

Punnett, R.C. (1936) The flaked sweet pea. *Journal of Genetics* 32, 171-177.

Radman, M. (1989) Mismatch repair and the fidelity of genetic recombination. *Genome* 3, 68-73.

Rasmusson, J. (1928) Genetically changed linkage-values in *Pisum*. *Hereditas* 10, 1-52.

Rawlins, D.J. and Shaw, P.J. (1988) Three-dimensional organization of chromosomes of *Crepis capillaris* by optical tomography. *Journal of Cell Science* 91, 401-414.

Rawlins, D.J. and Shaw, P.J. (1990) Three-dimensional organization of ribosomal DNA in interphase nuclei of *Pisum sativum* by *in situ* hybridization and optical tomography. *Chromosoma* 99, 143-151.

Rawlins, D.J. and Shaw, P.J. (1991) Localization of telomeres in plant interphase nuclei by *in situ* hybridization and 3D confocal microscopy. *Chromosoma* 100, 143-151.

Richards, E.J. and Ausubel, F.M. (1988) Isolation of a higher eukaryotic telomere from *Arabidopsis thaliana*. *Cell* 53, 127-136.

Rogers, S.O. and Bendich, A.J. (1987) Heritability in ribosomal RNA genes of *Vicia faba*. *Genetics* 117, 285-295.

Saitou, N. and Nei, M. (1987) The neighbor-joining method: a new method for reconstructing phylogenetic trees. *Molecular Biology and Evolution* 4, 406-425.

Salinas, J., Zerial, M, Filipski, J. and Bernardi, G. (1986) The distribution and nucleotide sequence organization in the mouse genome. *European Journal of Biochemistry* 160, 469-478.

Salinas, J., Matassi, G., Montero, L.M. and Bernardi, G. (1988) Compositional compartmentalization and compositional patterns in the nuclear genomes of plants. *Nucleic Acids Research* 16, 4269-4285.

Sansome, E.R. (1932) Segmental interchange in *Pisum sativum*. *Cytologia* 3, 200-219.

Sansome, E.R. (1933) Segmental interchange in *Pisum* II. *Cytologia* 5, 15-30.

Shirsat, A.H. (1988) A transposon-like structure in the 5' flanking sequence of a legumin gene from *Pisum sativum*. *Molecular and General Genetics* 212, 129-133.

Simpson, P.R., Newman, M.-A., and Davies, D.R. (1988) Detection of legumin gene DNA sequences in pea by *in situ* hybridization. *Chromosoma* 96, 454-458.

Simpson, P.R., Newman, M.-A., Davies, D.R., Ellis, T.H.N., Matthews, P.M. and Lee, D. (1990) Identification of translocations in pea by *in situ* hybridization with chromosome-specific DNA probes. *Genome* 33, 745-749.

Slatter, R.E., Dupree, P. and Gray, J.C. (1991) A scaffold-associated DNA region is located downstream of the pea plastocyanin gene. *Plant Cell* 3, 1239-1250.

Small, D., Nelkin, B. and Vogelstein, B. (1982) Nonrandom distribution of repeated DNA sequences with respect to supercoiled loops and the nuclear matrix. *Proceedings of the National Academy of Sciences USA* 79, 5911-5915.

Snoad, B. (1966) The chromosomes and linkage groups of *Pisum*. *Genetica* 37, 247-254.

Stahl, F.W. (1979) *Genetic Recombination. Thinking About It in Phage and Fungi*. W.H. Freeman & Co., San Francisco.

Stallings, R.L., Ford, A.F., Helson, D., Torney, D.C., Hildebrand, C.E. and Moyzis, R.K. (1991) Evolution of $(GT)_n$ repetitive sequences in mammalian genomes. *Genomics* 10, 807-815.

Szostak, J.W. and Wu, R. (1980) Unequal crossing over in the ribosomal DNA of *Saccharomyces cerevisiae*. *Nature* 284, 426-430.

Tartof, K.D. (1974) Unequal mitotic sister chromatid exchange as the mechanism of ribosomal RNA gene magnification. *Proceedings of the National Academy of Science, USA* 71, 1272-1276.

Therman, E. and Murashige, T. (1984) Polytene chromosomes in cultured pearoots (*Pisum*, Fabaceae). *Plant Systematics and Evolution* 148, 25-33.

Thompson, R.D. and Kirch, H.-H. (1992) The *s* locus of flowering plants: when self rejection is self-interest. *Trends in Genetics* 8, 381-387.

Thompson, W.F. and Murray, M.G. (1980) Sequence organization in pea and mung bean DNA and a model for genome evolution. In: Davies, D.R. and Hopwood, D.A. (eds), *The Plant Genome*. The John Innes Institute, Norwich, pp. 31-45.

Timmerman, G.M., Frew, T.J., Miller, A.L., Weeden, N.F. and Jermyn, W.A. (1993) Linkage mapping of *sbm-1*, a gene conferring resistance to pea seed-borne mosaic virus, using molecular markers in *Pisum sativum*. *Theoretical and Applied Genetics* 85, 609-615.

Treco, D. and Arnheim, N. (1986) The evolutionarily conserved repetitive sequence $d(TG.AC)_n$ promotes reciprocal exchange and generates unusual recombinant tetrads during yeast meiosis. *Molecular and Cellular Biology* 6, 3934-3947.

Turner, L, Hellens, R.P., Lee, D. and Ellis, T.H.N. (1993) Genetic aspects of the organization of legumin genes in pea. *Plant Molecular Biology* 22, 101-112.

van Driel, R., Humbel, B. and de Jong, L. (1991) The nucleus: a black box being opened. *Journal of Cell Biochemistry* 47, 311-316.

Van't Hof, J. (1966) Experimental control of DNA synthesizing and dividing cells in excised root tips of *Pisum*. *American Journal of Botany* 53, 970-976.

Van't Hof, J. and Lamm, S.S. (1991) Single-stranded replication intermediates of ribosomal DNA replicons of pea. *EMBO Journal* 10, 1949-1953.

Van't Hof, J. and Lamm, S.S. (1992) Site of initiation of replication of the ribosomal genes of pea (*Pisum sativum*) detected by two-dimensional gel electrophoresis. *Plant Molecular Biology* 20, 377-382.

Van't Hof, J., Hernández, P. Bjerknes, C.A., Kraszewska, E.K. and Lamm, S.S. (1987) Replication of the rRNA and legumin genes in synchronized root cells of pea (*Pisum sativum*): evidence for transient *Eco*RI sites in replicating rRNA genes. *Plant Molecular Biology* 8, 133-143.

Watson, J.C., Kaufman, L.S. and Thompson, W.F. (1987) Developmental regulation of cytosine methylation in the nuclear ribosomal RNA genes of *Pisum sativum*. *Journal of Molecular Biology* 193, 15-26.

Weeden, N.F. and Marx, G.A. (1987) Further genetic analysis and linkage relationships of isozyme loci in the pea. Confirmation of the diploid nature of the genome. *Journal of Heredity* 78, 153-159.

Weeden, N.F. and Wolko, B. (1990) Linkage map for the garden pea (*Pisum sativum*). In: O'Brien, S.J. (ed.), *Genetic Maps. Locus Maps of Complex Genomes*, 5th edition. Cold Spring Harbor Laboratory Press, Cold Spring Harbor, New York, pp. 6.106-6.112.

Weeden, N.F., Keen, B.E. and LaRue, T.A. (1990) Genetic analysis of *sym* genes and other nodule related genes in *Pisum sativum*. In: Gresshoff, P.H., Roth, L.E., Stacey, G.S. and Newton, W.E. (eds), *Nitrogen Fixation: Achievements and Objectives, Proceedings of the 8th International Congress on Nitrogen Fixation*. Chapman and Hall, London, pp. 323-330.

Wellensiek, S.J. (1925) Genetic monograph on *Pisum*. *Bibliographica Genetica* 2, 343-476.

Winge, Ö. (1936) Linkage in *Pisum*. *C. R. La. Carlsberg Copenhagen Series Physiology* 21, 271-393.

Wolko, B. and Weeden, N.F. (1990) Additional markers for chromosome 6. *Pisum Newsletter* 22, 71-74.

Yao, M.-C., Choi, J., Yokoyama, S., Austerberry, C.F. and Yao, C.H. (1984) DNA elimination in *Tetrahymena*: a developmental process involving extensive breakage and rejoining of DNA at defined sites. *Cell* 36, 433-440.

Yesufu, H.M.I., Hanley, A., Rinaldi, A. and Adams, R.L.P. (1991) DNA methylase from *Pisum sativum*. *Biochemical Journal* 273, 469-475.

Young, J.P.W. (1985) Linkage of *sym-2*, the symbiotic specificity locus of *Pisum sativum*. *Journal of Heredity* 76, 207-208.

Young, N.D., Miller, J.C. and Tanksley, S.D. (1987) Rapid chromosomal assignment of multiple genomic clones using primary trisomics. *Nucleic Acids Research* 15, 9339-9348.

The Plastid Genome 3

T.H.N. ELLIS
John Innes Institute, Colney Lane, Norwich, NR4 7UH UK

Introduction

Higher plants have three genomes: the nuclear genome, the mitochondrial genome and the plastid genome. These three interact with each other, so that particular combinations may provide genetic information not represented by either alone. Some of our knowledge of the pea nuclear genome has been reviewed in the previous chapter while this chapter describes some features of the pea plastid genome.

Very little is known about the structure of the pea mitochondrial genome, probably because cytoplasmic male sterility is of little relevance to pea breeding. However, the pea has been widely used as a model organism for plant biochemistry so that much is known about mitochondrial metabolism in pea. Several mitochondrial genes have been cloned and characterized: ATPase subunits α and 9 (Morikami and Nakamura, 1987a,b) and cytochrome oxidase subunit I (Kemmerer *et al.*, 1989; Kemmerer and Wu, 1990). Nuclear genes encoding several mitochondrial enzymes have been cloned and characterized, for example those encoding the glycine decarboxylase and serine hydroxymethyltransferase (Kim and Oliver, 1990; Macherel *et al.*, 1990; Turner *et al.*, 1992a,b, 1993). Despite the attention which the pea mitochondrion has received, the structure and organization of this genome remains undocumented. This may be a serious omission as the other two genomes hold many surprises.

In general, the plastid genomes of higher plants consist of circularly permuted DNA molecules with a monomeric size of about 150 kb (reviewed by Palmer, 1985; see Fig. 3.1). Circular DNA molecules that are integral multiples of the monomeric size have been found in many plant species, although there

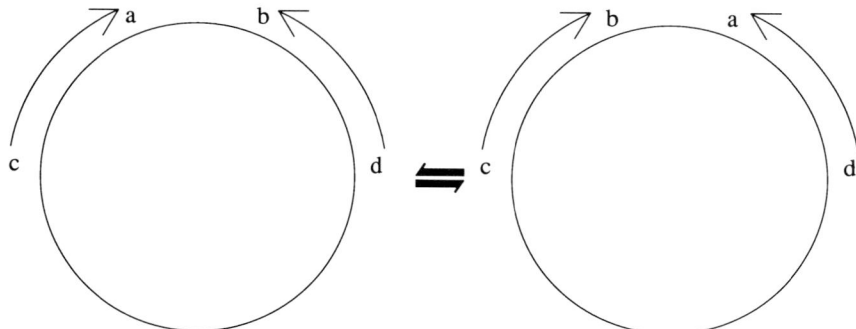

Fig. 3.1. Isomeric forms of plastid DNA. Homologous recombination between large inverted repeats in a circular DNA (long arrows) leads to the inversion of two segments of the DNA circle with respect to each other. The process is reversible.

has recently been some evidence to suggest that linear forms can also be found (Bendich and Smith, 1990; Collin and Ellis, 1991). Chloroplasts exist in many copies in higher plant cells, and within each chloroplast there are many copies of the plastid genome. The pattern of replication and segregation of DNA molecules and of the plastids that contain them is thus a complex process to understand. The pattern of 'sorting out' of different plastid types has been studied mostly in species which exhibit biparental inheritance of plastid genomes (Kirk and Tilney-Bassett, 1978).

The genome of the pea plastid was one of the first such DNAs to be studied in terms of its structural organization (Kolodner and Tewari, 1972, 1975a,b; see Fig. 3.2). It is a salutary lesson that it has proved to be so exceptional in structure.

Evolutionary Context

The vast majority of land plants have a plastid genome comprising two rather differently sized single-copy regions separated by two copies of an inverted repeat (IR) (Palmer, 1985; Ohyama *et al.*, 1986; Shinozaki *et al.*, 1986; Hiratsuka *et al.*, 1989). These genomes thus exist in two isomeric forms (Palmer, 1983; see Fig. 3.1). Variation on this theme generally takes the form of point mutations, small-scale insertions or deletions, or inversions within the single-copy regions (Palmer, 1985; Palmer and Stein, 1986; Wolfe *et al.*, 1987). Two groups of land plants are known to exhibit a major departure from this scheme; one is a group of gymnosperms in the Pinaceae (Strauss *et al.*, 1988; White, 1990), while the other is a group of legumes, including pea (Palmer and Thompson, 1982; Crouse *et al.*, 1986; Palmer *et al.*, 1987).

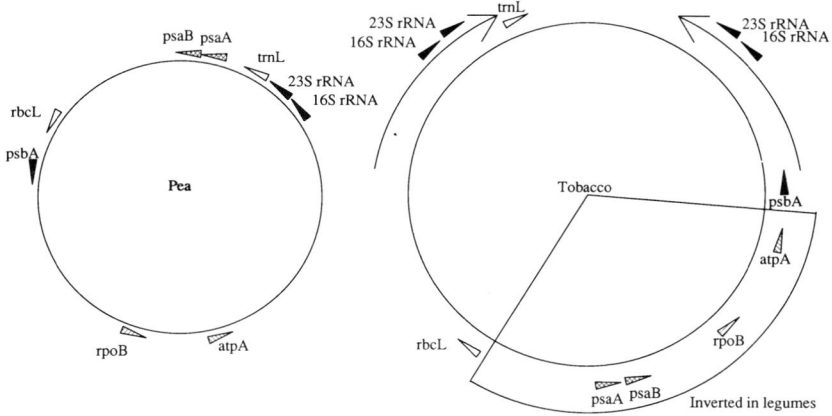

Fig. 3.2. Plastid genome organization. A comparison between the organization of the pea plastid genome and that of the tobacco is shown. The approximate position and direction of transcription of several genes is indicated. Data are taken from Hiratsuka et al. (1989) and Woodbury et al. (1988). The segment between *rbcL* and *psbA* of the tobacco genome is inverted in those legume plastid genomes which carry both copies of the large inverted repeat (reviewed in Doyle, 1987).

In both these examples one copy of the inverted repeat sequence has been lost. The structure of the resulting molecules is thus radically different from the standard plastid genome. Within both the groups which have lost one copy of the IR there has been extensive rearrangement of the organization of the plastid genomes (Palmer et al., 1988; Strauss et al., 1988). The observation of this pattern of variation in the legumes led to the suggestion that the IR organization stabilized the structure of the genome (Palmer and Thompson, 1982). However, an analysis of the structure of mutant plastid genomes in wheat and barley led to the suggestion that those genomes which have large inverted repeats are prone to breakage (Day and Ellis, 1984); that is, they are unstable because there are more opportunities for recombination in those genomes with a large inverted repeat. These two suggestions appear contradictory, but are probably not; this is because the broken genomes will be lost on an evolutionary time-scale. The apparent stability (or evolutionary conservation) of those genomes that have a pair of inverted repeats is not due to immutability of such molecules, but is a consequence of the phenotype arising from a major class of rearrangements.

The phylogeny of the legume plastid genomes that lack one copy of the IR has received much attention, and it has been argued that these legumes form a monophyletic group derived from a single common ancestor that lost one copy of the IR (Crouse et al., 1986; Palmer et al., 1987, 1988). This is potentially a powerful guide to resolving issues of phylogenetic relationships within the legumes, and the conclusions seem in reasonable agreement with

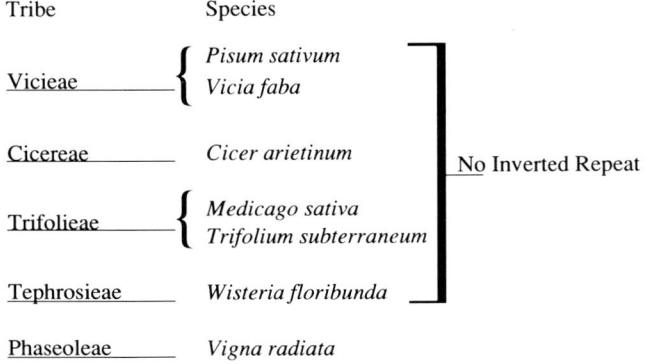

Fig. 3.3. Phylogenetic relationships. The taxonomic relationships between various legume species is shown in accordance with Polhill and Raven (1981). The structure of the *Cicer arietinum* chloroplast DNA is described by Chu and Tewari (1982). These relationships have been reviewed by Doyle (1987).

currently accepted legume taxonomy (Polhill and Raven, 1981; Doyle, 1987; Palmer *et al.*, 1987, 1988; see Fig. 3.3). Perhaps the most striking piece of evidence in favour of the monophyletic origin of these plastid genomes lacking one copy of the IR was the beautiful demonstration by Palmer *et al.* (1987) that the IR was lost from the same isomeric form of plastid DNA in the common ancestor to pea and *Vicia faba*. This is a stringent test of the proposal that a single DNA molecule that lost the IR was the common ancestor to all those legumes that share this structure. We have the corollary that within the monophyletic group of legume species, in which those species whose plastid genomes lack large inverted repeats are to be found, there should be no species with a large inverted repeat in their plastid genome.

Among the Vicieae, plastid genomes show frequent rearrangement. At least eleven inversions need to be proposed to account for the difference in the organization of the pea and *Vicia faba* plastid genomes (Palmer *et al.*, 1987, 1988). The Trifolieae have also exhibited rapid change in the lineage leading to *Trifolium subterraneum*. However, this high frequency of rearrangement is not seen in the comparison between alfalfa (*Medicago sativa*) or wisteria (*Wisteria floribunda*) and the IR-containing mung bean (*Vigna radiata*) genome (Palmer *et al.*, 1987, 1988). The alfalfa and wisteria genomes are colinear with that of mung bean except for the loss of one copy of the IR. This poses a problem: if all of the plastid genomes lacking one copy of the large inverted repeat are the descendants of a single molecule that lost one copy of the IR, it is necessary to propose that the rate of evolution of plastid genome organization in the lineage leading to alfalfa or wisteria has been much slower than in the others. This pattern of diversity of plastid genome structure suggests that plastid genomes that lack a pair of large inverted repeats may be able to exhibit evolutionary stability, countering the original arguments about the stability of IR-containing plastid DNA.

A much more accommodating model for the evolution of the plastid genomes of those legumes that lack a pair of large inverted repeats can be made if we propose a small alteration to the considerations above. The common ancestor to all plastid genomes that have lost one copy of the pair of inverted repeats could have been a molecule where one copy of the IR was prone to deletion. For example, if some other rearrangement brought a sequence near to the end of one of the IRs so that the latter was bounded by short direct repeats (see Michalowski et al., 1987; Wolfe, 1988), a recombination event between these would lead to deletion of one copy of the IR (Fig. 3.4a,b). Note that this deletion is possible in only one of the two isomeric forms (Day and Ellis, 1984). Only some of the progeny of such a molecule will lose the IR;

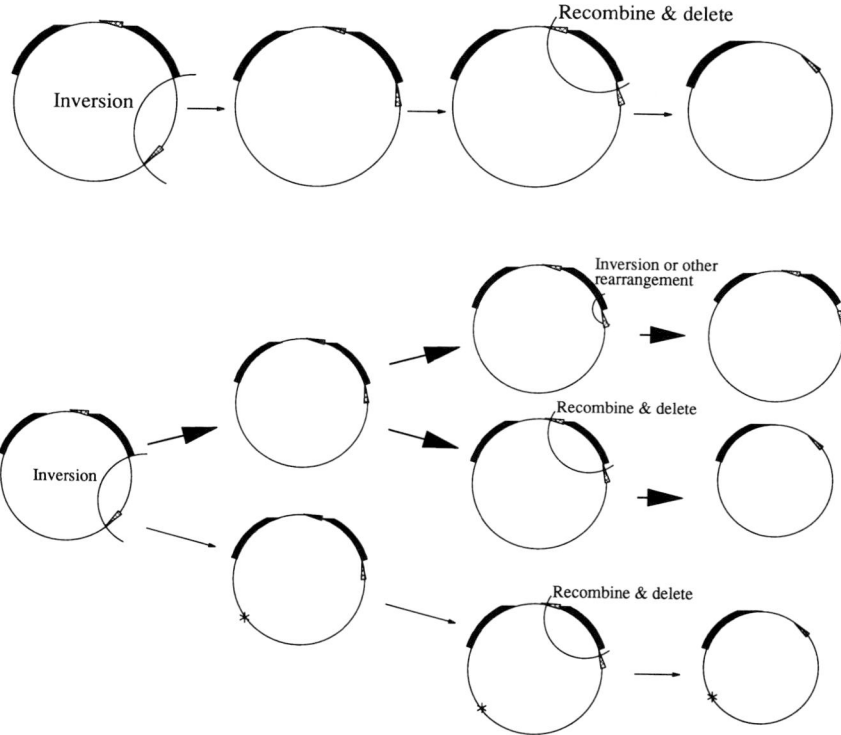

Fig. 3.4. Hypothetical evolutionary scheme. **(a)** Rearrangements leading to the loss of one copy of the IR. **(b)** This scheme shows how rearrangement in the organization of a chloroplast DNA could give rise to a complex pattern of relationships between genomes with and without large inverted repeats (filled shapes). Open triangles mark short repeated sequences. Other symbols represent lineage-specific markers.

others may not do so, either by chance or because further changes may have destroyed the organization that permits such events. Under this latter scheme the taxonomic interpretation of the distribution of plastid genome types in legumes is muddled, and more difficult to test than that proposed by Palmer and colleagues (Palmer and Thompson 1982; Palmer et al., 1987, 1988); it is a less restrictive and therefore less powerful hypothesis.

There is, at the moment, only one consequence of the scheme proposed here to commend it; we can suggest that the alfalfa or wisteria genome lost one copy of the IR more recently than that of pea, subterranean clover and *Vicia*. This interpretation suggests that close taxonomic allies of alfalfa and wisteria could well be found with IR-containing plastid genomes. This focuses attention on a restricted group of legume species. If such genomes can be found, then the monophyletic origin of plastid genomes lacking the large inverted repeats in legumes should be re-examined.

Gene Content

The coding capacity of the plastid genomes of land plants is highly conserved, as is exemplified by the comparison between the gene content of bryophyte and angiosperm genomes (Ohyama et al., 1986; Shinozaki et al., 1986; Hiratsuka et al., 1989). Among the angiosperms there are a few examples of loss of coding sequences in certain lineages. In the legumes the loss of *rpl*22 is a case in point (Gantt et al., 1991); this gene is not present in the pea plastid genome. The coding capacity of the pea plastid genome is not fully catalogued [although extensive transcription and transcript initiation maps have been generated (Woodbury et al., 1988, 1989)], so it is not yet possible to give a definitive statement about the conservation of coding capacity in this genome.

Replication

There are two aspects to the replication of plastid DNA. One is the replication that occurs in meristems and is associated with the proliferation of that organelle as cells divide. The second is the plastid DNA replication that accompanies the accumulation of chloroplasts. The copy number of plastid DNA, and of mitochondrial DNA, varies between different plant organs (roots and shoots), but the mitochondrial DNA does not increase along with plastid DNA in leaves (Lamppa and Bendich, 1985). Plastid DNA copy number in pea is strongly correlated with the expression of photosynthetic genes, but not with the amount of other transcripts (Sasaki et al., 1990). This suggests that the phytochrome-mediated plastid DNA accumulation (Sasaki et al., 1986) is an adaptive response to the pressure to generate an effective photosynthetic apparatus quickly. A pea mutant exhibiting light-independent photomorphogenesis has been described recently (Frances *et al.*, 1992). In some sense plastid DNA replication may be viewed as an analogue of rDNA amplification in frog oocyte (Rochaix *et al.*, 1974). There is a further parallel between these two systems; both involve a rolling circle mechanism (Rochaix *et al.*, 1974; Kolodner and Tewari, 1975b; Collin and Ellis, 1991). This replication mechanism is ideal for rapid amplification of DNA because it does not require multiple rounds of initiation.

In vitro analysis has shown that the pea chloroplast DNA polymerase has preferred templates for the initiation of DNA synthesis (Carrillo and Bogorad, 1988), suggesting that rolling circle replication alone does not account for plastid DNA replication; presumably other forms are also to be found (Kolodner and Tewari, 1975b). Two sites for the initiation of DNA synthesis (by the formation of D loops) have been mapped near to the rRNA spacer region in pea (Meeker *et al.*, 1988); this location is similar to one mapped in *Oenothera* (Chiu and Sears, 1992), and the involvement of these D loops in DNA replication has been demonstrated by their activity with a pea plastid DNA primase (Nielsen *et al.*, 1991).

Despite early attempts to characterize the mechanism of plastid DNA replication, the nature and mechanism of this process are still largely a mystery, and much work has focused on the identification of the region that acts as a replication origin (see above). Replicative intermediates of such large DNA molecules are hard to characterize, and it has been suggested that about 2% of plastid DNA is actively replicating (Lindbeck and Rose, 1990). Given that plastid DNA can be multimeric, it seems possible that all intact plastid DNA molecules are in some way involved in DNA replication.

Transmission Genetics

The replication of plastid DNA and its distribution among daughter plastids is a complex and ill-defined process (Kirk and Tilney-Bassett, 1978). However, the process of segregation of plastids between cells during microsporogenesis and especially their fate upon zygote formation, and the subsequent few cell divisions, are of fundamental importance for the transmission of plastids from one generation to the next. These processes have been investigated to some extent in pea (Polans *et al.*, 1990); a survey of several pea cultivars has shown that there is some genetic variation for the presence of plastid DNA in the generative cell of pollen (Corriveau *et al.*, 1989), suggesting that there is at least some opportunity for biparental transmission of plastids. Although pea cultivars exhibit considerable plastid DNA restriction fragment length polymorphism (Palmer *et al.*, 1985; Teeri *et al.*, 1985;) there is as yet no conclusive evidence for biparental transmission of plastids in this species.

Final Comments

The pea plastid genome provides an opportunity for the study of many aspects of its molecular genetics; the study of several of its attributes is well advanced despite the lack of a complete sequence for this molecule. The most exciting opportunities are raised by the characterization of its replication strategy and the possibility of limited biparental transmission. Taken together with the variability of this genome, we can see potential for a genetic dissection of the processes of replication and segregation of the pea plastid genome.

References

Bendich, A.J. and Smith, S.B. (1990) Moving pictures and pulsed-field gel electrophoresis show linear DNA molecules from chloroplasts and mitochondria. *Current Genetics* 17, 421–425.

Carrillo, N. and Bogorad, L. (1988) Chloroplast DNA replication *in vivo*: site specific initiation from preferred templates. *Nucleic Acids Research* 16, 5603–5621.

Chiu, W.-L., and Sears, B.B. (1992) Electron microscopic localization of replication origins in *Oenothera* chloroplast DNA. *Molecular and General Genetics* 232, 33–39.

Chu, N.M. and Tewari, K.K. (1982) Arrangement of the ribosomal RNA genes in chloroplast DNA of Leguminosae. *Molecular and General Genetics* 186, 23–32.

Collin, S. and Ellis, T.H.N. (1991) Evidence for the presence of hairpin chloroplast DNA molecules in barley cultivars. *Current Genetics* 20, 253–258.

Corriveau, J.L., Polans, N.O. and Coleman, A.W. (1989) Cultivar variability for the presence of plastid DNA in pollen of *Pisum sativum* L: implications for plastid transmission. *Current Genetics* 16, 47–51.

Crouse, E.J., Mubumbila, M., Stummann, B.M., Michalowski, C., Bonhert, H.J., Weil, J.-H., and Henningsen, K.W. (1986) Divergence of chloroplast gene organization in three legumes: *Pisum sativum*, *Vicia faba* and *Phaseolus vulgaris*. *Plant Molecular Biology* 7, 143–150.

Day, A. and Ellis, T.H.N. (1984) Chloroplast DNA deletions associated with wheat plants regenerated from pollen: possible basis for maternal inheritance of chloroplasts. *Cell* 39, 359–368.

Doyle, J.J. (1987) Variation at the DNA level: uses and potential in legume systematics. In: Stirton C.H. (ed.), *Advances in Legume Systematics Part 3*. Royal Botanic Gardens, Kew, pp. 1–30.

Frances, S., White, M.J., Edgerton, M.D., Jones, A.M., Elliot, R.C. and Thompson, W.F. (1992) Initial characterization of a pea mutant with light-independent photomorphogenesis. *Plant Cell* 4, 1519–1530.

Gantt, J.S., Baldauf, S.L., Calie, P.J., Weeden, N.F. and Palmer, J.D. (1991) Transfer of *rpl*22 to the nucleus greatly preceded its loss from the chloroplast and involved the gain of an intron. *EMBO Journal* 10, 3073–3078.

Hiratsuka, J., Shimada, H., Whittier, R., Ishibashi, T., Sakamoto, M., Mori, M., Kondo, C., Honji, Y., Sun, C.-R., Meng, B.-Y., Li, Y.-Q., Kanno, A.,

Nishizawa, Y., Hirai, A., Shinozaki, K. and Sugiura, M. (1989) The complete sequence of the rice (*Oryza sativa*) chloroplast genome: inter-molecular recombination between distinct tRNA genes accounts for a major plastid DNA inversion during the evolution of the cereals. *Molecular and General Genetics* 217, 185-194.

Kemmerer, E.C. and Wu, R. (1990) The *Pisum sativum* mitochondrial gene encoding cytochrome oxidase subunit I has an unusual transcription pattern. *Gene* 89, 157-162.

Kemmerer, E.C., Kao, T.-H., Deng, G.-R. and Wu, R. (1989) Isolation and nucleotide sequence of the pea cytochrome oxidase subunit I. *Plant Molecular Biology* 13, 121-124.

Kim, Y. and Oliver, D.J. (1990) Molecular cloning, transcriptional characterization, and sequencing of cDNA encoding the H-protein of the mitochondrial glycine decarboxylase complex in peas. *Journal of Biological Chemistry* 265, 848-853.

Kirk, J.T.O. and Tilney-Bassett, R.A.E. (1978) *The Plastids. Their Chemistry, Structure, Growth, Inheritance.* Elsevier, Amsterdam.

Kolodner, R. and Tewari, K.K. (1972) Molecular size and conformation of chloroplast DNA from pea leaves. *Journal of Biological Chemistry* 247, 6355-6364.

Kolodner, R. and Tewari, K.K. (1975a) Presence of displacement loops in the covalently closed circular chloroplast deoxyribonucleic acid from higher plants. *Journal of Biological Chemistry* 250, 8840-8847.

Kolodner, R. and Tewari, K.K. (1975b) Chloroplast DNA from higher plants replicates by both the Cairns and the rolling circle mechanism. *Nature* 256, 708-711.

Lamppa, G.K. and Bendich, A.J. (1985) Changes in mitochondrial DNA levels during development of pea (*Pisum sativum* L.). *Planta* 162, 463-468.

Lindbeck, A.G.C. and Rose, R.J. (1990) Thylakoid-bound chloroplast DNA from spinach is enriched for replication forks. *Biochemical and Biophysical Research Communications* 172, 204-210.

Macherel, D., Lebrun, M., Gagnon, J., Neuburger, M. and Douce, R. (1990) cDNA cloning, primary structure and gene expression for the H protein, a component of the glycine-cleavage system (glycine decarboxylase) of pea (*Pisum sativum*) leaf mitochondria. *Biochemical Journal* 268, 783-789.

Meeker, R., Nielsen, B. and Tewari, K.K. (1988) Localization of the replication origins in pea chloroplast DNA. *Molecular and Cell Biology* 8, 1216-1223.

Michalowski, C., Breuning, K.D. and Bonhert, H.J. (1987) Points of rearrangements between plastid chromosomes: location of protein coding regions on broad bean chloroplast DNA. *Current Genetics* 11, 265-274.

Morikami, A. and Nakamura, K. (1987a) Structure and expression of pea mitochondrial F_1 ATPase α-subunit gene and its pseudogene involved in homologous recombination. *Journal of Biochemistry* 101, 967-976.

Morikami, A. and Nakamura, K. (1987b) The pea mitochondrial ATPase subunit 9 gene is located upstream of the ATPase α-subunit gene. *Nucleic Acids Research* 15, 4692.

Nielsen, B.L., Rajasekhar, V.K. and Tewari, K.K. (1991) Pea chloroplast DNA primase: characterization and role in initiation of replication. *Plant Molecular Biology* 16, 1019-1034.

Ohyama, K., Fukuzawa, H., Kohchi, T., Shirai, H., Sano, T., Sano, S., Umesono, K., Shiki, Y., Takeuchi, M., Chang, Z., Aota, S-I., Inokuchi, H. and Ozeki, H.

(1986) Chloroplast gene organization deduced from complete sequence of liverwort *Marchantia polymorpha* chloroplast DNA. *Nature* 322, 572–574.

Palmer, J.D. (1983) Chloroplast DNA exists in two orientations. *Nature* 301, 92–93.

Palmer, J.D. (1985) Comparative organization of chloroplast genomes. *Annual Review of Genetics* 19, 325–354.

Palmer, J.D. and Stein, D.B. (1986) Conservation of chloroplast genome structure among vascular plants. *Current Genetics* 10, 823–833.

Palmer, J.D. and Thompson, W.F. (1982) Chloroplast DNA rearrangements are more frequent where a large inverted repeat is lost. *Cell* 29, 537–550.

Palmer, J.D., Jorgensen, R.A. and Thompson, W.F. (1985) Chloroplast DNA variation and evolution in *Pisum*: patterns of change and phylogenetic analysis. *Genetics* 109, 195–213.

Palmer, J.D., Osorio, B., Aldrich, J. and Thompson, W.F. (1987) Chloroplast DNA evolution among legumes: loss of a large inverted repeat occurred prior to other sequence rearrangements. *Current Genetics* 11, 275–286.

Palmer, J.D., Osorio, B. and Thompson, W.F. (1988) Evolutionary significance of inversions in legume chloroplast DNAs. *Current Genetics* 14, 65–74.

Polans, N.O., Corriveau, J.L. and Coleman, A.W. (1990) Plastid inheritance in *Pisum sativum* L. *Current Genetics* 18, 477–480.

Polhill, R.M. and Raven, P.H. (eds) (1981) *Advances in Legume Systematics Part 1*. Royal Botanic Gardens, Kew.

Rochaix, J.-D., Bird, A.P. and Bakken, A. (1974) Ribosomal RNA gene amplification by rolling circles. *Journal of Molecular Biology* 87, 473–487.

Sasaki, Y., Nakamura, Y. and Matsuno, R. (1986) Phytochrome-mediated accumulation of chloroplast DNA in pea leaves. *FEBS Letters* 196, 175–179.

Sasaki, Y., Morioka, S. and Matsuno, R. (1990) Correlation of plastid DNA copy number with plastid gene expression in various organs in mature pea plants (*Pisum sativum* L.). *Plant Cell Physiology* 31, 925–931.

Shinozaki, K., Ohme, M., Tanaka, M., Wakasugi, T., Hayashida, N., Matsubayashi, T., Zaita, N., Chunwongse, J., Obokata, J., Yamaguchi-Shinozaki, K., Ohto, C., Torazawa, K., Mang, B.Y., Kusuda, J., Takaiwa, F., Kato, A., Tohdoh, N., Shimada, H. and Sugiura, M. (1986) The complete nucleotide sequence of the tobacco chloroplast genome: its gene organization and expression. *EMBO Journal* 5, 2043–2049.

Strauss, S.H., Palmer, J.D., Howe, G.T. and Doeriksen, A.H. (1988) Chloroplast genomes of two conifers lack a large inverted repeat and are extensively rearranged. *Proceedings of the National Academy of Sciences USA* 85, 3898–3902.

Teeri, T.H., Saura, A. and Lokki, J. (1985) Insertion polymorphism in pea chloroplast DNA. *Theoretical and Applied Genetics* 69, 567–570.

Turner, S.R., Ireland, R., and Rawsthorne, S. (1992a) The cloning and characterisation of the P subunit of glycine decarboxylase from pea (*Pisum sativum*). *Journal of Biological Chemistry* 267, 5355–5360.

Turner, S.R., Ireland, R., and Rawsthorne, S (1992b) The cloning and characterisation of a cDNA encoding the L subunit of glycine decarboxylase from pea (*Pisum sativum*). *Journal of Biological Chemistry* 267, 7745–7750.

Turner, S.R., Hellens, R., Ireland, R., Ellis, N. and Rawsthorne, S. (1993) The organisation and expression of the genes encoding the mitochondrial glycine decarboxylase complex and the serine hydroxymethyltransferase in pea (*Pisum sativum*). *Molecular and General Genetics* 236, 402–408.

White, E.E. (1990) Chloroplast DNA in *Pinus moticola*. *Theoretical and Applied Genetics* 79, 119–124.

Wolfe, K.H. (1988) The site of deletion of the inverted repeat in the pea chloroplast DNA contains duplicated gene fragments. *Current Genetics* 13, 97–99.

Wolfe, K.H., Li, W.-H., and Sharp, P.M. (1987) Rates of nucleotide substitution vary greatly among plant mitochondria, chloroplast and nuclear DNAs. *Proceedings of the National Academy of Sciences USA* 84, 9054–9058.

Woodbury, N.W., Roberts, L.L., Palmer, J.D. and Thompson, W.F. (1988) A transcription map of the pea chloroplast genome. *Current Genetics* 14, 75–89.

Woodbury, N.W., Dobres, M. and Thompson, W.F. (1989) The identification and localization of 33 pea chloroplast transcript initiation sites. *Current Genetics* 16, 433–445.

Regulation of Gene Expression 4

P.M. GILMARTIN
Centre of Plant Biochemistry and Biotechnology, University of Leeds, Leeds, LS2 9JT, UK

Studies on peas have contributed greatly to our understanding of the regulatory DNA sequences that govern and modulate the temporal and spatial control of plant genes. Much of this chapter will be concerned with the pea *rbcS* genes, which code for the small subunit of ribulose-1,5-bisphosphate carboxylase (RUBISCO); these genes are differentially activated during development and respond to light.

Differential Gene Expression

In plants, as in all multicellular organisms, specific tissues and organs perform particular functions and play distinct roles. There is therefore a clear requirement for different structural proteins and enzymes at various times and in different places. The presence of all the coding information for every protein specified by the genome within each cell of a multicellular organism, coupled to the distinct structure and metabolic function of specific cell types, exemplifies the requirement for differential control of gene expression. The temporal and spatial regulation of gene activity not only determines the structure of the organism, but also ensures that the proteins required for complex metabolic pathways are present at the right time, in appropriate ratios, within the correct cell types. This basic requirement for differential gene regulation arises both during genetically programmed development and in response to changes in environmental stimuli.

The amount of information that exists within the tens of millions of the four nucleotides within the genome of a plant is vast. This information specifies not only the amino acid sequence of many thousands of proteins, but also the developmental programme of the organism and its ability to respond to changes in its environment. The protein-coding information is read in a triplet code from a defined start point to a defined end point. The regulatory information, which specifies the temporal and spatial control of gene expression, is also encoded within the linear array of the four nucleotides, but operates through the interactions of specific nucleotide sequences with regulatory macromolecules such as proteins or other nucleic acid sequences. The complexities of tissue-specific and environmentally stimulated patterns of gene expression arise out of the potentially vast array of regulatory sequences that can act to coordinate the expression of tens of thousands of genes.

Control Points of Gene Expression

Advances over the past several years in our understanding of the expression of eukaryotic genes in animals, fungi and plants have revealed that control over the abundance and activity of a particular gene product can be exerted at many levels. These can be simply partitioned into transcriptional, post-transcriptional, translational and post-translational control.

In eukaryotes, the process of transcribing DNA sequence into RNA sequence is performed by one of three enzymes, depending on the type of gene and its role. Genes encoding ribosomal RNA are transcribed by RNA polymerase I; protein-coding genes are transcribed by RNA polymerase II; genes encoding tRNA are transcribed by RNA polymerase III. The best characterized transcriptional regulatory elements are those involved in the control of genes transcribed by RNA polymerase II. The *cis*-acting regulatory sequences that modulate transcription of these protein-coding genes can be divided into two types: those that are required at specific positions in relation to the coding regions, and those whose position relative to the transcription start site is less critical. The TATA box is an example of the former type of control element, its position approximately 30 bp upstream of the transcription start point being highly conserved in virtually all eukaryotic genes. The TATA box is the binding site of the regulatory DNA-binding protein TFIID, which, in combination with other proteins, can form a transcription complex that includes RNA polymerase II (see Gasch *et al.*, 1990). The upstream region of a gene that contains the TATA box is generally referred to as the promoter.

Regulatory elements that can mediate their effect at a distance from the TATA box are referred to as enhancers, or upstream activation sequences. These regulatory elements can, through interactions with sequence-specific DNA-binding proteins, modulate transcription. Activation of transcription occurs through interactions between the enhancer-binding proteins (transcription factors) and the protein complex bound at the TATA box (transcription complex). Enhancers have been shown to be capable of mediating their effects independently of orientation relative to the coding region, and from various positions, both upstream and downstream, relative to the transcription start site.

Control of transcription is not exclusively mediated through gene up-regulation or activation by transcriptional enhancers. Patterns of gene expression can be modulated by a combination of both gene activation and gene suppression (see Simpson *et al.*, 1986; Kuhlemeier *et al.*, 1987a,b). The principle of transcriptional repression is similar to transcriptional activation, in that it arises out of interactions of DNA-binding proteins with their DNA target sequences. However, in contrast to transcriptional activation, transcription silencers and their associated proteins lead to reduced transcription rates either by preventing the binding of a transcriptional activator to its target regulatory sequence or by preventing stabilization of the transcription complex at the TATA box. The regulatory information that controls transcription of a gene can therefore comprise a combination of transcriptional enhancers and transcriptional silencers. Transcriptional enhancers are by far the more numerous of the characterized regulatory elements.

Since all cells of an organism contain the same complement of genes, and associated regulatory DNA elements, the specificity of transcriptional regulation arises out of the differential abundance or activity of the regulatory proteins in different cell types and following specific environmental cues. Following transcription, the primary RNA transcript undergoes a series of processing events that result in the formation of a mature mRNA molecule. These events include the removal of intervening sequences (intron splicing) and capping of the 5' terminus of the molecule by the addition of a methylated guanine residue. This 5' processing event not only protects the RNA molecule from degradation, but also plays a key role in the initiation of protein synthesis. The newly synthesized RNA molecule is also processed at the 3' end by cleavage and subsequent addition of a few hundred adenine residues (the poly-A tail), which are thought to be important for RNA stability.

The processing events that lead to removal of introns from the primary transcript also occur in the nucleus. Upon completion of splicing, the mature mRNA molecule is transported from the nucleus to the cytoplasm. Control over any of these RNA processing events can lead to alterations in mRNA abundance within the cytoplasm and therefore affect the availability of a transcript for translation. One further level of post-transcriptional control is mediated at the level of RNA stability. Transcripts derived from different genes may show differing half-lives, or the half-life of a specific transcript may be altered in response to a developmental or environmental cue. For example, it has been suggested that the gene encoding ferredoxin from pea (*Fed-1*) is controlled at least in part by post-transcriptional regulation (Elliot *et al.*, 1989).

In addition to transcriptional and post-transcriptional control of mRNA abundance, translational control can also influence the relative amounts of specific proteins present in distinct cell types, or induced by environmental stimuli (Berry *et al.*, 1990). Post-translational control events, such as glycosylation, phosphorylation, proteolytic processing, dimerization or protein turnover, can also contribute to the differential abundance and activity of a specific protein.

There are therefore many levels of control exerted over the abundance and activity of a gene product, but it is transcriptional regulation that appears to play the greatest role in the differential control of nuclear genes (Kuhlemeier, 1992). Some of the different levels at which gene expression can be modulated are shown in Fig. 4.1.

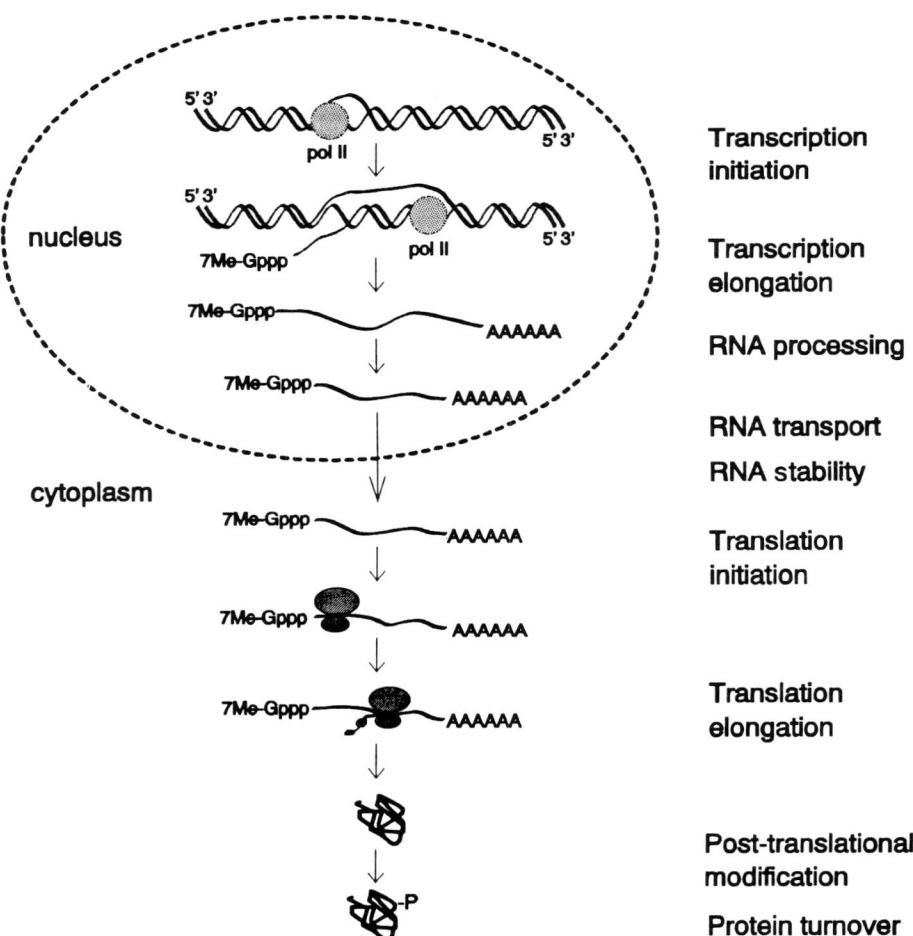

Fig. 4.1. Control points of gene expression. The various stages of gene expression from transcription to post-translational modification are shown and the points at which regulatory control may be exerted are indicated.

Analysis of Gene Expression Levels

There are numerous molecular biological tools available with which to dissect the regulatory mechanisms that control gene expression. Clearly, in understanding the mechanisms that lead to differential gene expression, an understanding is required of the level at which control operates. The situation is not always entirely straightforward, as control may be mediated at several levels.

Faced with the observation that the activity of a particular enzyme, for example ribulose-1,5-bisphosphate carboxylase (RUBISCO), is modulated by light, it is not immediately apparent at which level this control is mediated. The observation of different RUBISCO enzyme activities between light- and dark-grown leaves can be rationalized as follows. Since the enzyme activity is only detectable in the light, it is possible that the protein is present only in the light. It is also possible that the protein is present in both the light and the dark and yet shows enzyme activity only in the light. These two alternatives can be distinguished by studies of protein abundance using 1-D or 2-D polyacrylamide gel electrophoresis. The availability of antibodies specific to the protein provides a valuable increase in resolution in such studies. The presence of a particular protein both prior to and following a stimulus but different protein activity would suggest that control was mediated post-translationally. In the case of RUBISCO, light stimulates accumulation of the protein in leaves (Ellis, 1981).

Measurements of the mRNA levels derived from a gene can be used to further focus on the level of regulatory control. Measurements of mRNA levels can be made following separation of populations of mRNA molecules by gel electrophoresis and subsequent hybridization (Northern analysis) with a specific DNA probe (Bedbrook *et al.*, 1980). Similar information can be obtained by the techniques of S1 nuclease mapping and RNase protection studies. These two techniques provide enhanced resolution over Northern analysis, in that they permit distinctions to be made between mRNA derived from distinct, yet similar, genes (Fluhr *et al.*, 1986a). As we will see later, this is an essential technique for studies of gene expression in multigene families. Both S1 nuclease mapping and RNase protection rely on the ability of a single-stranded DNA or RNA probe to anneal to complementary transcripts; any base pair mismatching between the radio labelled probe and the mRNA molecule provides a target for S1 nuclease or RNase digestion. The resulting probe, trimmed to size by the nuclease, can be visualized on a polyacrylamide gel and provides a measure of the transcript abundance from a given gene. Observations resulting from such Northern analysis or nuclease digestion experiments can provide evidence for differential mRNA abundance.

Observations of differential protein abundance, but uniform mRNA levels following a particular stimulus would indicate that the control of the corresponding gene was likely to be translational or at the level of protein turnover. In contrast, differential abundance of an mRNA molecule would indicate either transcriptional or post-transcriptional control. Since measurements of transcript levels do not permit this distinction to be made, estimation of transcription rates must be made directly. The technique of run-off (or run-on) transcription provides the means with which to directly determine whether differential RNA abundance is modulated at the level of transcription.

The technique of run-on transcription relies on the fact that when nuclei isolated from a tissue are incubated in the presence of ribonucleotides, including ^{32}P-UTP, any transcripts that were initiated prior to nuclear isolation are elongated *in vitro*; reinitiation of transcription does not occur. The radiolabelled nascent transcripts can be detected by hybridization to a specific probe and comparative measurements made of the transcription of a given gene at different times. Differential mRNA abundance and yet constant transcription levels would indicate post-transcriptional regulation, whereas differential transcript levels (as measured by run-on transcription) provide direct evidence for transcriptional control of gene expression. As discussed in more detail below, the genes encoding the small subunit of RUBISCO are regulated at the level of transcription (Gallagher and Ellis, 1982). It is primarily this differential gene expression that gives rise to increased RUBISCO activity in the light.

One further means of demonstrating that regulation of gene expression is at the level of transcription is through the use of chimeric genes comprised of regulatory sequence information fused to reporter genes, such as those encoding chloramphenicol acetyl transferase (see Timko *et al.*, 1985a; Fluhr *et al.*, 1986b), β-glucuronidase (Jefferson *et al.*, 1987; Benfey and Chua, 1989) or luciferase (Millar *et al.*, 1992), followed by assay of the gene fusion in transgenic plants. These techniques have proved invaluable, not only as a means of determining the level of gene regulation, but also as a way of defining the regulatory DNA elements through which differential gene expression is modulated. Several studies, using genes derived from pea, have provided evidence that the primary level of gene control that results in the differential abundance of several proteins in response to light is that of transcription.

Location of Regulatory Sequence Information

Transcriptional control is mediated through *cis*-acting regulatory DNA elements whose effects are mediated by interactions with regulatory DNA-binding proteins. The location of the *cis*-regulatory DNA sequences within a gene can vary; most reside upstream of the coding regions of genes (Dean *et al.*, 1989), and the greatest emphasis in the study of regulatory DNA elements has been placed on these 5'-flanking *cis*-acting regulatory elements. Other studies have shown that the control sequences can be located beyond the 3' end, or downstream, of the coding region, or even within the transcribed region of a gene. This information may be contained within intron sequences that are spliced out following transcription, or could theoretically reside within the coding sequence of the gene itself. In this latter case the DNA would not only carry the information to specify the sequence of amino acids within the polypeptide product, but also, superimposed on this coding information, include the regulatory information that directs the activity of the gene.

In order for nucleic acid sequences to regulate events post-transcriptionally, they must be present in the RNA transcript. Such sequences, copied from the DNA into the messenger RNA, could mediate their effect at the level of mRNA stability or translational control. Furthermore, once the DNA sequence information is translated into a polypeptide, the sequence of amino acids encoded by the gene may specify additional levels of regulation, such as post-translational processing or protein turnover. Clearly the coordinated complexity of differential gene expression required to generate a plant and enable it to respond to changes in its environment arise out of the existence of a vast array of regulatory elements and multiple levels of control. Advances in molecular biology and associated techniques that permit the isolation and characterization of genes have provided insight into some of the regulatory sequences through which differential gene expression is controlled. These recent studies have in some cases identified specific DNA sequence elements that comprise the molecular switches through which gene expression is modulated.

Molecular Dissection of the Regulatory Elements of Pea *rbcS-3A*

Probably the most intensively studied genes from pea are the *rbcS* genes, which encode the small subunit of RUBISCO (Coruzzi *et al.*, 1984; Timko *et al.*, 1985b; Fluhr *et al.*, 1986a). All five members of the multigene family have been sequenced and their levels of expression established by S1 nuclease protection studies, which relied on sequence differences between the 3' non-coding regions of the five genes (Fluhr *et al.*, 1986a). All five genes are clustered on chromosome 5 (Polans *et al.*, 1985; see also Chapter 2) and encode identical mature polypeptides. The genes are expressed in a coordinate and organ-specific manner, but the contribution of each gene to the steady-state transcript pool ranges from less than 7% to 40%; *rbcS-3A* is the most highly expressed. There are also subtle differences in the relative levels of expression of the five genes in different tissues.

In order to localize and define the regulatory elements that mediate the light-responsive and organ-specific expression of pea *rbcS-3A*, extensive studies have been undertaken using chimeric promoter fusions assayed in transgenic plants. In the absence of a transgenic assay system for pea, the regulatory DNA elements have been studied in heterologous plant species. Some early work was done with transgenic petunia cells (Broglie *et al.*, 1981) and subsequently more extensive studies were undertaken using transgenic tobacco plants (see

Kuhlemeier et al., 1987a; Gilmartin et al., 1991). These studies demonstrated that the regulatory information which specifies light-responsiveness and organ-specificity is located in 410 bp upstream of the start of the coding sequence (Fluhr et al., 1986b; Kuhlemeier et al., 1987b). These initial studies were extended through the analysis of deletion derivatives of the −410 bp upstream sequence. A deletion to −166 retained sufficient regulatory information to confer both organ-specificity and light-responsiveness; a further deletion that removed an additional 17 bp and truncated the promoter to −149 was inactive. This observation highlighted the presence of a critical regulatory element between −166 and −149. These studies in transgenic tobacco plants revealed not only that sufficient regulatory information to specify light-responsiveness and organ-specificity resided in a sequence of only 166 bp 5′ to the transcription start site, but also that the tobacco regulatory machinery could recognize and decipher the regulatory information encoded within the pea gene. This observation argues for some universality in the regulatory elements that modulate light-responsive transcription and demonstrates the power of heterologous expression systems.

Early studies using run-on transcription assays did not differentiate between the contribution of individual members of the *rbcS* family to the overall expression (Gallagher and Ellis, 1982). However, from these studies it was apparent that the regulation of the *rbcS* genes was predominantly transcriptional. Subsequent observations using transgenic tobacco plants demonstrated that the upstream sequences of the *rbcS* genes could make reporter genes respond to light (Herrera-Estrella et al., 1984; Nagy et al., 1985; Fluhr et al., 1986b). These observations confirmed that transcriptional control was the predominant regulatory mechanism in *rbcS* gene expression and demonstrated that sufficient regulatory information for light-responsive and organ-specific expression was located within a very short region of upstream sequence. Further studies, using chimeric promoters derived from *rbcS-3A* upstream regions, demonstrated that the regulatory information which specified the light-responsiveness and organ-specificity of this gene was repeated within the upstream region (Kuhlemeier et al., 1988; Davis et al., 1990). Multiple elements present upstream of the gene can confer similar expression patterns; there is therefore a redundancy of regulatory information. These studies, using upstream fragments of *rbcS-3A* fused to the heterologous cauliflower mosaic virus (CaMV) 35S promoter, as well as to an inactive −50 deleted *rbcS-3A* gene, revealed that sequences located between −410 and −330, between −330 and −166 and between −166 and −50 all contained light-responsive elements (LREs) that could direct light-responsive and organ-specific expression. In addition, a further light-responsive element is located between −50 and +15 of *rbcS-3A* (Kuhlemeier et al., 1989). This last element is too weak to observe unless additional enhancer activity is provided to uncover the action of the molecular light switch. Earlier studies had demonstrated that a 33 bp region spanning the TATA element of pea *rbcS-E9* can confer very low levels of light-responsive expression in transgenic callus tissue (Morelli et al., 1985). There are therefore multiple light-responsive elements involved in modulating *rbcS* gene expression.

The redundancy of function observed within the *rbcS-3A* promoter can be considered in two ways. Identical regulatory elements that mediate these patterns of expression could be located in all of these four regions; alternatively, there may be more than one regulatory element that can specify light-responsive and leaf-specific expression. As we will see, it is a combination of these possibilities that provides the multiple light-responsive elements of pea *rbcS-3A*. Not only are there several distinct light-responsive elements within the upstream regions of *rbcS-3A*, but such regulatory elements can occur more than once.

As the expression of *rbcS* genes is light-responsive and organ-specific, it was of interest to establish whether distinct regulatory elements mediate these two aspects of expression, or whether both responses are mediated by a single switch and are immutably coupled. Most of the upstream light-responsive elements of *rbcS-3A* also mediate leaf-specific expression patterns, but it is clear that light-regulation and organ-specificity can, under certain circumstances, be uncoupled (Simpson et al., 1986; Kuhlemeier et al., 1989; Lam et al., 1989). Such dissociation of light-responsiveness and organ-specificity can also be seen from work on the *Nicotiana plumbaginifolia rbcS-8B* gene (Poulsen and Chua, 1988). In this example, an upstream regulatory element was identified that confers leaf-specificity but is not light-responsive.

Studies of chimeric *rbcS* promoters in transgenic plants have therefore led to the definition of some of the *cis*-acting regulatory elements through which light-responsive transcription is modulated. In parallel to the definition of *cis*-acting regulatory elements, studies to identify the *trans*-acting nuclear proteins that specifically interact with the DNA regulatory elements have been undertaken. Such studies, using nuclear extracts from pea, as well as studies with numerous other plants, have revealed the identity of several nuclear proteins that specifically interact with regulatory elements of *rbcS-3A*.

DNA–Protein Interactions Within the Light-responsive Elements of Pea *rbcS-3A*

From extensive studies of pea *rbcS-3A*, and from comparisons with numerous other light-regulated genes, it is apparent that there are several target sequences for sequence-specific DNA-binding proteins located between -410 and $+1$ of the upstream region (see Gilmartin et al., 1990). Figures 4.2 and 4.3 summarize the functionally defined regions of the upstream regions of pea *rbcS-3A* and indicate the locations of defined factor-binding sites. Some of these conserved DNA sequence motifs are also present upstream of other light-responsive genes, although there appear to be extensive differences between

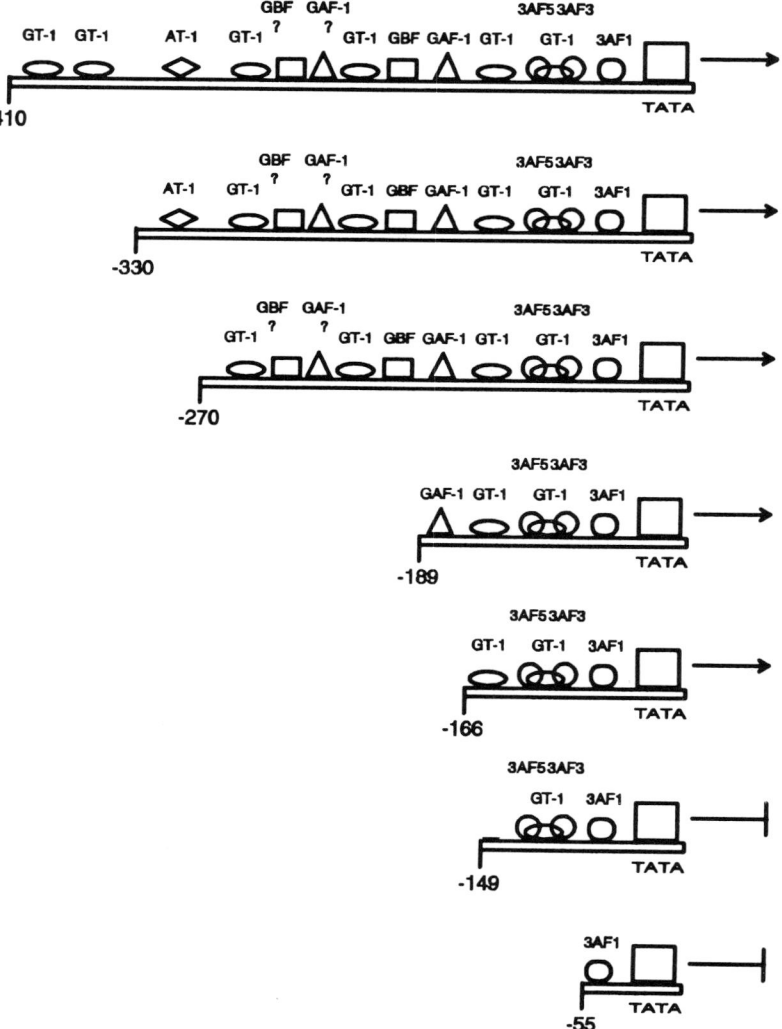

Fig. 4.2. Deletion analysis of pea *rbcS-3A*. The deletion endpoints of *rbcS-3A* used to define the location of light-responsive elements are shown (Kuhlemeier *et al.* 1987b). The nuclear DNA-binding proteins subsequently identified are indicated. Those indicated by a question mark (?) indicate binding sites predicted by homology to defined sequences. The ability of a particular deletion derivative to drive transcription is indicated by an arrow.

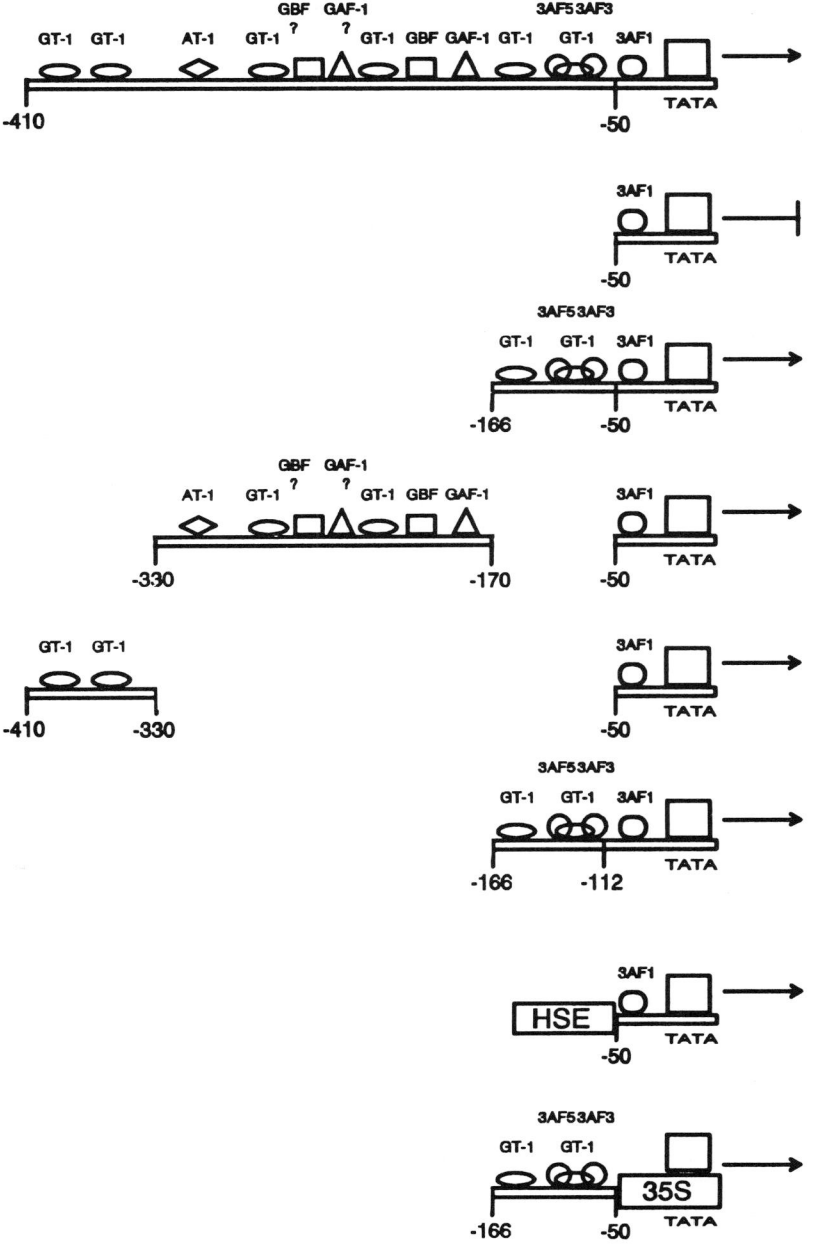

Fig. 4.3. Location of light-responsive elements upstream of pea *rbcS-3A*. The full −410 promoter and the binding sites for defined DNA-binding proteins are shown. The chimeric promoter elements assayed to define the light-responsive elements within this upstream sequence are shown. The heat-shock element (HSE) is shown and the cauliflower mosaic virus 35S promoter is indicated (35S).

the upstream sequence of *rbcS* and *cab* sequences from pea. This observation is a little surprising considering their somewhat similar expression profiles. However, from the studies of *rbcS-3A*, it is apparent that similar expression patterns can be derived from distinct combinations of regulatory proteins and their cognate binding sites. The first such sequence-specific protein-binding sites identified upstream of pea *rbcS-3A* were those for the protein GT-1 (Green *et al.*, 1987). This protein, so named because of its ability to interact with sequences containing GT residues, has been found in several plant species including tobacco; the binding activity of GT-1 is present in plants grown in both the light and the dark.

Analyses of the binding-site specificity of GT-1 have demonstrated that a pair of GT-1-binding sites are located within each of the three upstream elements of *rbcS-3A* shown to be capable of driving transcription from the -50 deleted promoter, namely -410 to -330, -330 to -170, and -166 to -50 (Green *et al.*, 1988). The six *rbcS-3A* GT-1-binding sites all differ slightly in their nucleotide sequence but, despite this, they each apparently interact with the same nuclear protein. This observation suggests that GT-1 can interact with a series of related and yet distinct sequences.

The best characterized of these light-responsive elements is the -166 to -50 region, which confers light-responsive and organ-specific expression when fused to either the -50 *rbcS-3A* promoter or a heterologous CaMV 35S promoter (Kuhlemeier *et al.*, 1989). Within this element are two GT-1-binding sites, box II (GTGTGGTTAATATG) and box III (ATCATTTTCACT) (Green *et al.*, 1987). Further studies on this element demonstrated that the 58-bp region containing boxes II and III was sufficient to confer light-responsive transcription (Kuhlemeier *et al.*, 1987a). Binding-site replacement studies, in which either of the entire box II or box III elements was replaced by a random sequence, revealed a functional role for the two GT-1-binding sites within the -166 deleted promoter (Kuhlemeier *et al.*, 1988). These studies demonstrated that both box II and box III are required for GT-1 binding *in vitro*, and are also required for transcriptional activity *in vivo*. Further studies *in vitro* defined the core binding site of box II as GTGT*GGTTAA*TATG, the middle six nucleotides (in italic) being critical for interaction with GT-1 (Green *et al.*, 1988). Furthermore, mutation of the double GGs, originally identified as critical for binding by methylation interference studies, within the -166 promoter resulted in a dramatic reduction in transcriptional activity *in vivo* (Kuhlemeier *et al.*, 1988).

A correlation between GT-1-binding and transcriptional activity has therefore been established. These studies demonstrated a requirement for two GT-1-binding sites for transcriptional activity, suggesting that interactions between nuclear proteins bound to these elements were essential for the observed transcriptional response. Construction of mutant promoter elements in which the spacing between box II and box III was altered demonstrated that a functional spacing requirement existed between these two elements;

increasing the spacing by only 2 bp resulted in a dramatic reduction in transcriptional activity. However, this change in spacing between the two elements did not affect the affinity of the promoter for GT-1. If the two elements were moved closer together within the promoter by deleting sequences between them, transcriptional activity was not dramatically affected until 10 bp had been removed, suggesting that a functional interaction could still occur between these two GT-1-binding sites up until this point. Since no helical turn dependence on expression was observed, this suggested that proteins other than GT-1 could be involved in mediating the observed transcriptional response through this promoter region (Gilmartin and Chua, 1990a).

Concurrent studies had also established that the contribution of box II and box III to the observed transcriptional response was not equal. A promoter in which box II replaced box III was as active as the wild type; conversely a promoter in which box III replaced box II had a 20-fold reduction in activity *in vivo*. Furthermore, the affinity of GT-1 for each of these elements is significantly different. GT-1 has a much higher affinity for box II than it does for box III. These combined data suggest that the box II and box III elements contribute differently to the transcriptional activation of *rbcS-3A* (Gilmartin and Chua, 1990b).

Detailed studies of the role of box III in this response demonstrated not only that box III is the target site for GT-1, but that two additional proteins interact with the 5' and 3' end of this element. These proteins, 3AF5 (*rbcS-3A* factor 5') and 3AF3 (*rbcS-3A* factor 3'), bind to the sequences AAACTTT*AT-CATT* and *TCAC*TATCTAA respectively, which overlap with the previously defined GT-1-binding site, box III (ATCATTTTCACT); overlapping regions are in italic. The phosphorylation state of these two novel proteins was shown to differ in nuclear extracts prepared from light-grown and dark-adapted plants and the nucleotide sequences within these binding sites were shown to be required for a transcriptional response *in vivo*. These combined observations suggest that the short −166 deleted promoter, whilst much simpler in structure than the full promoter, still retains a number of different DNA regulatory elements that are essential for the transcriptional response to light (Sarokin and Chua, 1992).

From these loss-of-function experiments it is possible to conclude that the binding sites for GT-1, 3AF3 and 3AF5 within the −166 promoter are required for a transcriptional light response. However, it is possible that mutation of a specific element and the subsequent loss of transcriptional activity could be due to the ablation of a purely positive element. Such a positive element could be involved in activating transcription, but not necessarily involved in the act of responding to the light signal transduction pathway. This question was resolved for the GT-1-binding site box II by gain-of-function experiments. In these experiments a synthetic tetramer comprising four tandem box II elements was ligated to a −90 deleted CaMV 35S promoter fused to the β-glucuronidase reporter gene *uidA* (Lam and Chua, 1990). This construct was introduced into transgenic tobacco and the chimeric constructs were assayed for their ability to respond to light. These studies demonstrated that addition of a tetramer of GT-1-binding sites to a promoter that was inactive in leaves resulted in light-responsive expression in chloroplast-containing cells. The GT-1 binding site box II is therefore a regulatory component of the light-responsive element and not merely a quantitative element.

The isolation of cDNA sequences that encode a protein with binding specificity for the box II element provides a possible route through which to address the role of this protein in the modulation of light-responsive transcription. The cDNA sequences were isolated from a tobacco cDNA expression library by South-Western screening (Gilmartin *et al.*, 1992; Perisic and Lam, 1992). This procedure relies on the ability of a recombinant protein, expressed in bacterial cells from a cDNA sequence, to interact with its radiolabelled DNA-binding site in a filter binding assay. These studies identified a cDNA that encodes a DNA-binding protein with similar sequence specificity to nuclear GT-1. In addition, studies with antibodies demonstrated that the recombinant and nuclear proteins were antigenically related. The predicted sequence of this protein indicates that the DNA-binding motif is of an unusual helix-helix-turn-helix configuration. Interestingly, the only known related protein sequence is that for the rice GT-binding protein GT-2, which interacts with the rice phytochrome promoter and appears to be involved in dark-dependent expression of this promoter (Dehesh *et al.*, 1990). The finding that related polypeptide sequences are present within DNA-binding proteins implicated in up-regulation of *rbcS* genes in the light and up-regulation of phytochrome genes in the dark demonstrates the complex nature of, and subtle differences in, the regulatory machinery through which light-responsive transcription is controlled. The presence of GT-1 in both the light and the dark and the constitutive expression of the gene encoding it suggest that its role in light-responsive transcription may arise through interactions with other proteins, or through post-translational modification of the protein for specific activity in the light.

One further sequence-specific DNA-binding protein that interacts with pea *rbcS-3A* and has been characterized in some detail is the tobacco protein 3AF1 (*rbcS-3A* factor 1) (Lam *et al.*, 1990), which binds to a region between -51 and -31 of *rbcS-3A*. This region, which is directly adjacent to the TATA box, is contained within the -55 to $+15$ region previously defined as containing a light-responsive element. However, studies in transgenic tobacco plants with a tetramer of the 3AF1-binding site fused upstream of a -90 CaMV 35S promoter (linked to the gene encoding β-glucuronidase, *uidA*), resulted in expression in most cell types, and in both the light and the dark. These data therefore suggest that the binding site for 3AF1 may play a role in basal expression of *rbcS-3A* but is unlikely to contribute to the light-responsive modulation of transcription. A tobacco cDNA sequence encoding a protein with specificity for the 3AF1-binding site has been isolated. This cDNA encodes a putative Zn-finger protein that appears by Northern and Southern blotting analyses to be encoded by a small multigene family.

A tobacco DNA-binding protein GAF-1 (GA binding factor 1) has been identified that interacts with a sequence motif present between nucleotides −184 and −176 of pea *rbcS-3A* (Gilmartin *et al.*, 1990). This motif (ATGATAAGG) is highly conserved in virtually all light-responsive genes and appears to be a member of the GATA family of sequences. These GATA elements occur frequently in the upstream regions of *rbcS* and *cab* genes. Several proteins have been identified that can interact with such sequences; GAF-1 is one such example, but was originally noted as being distinct from these other GATA-binding proteins by its presence in nuclear extracts prepared from the leaves of light-grown, as opposed to dark-adapted, plants. Whilst the basis for this differential binding is not yet clear, it is reminiscent of a binding protein LRF-1 from *Lemna* that shows similar light–dark binding differences (Buzby *et al.*, 1990). This observation may suggest a regulatory role for GAF-1 in light-responsive gene expression. The binding specificity of GAF-1 appears distinct from several other GATA-binding proteins that have been identified, suggesting the existence of families of distinct GATA-binding proteins with related binding specificities. The role of the GAF-1-binding site within the *rbcS-3A* promoter has not been defined, but it is interesting to note that mutation of this sequence within the *Arabidopsis rbcS-1A* promoter results in a dramatic reduction in transcriptional activity (Donald and Cashmore, 1990).

An additional DNA-binding activity that has been shown to interact with the upstream regulatory elements of pea *rbcS* genes is GBF (G-box factor), the G-box-binding protein (Giuliano *et al.*, 1988). This protein, first identified in tomato and *Arabidopsis* nuclear extracts, was shown to interact with the G-box (CACGTG) element within pea *rbcS3.6*. Subsequent studies of this DNA-binding activity have led to the identification of similar binding sites in numerous genes, both light-responsive and light-insensitive, from a range of plant species. The role of this element in all these different contexts has not been defined. Several cDNA sequences that encode related GBF proteins have been identified from a range of species (Schindler *et al.*, 1991, 1992). These observations again suggest the presence of a small multigene family of related DNA-binding proteins. Studies *in vitro* have shown that the binding-site specificity of these different proteins is determined by the sequences flanking the core CACGTG motif (Williams *et al.*, 1992). It is therefore likely that the role of a G-box element in a particular gene will be dependent not only on the sequence context within which the core element is located, but also on which member of the GBF family can interact with that element in that context.

Studies with pea nuclear extracts have revealed a DNA-binding protein termed AT-1 (after its ability to interact specifically with an AT-rich sequence, AATATTTTTATT, upstream of the pea *rbcS3.6* gene). This protein can also interact with similar elements in other genes (Datta and Cashmore, 1989). The particularly interesting feature of this DNA-binding protein is the modulation of its binding activity by phosphorylation. AT-1 can interact with its target sequence only in a dephosphorylated form; phosphorylation of AT-1 results in a dramatic reduction in its ability to interact with its target sequence. This modulation of binding activity contrasts with that observed for 3AF3, the protein that interacts with the −166 *rbcS-3A* promoter, since in the case of 3AF3 it is dephosphorylation of the protein that leads to a loss of DNA binding (Sarokin and Chua, 1992).

These studies of DNA-binding proteins that have specificity for the light-responsive elements of the pea *rbcS* gene demonstrate that multiple DNA–protein complexes can form between sequence motifs and numerous sequence-specific DNA-binding proteins. Some of these sequences have been implicated in specific roles for *rbcS* gene expression; a role for others has yet to be defined. The example given by the pea *rbcS* genes demonstrates the complexities of transcriptional regulation. The observations also introduce the possibility that putative transcriptional regulators are themselves post-translationally regulated, such that their ability to interact with the *cis*-acting regulatory elements of the *rbcS* genes can be modulated. This provides a clear example of how multiple controls can be exerted over the transcription of a specific gene. There must be an alternative means of regulation to transcriptional control at some point in a regulatory cascade. If a transcriptional regulator is itself transcriptionally regulated, it is difficult to imagine how differential control could be generated. These observations also provide a possible insight into the way in which a signal transduction pathway can terminate on a specific DNA element and activate transcription, through the biochemical modification of a regulatory protein. When the multiplicity of transcriptional controls is considered in parallel to the possibilities, with some genes, of post-transcriptional, translational and post-translational control, a picture of immense complexity and yet delicate balance emerges.

Future Prospects

From the above it is apparent that *cis*-acting sequences that mediate various developmental and environmental responses can be dissected to reveal the individual regulatory elements responsible for a specific aspect of a complex expression pattern. These types of molecular analyses, using cloned genes and assays of their mutant derivatives in transgenic plants, have provided information on the DNA sequences that comprise the molecular switches through which plant genes are regulated. Through the complexities of coordinated and differential gene regulation, plants possess the ability to develop to a predefined plan and yet maintain the plasticity to respond to rapid and dramatic changes in their environment.

Studies of several genes isolated from pea, most notably those encoding the small subunit of RUBISCO, the chlorophyll-a/b-binding proteins (Simpson *et al.*, 1986; Arguello *et al.*, 1992), glutamine synthetase (Tingey *et al.*, 1987; Brears *et al.*, 1991), asparagine synthetase (Tsai and Coruzzi, 1990), ferredoxin (Elliot *et al.*, 1989; Dickey *et al.*, 1992), plastocyanin (Dupree *et al.*, 1991) and phytochrome (Sato, 1988; Komeda *et al.*, 1991), have contributed significantly to our understanding of molecular aspects of plant gene regulation. I have discussed several aspects of gene regulation, but have focused primarily on transcriptional control in relation to leaf-specific and light-responsive expression of pea *rbcS* genes.

An understanding of the mechanism through which genes are turned on and off is a prerequisite to manipulating plants for agricultural purposes. The introduction of novel genes into plants, either to alter their metabolism, their developmental programmes or their responses to environmental stimuli, or to provide resistance to pests and pathogens, requires that the foreign genes be expressed at appropriate times and in appropriate tissues. In order to achieve this, suitable regulatory elements are required that will direct gene expression in the desired manner. Such information can be gained by analysing and characterizing the regulatory elements from a wide range of plant genes. Such studies are essential to determine not only what regulatory sequences are required for developmental or environmental regulation, but also at what level in the hierarchy of regulatory mechanisms they act.

From studies using pea genes in heterologous transformation systems, it is clear that similar transcriptional regulatory mechanisms exist in transgenic hosts such as petunia and tobacco. Many of the underlying principles of gene regulation are common to different plant, and even non-plant, species. It therefore follows that the regulatory elements used to drive the expression of a modified or foreign gene in pea need not necessarily originate from peas. On the other hand, some studies using heterologous transgenic systems have demonstrated differences in the expression patterns of introduced genes compared with other plants. Examples of divergent expression patterns in different transgenic hosts include those conferred by the CaMV 35S promoter in transgenic petunia and tobacco plants (Benfey and Chua, 1990), and by pea ferredoxin in transgenic tobacco (Gallo-Meagher *et al.*, 1992). Such differences necessitate the detailed analysis of tissue-specific and environmentally responsive regulatory elements used to express transgenes at specific times and places.

From our increased understanding of the complex regulatory DNA elements through which gene expression is modulated it should be possible to generate synthetic regulatory sequence elements that direct gene expression in novel patterns. Some such studies have already demonstrated the modular nature of promoter elements and the ability to construct chimeric sequences whose activity is derived from a combination of the unique elements. One such example is the introduction of the as-1 element derived from the CaMV 35S promoter into the pea *rbcS-3A* gene (Lam *et al.*, 1989). The as-1 element is the target sequence of the nuclear factor ASF-1, initially identified in whole-cell extracts from pea. The as-1 element was implicated as the root-specific regulatory element of the CaMV 35S promoter by mutational analysis. Introduction of this short 21-bp DNA sequence into the leaf-specific and light-responsive pea *rbcS-3A* promoter resulted in a chimeric promoter that retained light-responsiveness in leaves and yet lost leaf-specificity. The root-specificity conferred by the as-1 element was superimposed on the previous leaf-specific expression of *rbcS-3A* such that the gene was active in leaves and roots (Lam *et al.*, 1989).

Similar studies from the combination of a heat-shock element (HSE) with the *rbcS-3A* upstream sequence generated a novel expression pattern whereby the heat-shock response of this chimeric promoter was rendered light-responsive and organ-specific (Kuhlemeier *et al.*, 1989). These studies with pea *rbcS-3A* demonstrate modular aspects of promoter architecture. However, extensive studies on the regulatory elements of the CaMV 35S promoter indicate that the individual *cis*-regulatory elements can provide novel and unexpected expression patterns depending on their combination. It is clear that many cellular processes are common to all cell types and that the products of certain genes are required by all cell types at some stage of their lives. However, the large number of distinct cell types within a plant, with explicit roles and functions, also demands the presence of cell-specific gene products.

The regulatory code is not simply additive, it is also interactive. It is this non-additive information that will require the greatest effort to unravel. However, an understanding of the *cis*-regulatory elements that modulate gene expression at the level of transcription and beyond should provide the tools with which to combine modular regulatory elements to construct sequences for the expression of foreign genes in novel combinations of cell types, or in response to combinations of environmental stimuli. For the genetic manipulation of pea, and indeed of any crop plant, it will be necessary to have not only a reliable and efficient transformation system (see Chapter 10), but also a detailed understanding of the regulatory DNA sequences that provide the exquisite precision of differential gene regulation.

References

Arguello, G., Garcia-Hernandez, E., Sanchez, M., Gariglio, P., Herrera-Estrella, L. and Simpson, J. (1992) Characterization of DNA sequences that mediate nuclear protein binding to the regulatory region of the *Pisum sativum* (pea) chlorophyll a/b binding protein gene AB80: identification of a repeated heptamer motif. *Plant Journal* 2, 301–309.

Bedbrook J.R., Smith, S.M. and Ellis, R.J. (1980) Molecular cloning and sequencing of cDNA encoding the precursor to the small subunit of chloroplast ribulose-1, 5-bisphosphate carboxylase. *Nature* 287, 692–697.

Benfey, P.N. and Chua, N.-H. (1989). Regulated genes in transgenic plants. *Science* 244, 174–181.

Berry, J.O., Breiding, D.E. and Klessig, D.F. (1990) Light-mediated control of translational initiation of ribulose-1,5-bisphosphate carboxylase in amaranth cotyledons. *Plant Cell* 2, 795–803.

Brears, T., Walker, E.L. and Coruzzi, G.M. (1991) A promoter sequence involved in cell-specific expression of the pea glutamine synthetase *GSA3* gene in organs of transgenic tobacco and alfalfa. *Plant Journal* 1, 235–244.

Broglie, R., Bellemare, G., Bartlett, S.G., Chua, N.-H. and Cashmore, A.R. (1981) Cloned DNA sequences complementary to mRNAs encoding precursors to the small subunit of ribulose-1,5-bisphosphate carboxylase and a chlorophyll a/b binding polypeptide. *Proceedings of the National Academy of Sciences USA* 78, 7304–7308.

Buzby, J.S., Yamada, T. and Tobin, E.M. (1990) A light-regulated DNA binding activity interacts with a conserved region of a *Lemna gibba rbcS* promoter. *Plant Cell* 2, 805–814.

Coruzzi, G., Broglie, R., Edwards, C. and Chua, N.-H. (1984) Tissue-specific and light-regulated expression of a pea nuclear gene encoding the small subunit of ribulose-1,5-bisphosphate carboxylase. *EMBO Journal* 3, 1671–1679.

Datta, N. and Cashmore, A.R. (1989) Binding of a pea nuclear protein to promoters of certain photoregulated genes is modulated by phosphorylation. *Plant Cell* 1, 1069–1077.

Davis, M.C., Yong, M.-H., Gilmartin, P.M., Goyvaerts, E., Kuhlemeier, C., Sarokin, L. and Chua, N.-H. (1990) Minimal sequence requirements for the regulated expression of *rbcS-3a from Pisum sativum* in transgenic tobacco plants. *Photochemistry Photobiology* 52, 43–50.

Dean, C., Favreau, M., Bond-Nutter, D., Bedbrook, J. and Dunsmuir, P. (1989) Sequences downstream of translation start regulate quantitative expression of two petunia *rbcS* genes. *Plant Cell* 1, 201–208.

Dehesh, K., Bruce, W.B. and Quail, P.H. (1990) A *trans*-acting factor that binds to a GT-motif in a phytochrome gene promoter. *Science* 250, 1397–1399.

Dickey, L.F., Gallo-Meagher, M. and Thompson, W.F. (1992) Light regulatory sequences are located within the 5′ portion of the Fed-1 message sequence. *EMBO Journal* 11, 2311–2317.

Donald, R.G.K. and Cashmore, A.R. (1990) Mutation of either G box or I box sequences profoundly affects expression from the *Arabidopsis rbcS-1A* promoter. *EMBO Journal* 9, 1717–1726.

Dupree, P., Pwee, K.-H. and Gray, J.C. (1991) Expression of photosynthesis gene-promoter fusions in leaf epidermal cells of transgenic tobacco plants. *Plant Journal* 1, 115–120.

Elliot, C.R., Dickey, L.F., White, M.J. and Thompson, W.F. (1989) *Cis*-acting elements for light regulation of pea ferredoxin I gene expression are located within transcribed sequences. *Plant Cell* 1, 691–698.

Ellis, R.J. (1981) Chloroplast proteins: synthesis, transport and assembly. *Annual Review of Plant Physiology* 32, 111–137.

Fluhr, R., Moses, P., Morelli, G., Coruzzi, G. and Chua, N.-H. (1986a) Expression dynamics of the pea *rbcS* multigene family and organ distribution of the transcripts. *EMBO Journal* 5, 2063–2071.

Fluhr, R., Kuhlemeier C., Nagy, F. and Chua, N.-H. (1986b) Organ-specific and light-induced expression of plant genes. *Science* 232, 1106–1112.

Gallagher, T.F. and Ellis, R.J. (1982) Light-stimulated transcription of genes for two chloroplast polypeptides in isolated pea leaf nuclei. *EMBO Journal* 1, 1493–1498.

Gallo-Meagher, M., Sowinski, D.A. and Thompson, W.F. (1992) The pea ferredoxin 1 gene exhibits different light responses in pea and tobacco. *Plant Cell* 4, 383–388.

Gasch, A., Hoffman, A., Horikoshi, M., Roeder, R.G. and Chua, N.-H. (1990) *Arabidopsis thaliana* contains two genes for TFIID. *Nature* 346, 390–394.

Gilmartin, P.M. and Chua, N.-H. (1990a) Spacing between GT-1 binding sites within a light-responsive element is critical for transcriptional activity. *Plant Cell* 2, 447–455.

Gilmartin, P.M. and Chua, N.-H. (1990b) Localization of a phytochrome responsive element within the upstream region of pea *rbcS-3A*. *Molecular and Cell Biology* 10, 5565–5568.

Gilmartin, P.M., Sarokin, L., Memelink, J. and Chua, N.-H. (1990) Molecular light switches for plant genes. *Plant Cell* 2, 369–378.

Gilmartin, P.M., Memelink, J. and Chua, N.-H., (1991) Dissection of the light-responsive elements of pea *rbcs-3A*. In: Thomas B. and Johnson, C.B. (eds), *Phytochrome Properties and Biological Action*, NATO asi series vol. 50. Springer-Verlag, Berlin, pp. 141–155.

Gilmartin, P.M., Memelink, J., Hiratsuka, K., Kay, S.A. and Chua, N.-H. (1992) Characterization of a gene encoding a DNA binding protein with specificity for a light-responsive element. *Plant Cell* 4, 839–849.

Giuliano, G., Pichersky, E., Malik, V.S., Timko, M.P., Scolnik, P.A. and Cashmore, A.R. (1988) An evolutionarily conserved protein binding sequence upstream of a plant light-regulated gene. *Proceedings of the National Academy of Sciences USA* 85, 7089–7093.

Green, P.J., Kay, S.A. and Chua, N.-H. (1987) Sequence-specific interactions of a pea nuclear factor with light-responsive elements upstream of the *rbcS-3A* gene. *EMBO Journal* 6, 2543–2549.

Green, P.J., Yong, M.-H., Cuozzo, M., Kano-Murakami, Y., Silverstein, P. and Chua, N.-H (1988) Binding site requirements for pea nuclear protein factor GT-1 correlate with sequences required for light-dependent transcriptional activation of the *rbcS-3A* gene. *EMBO Journal* 7, 4035–4044.

Herrera-Estrella, L., Van Den Broek, G., Maenhaut, R., Van Montagu, M., Schell, J., Timko, M. and Cashmore, A.R (1984) Light-inducible and chloroplast-associated expression of a chimeric gene introduced into *Nicotiana tabacum* using a Ti plasmid vector. *Nature* 310, 115–120.

Jefferson, R.A., Kavanagh, T.A. and Bevan, M.W. (1987) Beta glucuronidase as a sensitive and versatile fusion marker in higher plants. *EMBO Journal* 6, 3901–3910.

Komeda, Y., Yamashita, H., Sato, N., Tsukaya, H. and Naito, S. (1991) Regulated expression of a gene-fusion product derived from the gene for phytochrome I from *Pisum sativum* and the *uidA* gene from *E. coli* in transgenic *Petunia hybrida*. *Plant Cell Physiology* 32, 737–743.

Kuhlemeier, C. (1992) Transcriptional and post-transcriptional regulation of gene expression in plants. *Plant Molecular Biology* 19, 1–14.

Kuhlemeier, C., Green, P.J. and Chua, N.-H. (1987a) Regulation of gene expression in higher plants. *Annual Review of Plant Physiology* 38, 221–257.

Kuhlemeier, C., Fluhr, R., Green, P. and Chua, N.-H. (1987b) Sequences in the pea *rbcS-3A* gene have homology to constitutive mammalian enhancers but function as negative regulatory elements. *Genes and Development* 1, 247–255.

Kuhlemeier, C., Cuozzo, M., Green, P., Goyvaerts, E., Ward, K. and Chua, N.-H. (1988) Localization and conditional redundancy of regulatory elements in *rbcS-3A*, a pea gene encoding the small subunit of ribulose-bisphosphate carboxylase. *Proceedings of the National Academy of Sciences USA* 85, 4662–4666.

Kuhlemeier, C., Strittmatter, G., Ward, K. and Chua, N.-H. (1989) The pea *rbcS-3A* promoter mediates light responsiveness but not organ specificity. *Plant Cell* 1, 471–478.

Lam, E., and Chua, N.-H. (1990) GT-1 binding site confers light-responsive expression in transgenic tobacco. *Science* 248, 471–474.

Lam, E., Benfey, P.N., Gilmartin, P.M., Fang, R.-X. and Chua, N.-H. (1989) Site-specific mutations alter *in vitro* factor binding and change promoter expression pattern in transgenic plants. *Proceedings of the National Academy of Sciences USA* 86, 7890–7894.

Lam E., Kano-Murakami, Y., Gilmartin, P.M., Niner, B. and Chua, N.-H. (1990) A metal-dependent DNA-binding protein interacts with a constitutive element of a light-responsive promoter. *Plant Cell* 2, 857–866.

Millar A.J., Short, S., Chua, N.-H. and Kay, S.A. (1992) Firefly luciferase as a reporter of regulated expression in higher plants. *Plant Molecular Biology Reporter* 10, 320–328.

Morelli, G., Nagy, F., Fraley, R.T., Rogers, S.G. and Chua, N.-H. (1985) A short conserved sequence is involved in the light-inducibility of a gene encoding ribulose-1, 5-bisphosphate carboxylase small subunit of pea. *Nature* 315, 200–204.

Nagy, F., Morelli, G., Fraley, R.T., Rogers S.G. and Chua, N.-H. (1985) Photoregulated expression of a pea *rbcS* gene in leaves of transgenic plants. *EMBO Journal* 4, 3063–3068.

Perisic, O. and Lam, E. (1992) A tobacco DNA binding protein that interacts with a light-responsive box II element. *Plant Cell* 4, 831–838.

Polans N.O., Weeden, N.F. and Thompson, W.F. (1985) Inheritance, organization and mapping of the *rbcS* and *cab* multigene families in pea. *Proceedings of the National Academy of Science USA* 82, 5083–5087.

Poulsen, C., and Chua, N.-H. (1988) Dissection of 5′ upstream sequences for selective expression of the *Nicotiana plumbaginifolia rbcS-8B* gene. *Molecular and General Genetics* 214, 16–23.

Sarokin L.P. and Chua, N.-H. (1992) Binding sites for two novel phosphoproteins, 3AF5 and 3AF3, are required for *rbcs-3A* expression. *Plant Cell* 4, 473–483.

Sato, N. (1988) Nucleotide sequence and expression of the phytochrome gene in *Pisum sativum*: differential regulation by light of multiple transcripts. *Plant Molecular Biology* 11, 697–710.

Schindler, U., Ecker, J.R. and Cashmore, A.R. (1991) An *Arabidopsis thaliana* leucine zipper protein that binds to the G-box promoter sequences. In: Thomas, B. and Johnson, C.B. (eds), *Phytochrome Properties and Biological Action*, NATO asi series vol. 50, Springer-Verlag, Berlin, pp. 157–165.

Schindler, U., Menkens, A.E., Beckmann, H., Ecker, J.R. and Cashmore, A.R. (1992) Heterodimerization between light-regulated and ubiquitously expressed *Arabidopsis* GBF bZIP protein. *EMBO Journal* 11, 1261–1273.

Simpson, J., Schell, J., Van Montagu, M. and Herrera-Estrella, L. (1986) Light-inducible and tissue specific pea *lhcp* gene expression involves an upstream element combining enhancer-and silencer-like properties. *Nature* 323, 551-554.

Timko, M.P., Kausch A.P., Hand, J.M., Cashmore, A.R., Herrera-Estrella, L., Van Den Broeck, G. and Van Montagu, M. (1985a) Structure and expression of nuclear genes encoding polypeptides of the photosynthetic apparatus. In: K.E. Steinbeck, S. Banitz, C.J. Amtzen and L. Bogorad (eds), *Molecular Biology of the Photosynthetic Apparatus*. Cold Spring Harbor Laboratory Press, Cold Spring Harbor, New York, pp. 381-396.

Timko, M.P., Kausch A.P., Castresana, C., Fassler, J., Herrera-Estrella, L., Van Den Broeck, G., Van Montagu, M. and Cashmore, A.R. (1985b) Light regulation of plant gene expression by an upstream enhancer-like element. *Nature* 318, 579-582.

Tingey, S.V., Walker, E.L. and Coruzzi, G.M. (1987) Glutamine synthetase genes of pea encode distinct polypeptides which are differentially expressed in leaves, roots and nodules. *EMBO Journal* 6, 1-9.

Tsai, F.-Y. and Coruzzi, G.M. (1990) Dark-induced and organ-specific expression of two asparagine synthase genes in *Pisum sativum*. *EMBO Journal* 9, 323-332.

Williams, M., Foster, R. and Chua, N.-H. (1992) Sequences flanking the hexameric G-box core CACGTG affect the specificity of protein binding. *Plant Cell* 4, 485-496.

Genetic and Developmental Analysis of the Seed

5

T.L. WANG AND C.L. HEDLEY
John Innes Institute, Colney Lane, Norwich, NR4 7UH, UK

Seed Development

Seed growth analysis

In general, the processes leading to the development of the mature seed are less well researched than those concerned with germination. The main reason for this disparity is the difficulty in producing the large numbers of synchronously developing seeds required to study growth and development compared with the availability of uniform samples of dry mature seeds. Peas, however, have been a favoured species with regard to both germination and developmental studies and a reasonable amount of information has accumulated on both aspects. Unfortunately, much of the data on pea seed development have been obtained without any regard to the genotypes used and the environments in which the plants have been grown. Researchers have used many different cultivars and genotypes with the result that few pieces of work can be satisfactorily correlated. Despite the comment of Dure (1975) that 'the development pattern of legume seed formation is so similar among the species studied, so noncontradictory, that it is possible to diagram the sequence of many of the major events', we shall see that there is much useful variation even within *Pisum*. Such schemes remain, nevertheless, useful guides but no more since they cover only 'major' events. As more is learnt, variation between genotypes can be identified at most levels of analysis.

Differences in the pattern of seed development have been reported that may be genetic, environmental or due to an interaction between these two components. Carr and Skene (1961) interpreted growth data derived by Bisson and Jones (1932) and McKee *et al.* (1955) on two *Pisum sativum* varieties,

Dwarf Telephone and Canner's Perfection, as having biphasic growth patterns, each growth phase separated by a 'lag' phase. This lag occurred at $c.$ 80% and 90% of their maximum fresh weights respectively (920 mg and 470 mg). In contrast, lags in the development of two *Pisum arvense* lines, New Zealand Maple (Burrows and Carr, 1970) and Black-eyed Susan (Flinn and Pate, 1968), occurred relatively early in development. In both of these cases it was suggested that the lags correspond to the time when the endosperm disappears and the embryo completely fills the embryo sac (see later). In other growth studies, e.g. on Victory Freezer (Bain and Mercer, 1966) and Progress No. 9 (Frydman *et al.*, 1974), no apparent lags in development were found.

In each of these studies the plants were grown in different environments and there was little attempt to minimize or measure the variation between plants or seed samples. In addition, no reference was made to the presence of mutations in any major genes which may affect seed development, such as those at the *rugosus* loci (see later). One of the first studies to take into account possible variation between plants and to rigorously control the environment throughout development of the plants was reported by Eeuwens and Schwabe (1975), who used the variety Alaska. Plants were grown in a controlled environment and only a single seed from each of 50 pods of similar age was taken for each point of the analysis. The growth pattern was interpreted as a double sigmoid curve separated by a lag phase, the lag correlating with a decline in the growth of the testa and with the filling of the embryo sac by the embryo.

Using the biphasic growth curve as a basic model for seed growth and information from the literature (e.g. Bisson and Jones, 1932; Rowan and Turner, 1957; Turner *et al.*, 1957; Bain and Mercer, 1966; Flinn and Pate, 1968; Smith, 1973), Pate (1975) developed a general schematic representation of the cardinal events during pea seed maturation (Fig. 5.1). As with the data on the pattern of seed growth, there was no uniformity either in environment or in genotype between these reports and so the resulting scheme is only a general statement for the species. In this scheme Pate suggested that storage product synthesis is not initiated in the cotyledons until the phase of cell division is virtually complete. Some distinction was made in the timing of these events between genotypes characterized as either *P. arvense* or *P. sativum*, which were thought to be different species but are now all included within *P. sativum*. The problems mentioned above about correlating information from different sources, however, did not allow true genetic differences in the pattern of pea seed development to be highlighted.

Genetic variation for seed growth

The first attempt to compare the developmental patterns of a range of pea genotypes was reported in 1980 by Hedley and Ambrose. In this study the seed from three round-(RR) and three wrinkled-(rr) seeded accessions from the John Innes gene bank were compared using plants grown in a controlled environment maintained at 15°C day/night and 16-hour days. The seed size of the

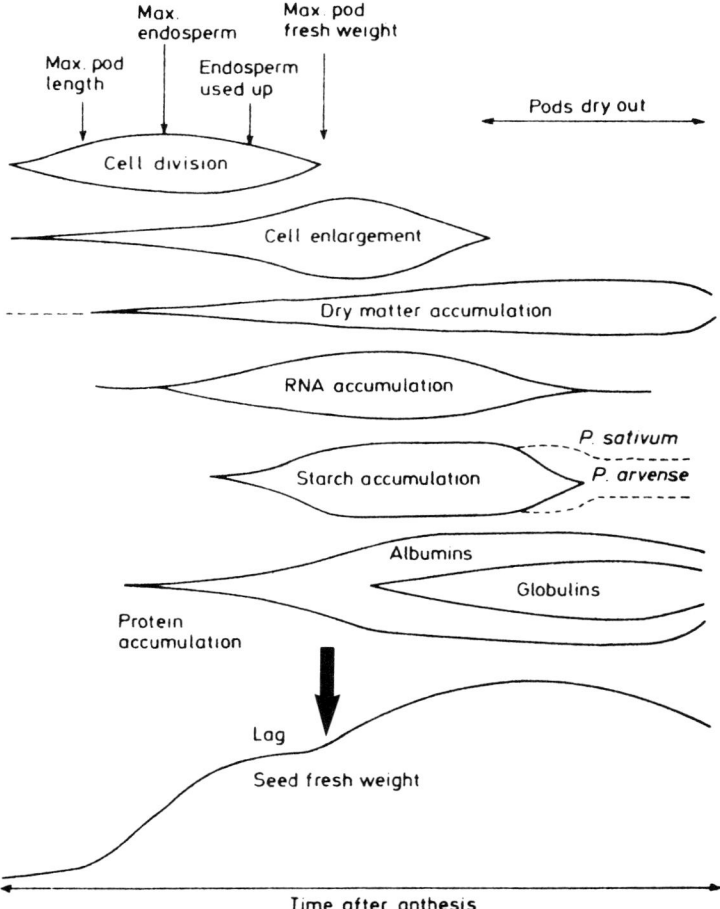

Fig. 5.1. Schematic representation of the cardinal events in maturation of *Pisum* species. The common time-scale is the growth curve for seed fresh weight. All events depicted refer to rates within the developing cotyledons, divergent lines denoting periods of increasing rate, convergent lines times of decreasing rate. Differences between *P. sativum* and *P. arvense* are indicated. (This represents variation within the species, *P. sativum*, since the two are now considered the same species; see Chapter 1.) The solid arrow marks the approximate time when seeds of *P. sativum* are harvested for quick freezing. (Redrawn from Pate (1975), with permission of Cambridge University Press.)

genotypes ranged from 125 mg to 400 mg dry seed weight and contained both cultivars (Alaska and Trapper) and other lines. To reduce variation within a genotype, a strictly controlled growth system was adopted where shoot growth was restricted to the main stem and only four pods were allowed to set. Only three central seeds from the second flowering node were chosen and replication was achieved by using five to ten plants for each sampling date. Three rapid phases of seed growth separated by two lag phases were identified. Genetic variation was found for the slopes of the growth phases and for the timing of

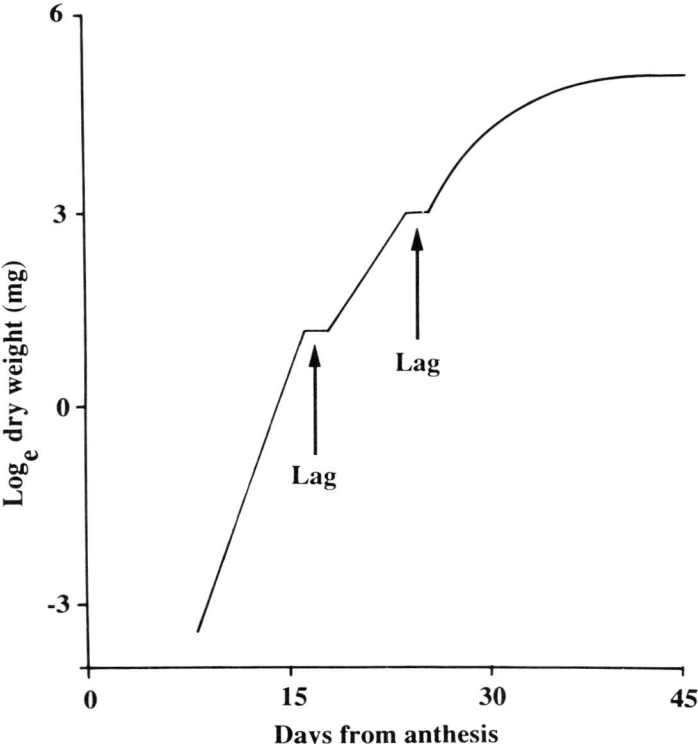

Fig. 5.2. Seed growth curve of *Pisum sativum* cv. Trapper showing two lags (arrowed) separating the three phases. (Based on data from Hedley and Ambrose, 1980.)

the lags. The initial growth phase of the seeds could be accounted for by changes in the testa and endosperm, the absolute growth rates of both reaching a maximum at the end of this phase. The first lag in seed growth corresponded to a rapid decline in the growth of the testa and endosperm. The second rapid growth phase related to very high relative rates of growth for the embryos. The time when the level of the liquid endosperm declined to a minimum and the embryo made contact with the testa related to the second lag phase. The third phase of rapid seed growth was due mainly to development of the embryo, the rate finally declining as the seed matured and began to dry (Fig. 5.2).

The complexity observed within this study may explain some of the problems in correlating information between the earlier seed growth analyses described above, in particular, the differences in the timing of the lag phases. It is evident that some studies observed the first and others the second of the two phases of reduced growth.

As well as providing a model for pea seed growth, Hedley and Ambrose

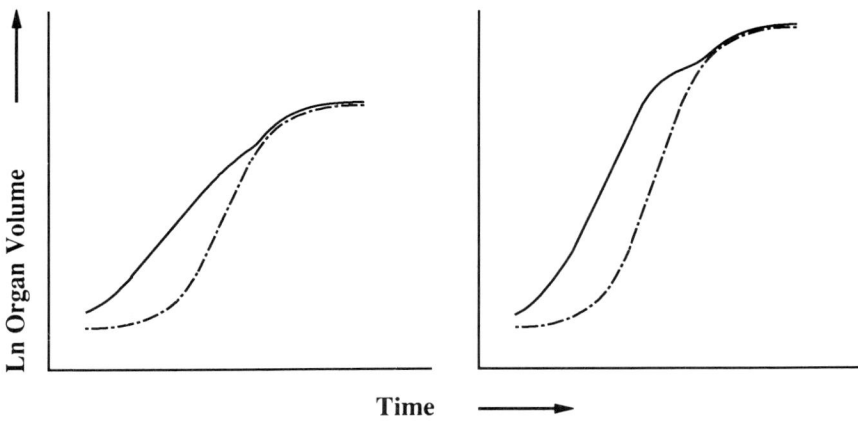

Fig. 5.3. Diagrammatic representation of the relationship between the growth of the embryo sac (solid line) and embryo (broken line) in hypothetical small-seeded (left) and large-seeded (right) pea lines.

(1980) demonstrated that final seed size was positively correlated with the maximum volume of endosperm. This in turn was dependent on the relationship between the growth rates of the embryo and the embryo sac. In large-seeded genotypes the relative rates of growth of the embryo sac and embryo were similar, while in small seeds they were very different (Fig. 5.3). This model can be used to explain the seed size effect resulting from reciprocal crosses between large- and small-seeded lines (Davies, 1975). In such crosses the reciprocal F_1 hybrid seeds resemble the maternal parent in size, even though the embryos from the two crosses are genetically identical (Fig. 5.4).

Maternal/embryo interactions

It is not known how the development of the testa influences the size of the developing embryo and, in particular, the number and size of cells in it at maturity (see later). What is known, however, is that all the nutrients reaching the developing embryo must pass through the vascular system within the testa. A model for the transfer of nutrients from the mother plant to the developing embryo has been proposed for soyabean (Thorne, 1980) and can probably be applied to pea. This model suggests a symplastic route for nutrients from the mother plant to the sieve tubes of the testa. Nutrient flow is then dependent on diffusion, when the endosperm is present, and apoplastic transport through transfer cells on the inner layer of the testa and outer layer of the embryo, once the endosperm has been absorbed. Although there is not a continuous symplastic path between the maternal tissue and the embryo, the mother plant still controls the flow of nutrients up to the point where it enters the

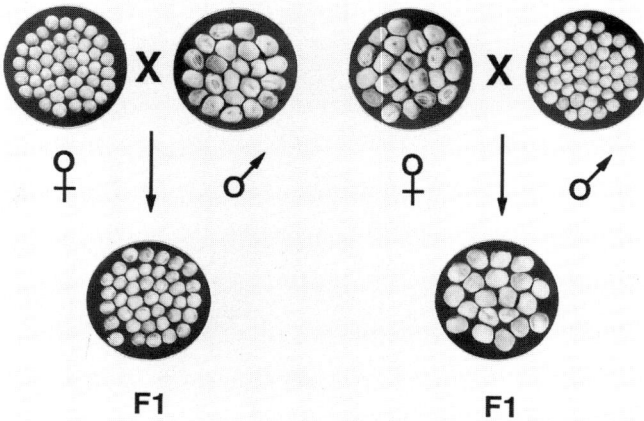

Fig. 5.4. A reciprocal cross between two genotypes of pea (cvs. Trapper and Maro) showing that the F_1 resembles more closely the size of the maternal parent.

cotyledons. The maternal tissues including the testa can affect, therefore, the rate of development of the embryo and possibly the quantity, quality and rate of deposition of the storage products, starch, protein and lipid.

Many of the studies relating to the interaction between testa and embryo have used an 'open seed' approach (Wolswinkel and Ammerlaan, 1984). This entails surgical removal of the embryo from within the embryo sac of developing seeds while still within the growing pod. An embryo sac 'cup' is left connected to the pod wall by the funiculus. The endosperm within this 'cup' is then replaced with buffer and the transfer of nutritional or other components determined by chemical analysis of the buffer solution after a period of time. Using this technique it has been shown that osmotic pressure (Wolswinkel and Ammerlaan, 1984), potassium (Wolswinkel and Ammerlaan, 1985) and a sulphhydryl-modifying reagent (PCMBS; Wolswinkel, 1985) affect the transport of sugars and amino acids into the embryo sac.

The testa has a positive role in the transfer of nutrients to the embryo. It has been shown that amino acids are metabolized in the testa cells (Murray, 1987; Rochat and Boutin, 1991) and that arginine is synthesized in large amounts prior to transfer to the developing embryo (Rochat and Boutin, 1992). Likewise, starch is accumulated early in the development of the testa for later remobilization to the developing embryo, the process corresponding to high activities within the testa of sucrose synthase and ADP glucose pyrophosphorylase, enzymes involved in sucrose–starch metabolism (Rochat and Boutin, 1992).

It is only very recently that genetic variants have been used to investigate the interaction between the testa and embryo during seed development. The *rugosus* loci, *r* and *rb*, discussed in detail later, are known to affect the content

Table 5.1. Effect of *rugosus* loci on the starch content of seed tissues.

Genotype	Embryo		Testa	
	%	mg	%	mg
RRRbRb	12	3	20	7
rrRbRb	6	2	16	6
RRrbrb	8	1	4	1
rrrbrb	5	1	4	1

Starch data are in percent organ dry weight or total weight per organ. Measurements were obtained from seeds at the point of maximum accumulation of starch in the testa.

and composition of starch in the dry seed. Recently, however, it has been shown that a mutation at the *rb* locus substantially reduces the amount of starch accumulated in the testa during development, whereas a mutant at the *r* locus appears to have little effect on the testa (C.Rochat, personal communication; Table 5.1); the double mutant, *rrrbrb*, has a similar effect to *rb*. Reciprocal crosses between the *rb* mutant (*rbrb*) and the wild type (*RbRb*) will give seeds which have genetically identical embryos (*rbRb* and *Rbrb*) but either mutant or wild type testas; such seed can be used to examine the role of the testa in more detail.

In addition to mutants affecting the embryo, there are those that affect testa characters: for example, testa thickness (*ep1*, *ep2*), attachment of the funiculus to the pod wall (*def*), size of the hilum (*Him*, *ar*) and seed shape (*di*, *mifo*, *foe*) (Hedley and Wang, 1987). The effect of these mutations on the development of the embryo, or on the interaction between the testa and embryo, is not known. An experimental approach using embryo and testa mutants combined with reciprocal crosses is a powerful analytical tool for understanding this interaction and the role of the testa in seed development.

Embryo Development

The development of the pea embryo, like the seed, can be divided into stages in a similar way to that undertaken in many other species (e.g. cotton; Dure, 1975). However, such staging systems have been rather empirical and often based on gross morphological changes rather than any underlying processes common to embryo development in many species. Hence, it is frequently difficult to make comparisons with other species. More recently, molecular markers have been isolated that can be associated with different steps in development (e.g. in cotton; Dure, 1985; Hughes and Galau, 1989), although the majority are linked to maturation or germination. Hopefully, as such

Embryo morphogenesis

The embryo of *Pisum sativum* L. develops in a typical solanad manner (Johri, 1984). The fertilized egg divides transversely into two approximately equal apical and basal (nearest the micropyle) cells. The apical cell divides again, the upper cell forming the embryo proper while the lower cell forms the neck of the suspensor; the remainder of the suspensor is formed from the original basal cell, which divides again to produce a five-cell proembryo (Cooper, D.C., 1938). The suspensor cells become elongated and multinucleate by a series of synchronous nuclear divisions, and are believed to channel all the nutrients to the young embryo. The evidence for this, however, is circumstantial (Natesh and Rau, 1984; Raghavan, 1986; Johri *et al.*, 1992) in that the bases of the suspensor cells contain numerous convolutions which are believed to increase the surface area for absorption of nutrients, rather like transfer cells (Gunning and Pate, 1969; Pate and Gunning, 1972). At later stages, the abaxial surface of the cotyledons and the inner surface of the testa become transfer cells and possess similar morphology. Early on, they too contain invaginations into the endospermic cytoplasm (Marinos, 1970). The nutrient flow into the embryo has to proceed via these transfer processes since there is no cytoplasmic continuity between embryo and mother plant, as mentioned earlier. Only the suspensor and embryo contain plasmodesmatal connections (Raghavan, 1986). The suspensor of the pea remains relatively small and undistinguished, never reaching the proportions of those in some other legume species such as *Phaseolus* (Yeung and Clutter, 1978). The elongation of the suspensor cells, however, moves the neck cells and embryo into the bulk of the endosperm and away from the micropyle.

The upper cell of the zygote which forms the embryo proper continues to divide to produce a globular embryo (Fig. 5.5A). The neck cells enlarge to form a large central vacuole surrounded by a multinucleate cytoplasm (Fig. 5.5A) but shortly afterward the whole suspensor starts to degenerate. The embryo consists of two main regions at this stage – the central region or ground meristem and an outer zone or protoderm. The ground meristem is believed to give rise to the procambium. Further divisions change the overall structure so that it becomes heart-shaped (Fig. 5.5B) delineating the cotyledons (Reeve, 1948). The direct progenitor cells of these cotyledons are difficult to determine unlike, for example, in rape where the presence of fewer cells with different shapes makes identification easy (Tykarska, 1979). In pea, the cotyledons form the main storage organs of the seed. At the cleft of the heart the shoot meristem will form while vertically below this, at the base of the heart, the

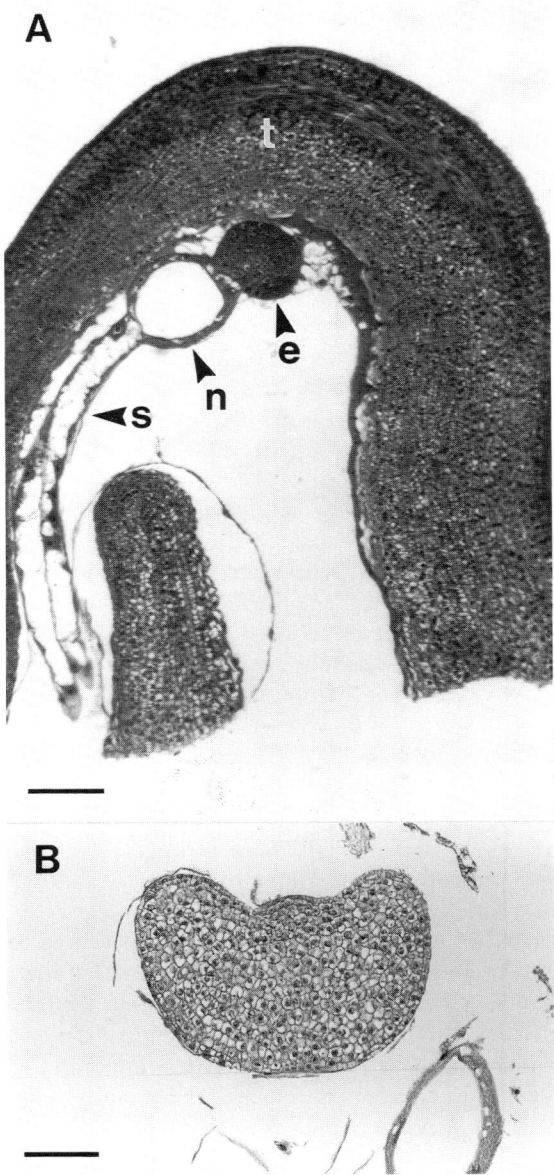

Fig. 5.5. Two stages in the morphogenesis of the pea embryo. (A) A globular-stage embryo (e) attached to suspensor consisting of a bulbous neck cell (n) and two elongated cells (s), all within the testa (t). Bar = 125 μm. (B) Early heart-stage embryo showing incipient cotyledons and acquisition of bipolar symmetry. Bar = 100 μm.

root meristem forms. Reeve (1948) indicated that the apical meristem arises first, by periclinal divisions, frequently in pairs, and a true and full protoderm cannot be distinguished until after this.

The endosperm is of the nuclear type (Johri et al., 1992) and is formed by the fusion of the two maternal polar nuclei from the central cell of the egg sac with the second male gamete nucleus. It is thus triploid. Divisions continue to form a multinucleate, partially cytoplasmic and partially liquid, tissue. It is not cellularized, although there is evidence for cellulose strands bridging testa and embryo (Marinos, 1970). The endosperm will eventually disappear as the embryo grows and, as mentioned above, the outer surface of the latter and inner one of the former will come to lie against each other and differentiate into transfer cells.

The cotyledons grow by cell division and cell expansion, the relative duration of the two processes depending much on genotype (Hedley and Smith, 1985). They consist largely of parenchymatous storage tissues with an outer epidermal layer and internal provascular strands. These strands do not develop sufficiently to be considered true vascular tissue until the pea germinates (Smith and Flinn, 1967). For most of the cotyledons' development, cell divisions are confined to the outer layers of the parenchyma, the epidermis and the provascular areas (Hauxwell et al., 1990). The cotyledons do not become photosynthetic as in *Glycine max* and *Phaseolus vulgaris* and remain underground following hypogeal germination. Divisions, of course, continue during the formation of the root and shoot axes. These will become quiescent once they are well developed. The embryo shoot apex contains several nodes at maturity and, in some early-flowering genotypes of pea (e.g. JI 141), all the vegetative nodes and some floral meristems will be present within the seed. This contrasts strongly with, for example, brassicas such as rape and *Arabidopsis*, where the shoot apex remains as either a simple dome of cells or a dense patch of meristematic cells between the cotyledons (Tykarska, 1979). The major organ-forming events, the embryology of the pea, however, occur very early in development and are essentially complete once the cotyledons have formed and start to expand. Marinos (1970) separated embryo development in pea into 25 developmental stages, with most embryo growth covered by the last three. Stages 23–25, therefore, represent no change in development but rather an increase in growth. The development of the embryo could thus be considered as consisting of two phases: organ formation and organ maturation, where the majority of cotyledon development would be part of the latter and thus a maturation phenomenon.

Cellular development

Although many biochemical and physiological measurements are performed on whole cotyledons, the cotyledon is not a uniform tissue; there are distinct epidermal and transfer cells, regions of elongated cells showing high frequency of division (the provascular tissue) and the storage parenchyma itself, in which the starch, protein and lipid are stored. Even within this parenchyma, the cells are not uniform throughout development. In a detailed analysis of the cell population of a round-seeded genotype (BC1/RR) using image analysis of cell areas, Ambrose *et al.* (1987) showed that there was a change in the population structure during development. Early in development of BC1/RR (0.3 mg fresh weight), the cotyledons consist of a single population of cells that are predominantly the size of dividing cells. As the cotyledon grows (30 mg) the population shifts, still remaining as a single distribution, to one with higher maximum cell sizes. By 100 mg, the population becomes bimodal, a characteristic that becomes more pronounced so that by 400 mg the maximum cell area can be 100 times that of the original population of meristematic-like cells (Fig. 5.6). These large cells are packed with starch. The switch between

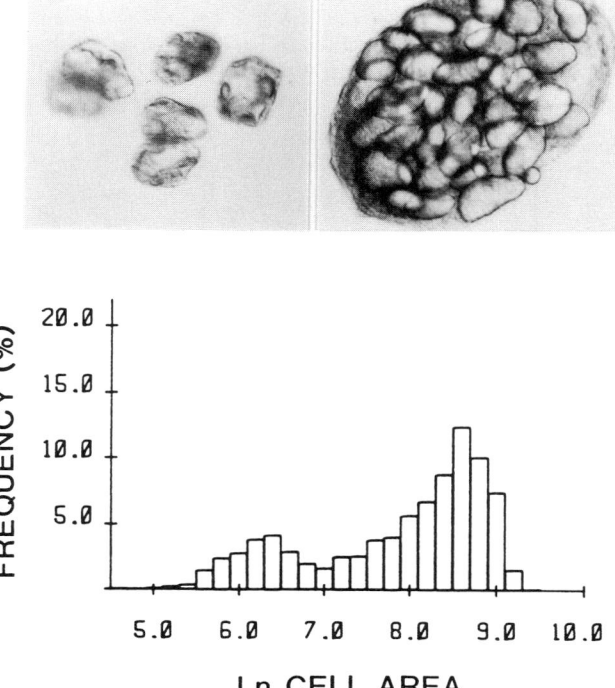

Fig. 5.6. A cotyledonary cell population (at 320 mg fresh weight) showing bimodal distribution of cell areas. Above each population is a photograph of representative cells fixed using chromic acid. (Based on data from Ambrose *et al.*, 1987.)

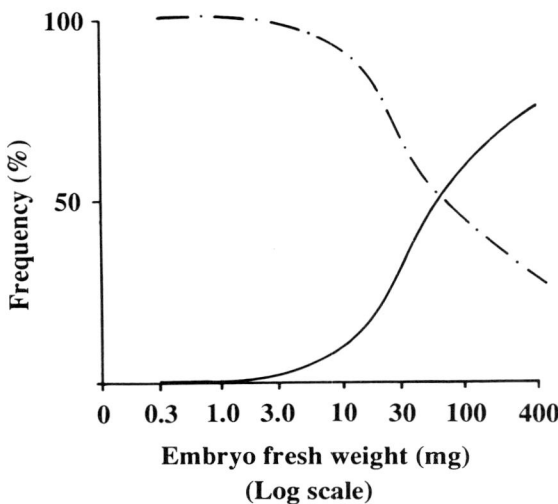

Fig. 5.7. A graph showing the change during embryo development in the two cotyledonary cell populations as a proportion of the total. Population cell areas: broken line = < 1260 μm^2; solid line = > 1260 μm^2. (Based on data from Ambrose et al., 1987.)

the populations, when the large-cell population becomes more significant, occurs at c. 50 mg fresh weight (Fig. 5.7). Measurements of cell size can be used to distinguish two genotypes, since the population structure can differ between the genotypes at given embryo fresh weights, as we shall see in the section on *rugosus* loci. The cell population structure can also provide a marker of the stage of development an embryo has reached or a cellular basis against which to measure, for example, enzyme activities or gene expression.

Simple measurements of cotyledonary volume and cell numbers can be used also to distinguish genetic variation, since the relationship between the two parameters varies considerably between genotypes (Hedley and Smith, 1985). The pea cotyledon develops essentially in two phases reflecting the development of its constituent cells. Initially, the cells will undergo a period of cell division to increase the cell population of the cotyledon. It is this population which will dictate ultimately the size of the seed since there is a linear relationship between cell number and seed dry weight (Davies, 1975). Thus, in general, a small-seeded line has fewer cells than a large-seeded line. The cells then expand, the DNA endoreduplicates (see below) and storage products are deposited. The timing of the switch between cell division and expansion, however, varies between genotypes (Hedley and Smith, 1985). Some genotypes, either small-seeded (e.g. JI 181) (Corke *et al.*, 1987) or large-seeded (e.g. JI 813), have an initial phase which is largely one of cell division and is followed by a clear switch into cell expansion. This type of development is

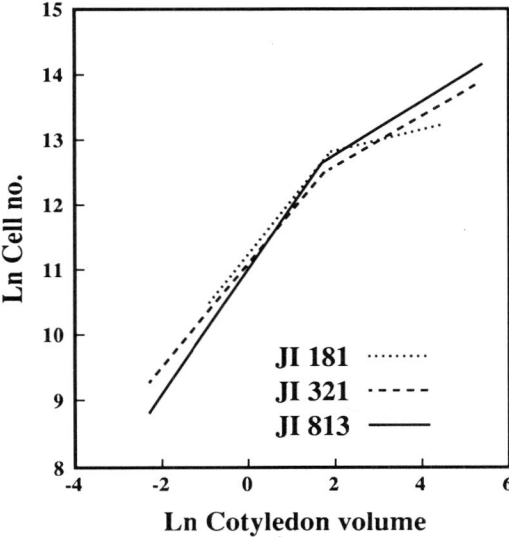

Fig. 5.8. A line diagram illustrating the differences in relationship between cell number and cotyledon size (volume) for three pea genotypes.

easily distinguished from that of a cotyledon where the switch is less distinct (e.g. JI 321; cv. Alaska), by measuring the volume of the cotyledon and the number of cells it contains (Fig. 5.8). In some other types, it is not possible to distinguish phases by such measurements (Hedley and Smith, 1985). It is not known how these developmental patterns relate to the potential of different genotypes to accumulate storage products.

There is a strong influence of maternal tissues on the final size of the seed (Davies, 1975) and on ribosomal RNA content (Davies and Brewster, 1975). Some specific genes are also known to affect seed development through maternal influences. Both Murfet (1985) and Reid (1989) have shown that the genes *ar* (Tedin, 1920) and *n* (Wellensiek, 1925) reduce final seed weights. Both probably operate in different ways, since *nn* plants have a smaller pod lumen which prevents the seeds achieving their potential seed size, whereas *arar* testas have defective hilums. It is considered, therefore, that these genes exert their effects physically by restricting seed inflation, in the case of *n*, or the flow of nutrients to the seed, in the case of *ar*. Such maternal restraints fit with observations in many reciprocal crosses where the seed more closely resembles that of the mother plant (see Fig. 5.4).

Endoreduplication in cotyledons

A peculiarity of many legumes is the vast increase in DNA that occurs in the cotyledon cells via endoreduplication. During development of the cotyledon, the DNA content can rise from the normal diploid level of 2C/4C (the levels in the G1/G2 phases of the cell cycle) to as much as 128C. Peas are no exception to this and have been shown to contain cells up to 64C (Scharpé and van Parijs, 1973; Smith, 1973; Davies, 1976). The increase occurs gradually, rather than at any particular stage, so that cells with low DNA contents are present in the embryo at the same time as those with high values. Scharpé and van Parijs (1973) and Le Gal *et al.* (1984) also noted a gradient in C values across the embryo using isolated sections, the highest values occurring towards the adaxial (inner) surface. During this increase, starch, protein and lipid are laid down in the cotyledons. Although not normally considered as such, the DNA accumulated could have a storage function since the transcriptional activity of, for example, storage protein genes is not related to the increase in available DNA template (Raghavan, 1986) nor is the endoreduplication selective – the whole genome is amplified (Millerd and Whitfield, 1973; Cullis and Davies, 1975; Dhillon and Miksche, 1983).

Cellular development and storage protein deposition

Recent investigations have shown that the accumulation of storage protein is initiated relatively early in cotyledon expansion when cells have a low DNA content. Corke *et al.* (1987) related the DNA content of individual enzymically separated cotyledon cells to their accumulation of the storage protein, vicilin (see Chapter 6). The genotype chosen, JI 181, was one which showed a clear switch from cell division to cell expansion in its cotyledons. Fluorescent and immunocytochemical staining techniques were employed to quantify the DNA, using the fluorochrome 4′,6-diamidino-2-phenylindole (DAPI), and to identify vicilin and the cytoskeletal protein tubulin, by indirect immunofluorescence. In contrast to previous biochemical measurements (Scharpé and van Parijs; 1973; Smith, 1973; Davies, 1976; Le Gal 1984), this technique allowed measurements to be made in the same cell rather than on the average basis per seed, cotyledon or part of a cotyledon. By integrating total fluorescence from the nucleus and relating this to the presence of the storage protein and the state of the cytoskeleton, Corke *et al.* (1987) were able to determine the minimum DNA levels and embryo fresh weight at which storage protein in individual cells could be detected and to identify cells in which deposition occurred. Previous estimates, based on conventional methods, had indicated that storage products including protein were only detectable when cotyledon cells contained DNA levels of 32C (Scharpé and Van Parijs, 1973). The 'triple fluorescence' technique revealed, however, that some cells with DNA contents as low as 5C contained vicilin. By the time the cells had increased their DNA contents to 32C, they all contained vicilin.

Examination of the spatial distribution of cells containing vicilin in serial sections of the pea embryo did not reveal any particular pattern of deposition. It was clear that the deposition was neither synchronous nor apparently confined to any particular region of the cotyledon (Corke *et al.*, 1990a). In such sections, it was not possible to correlate deposition with DNA contents or any other parameter. Extending the studies on these spatial patterns, however, to the detection of storage protein mRNA using *in situ* hybridization, Hauxwell *et al.* (1990) showed very specific patterns of gene expression. Tritium-labelled sense and antisense sequences corresponding to the cDNA clones for a 47,000 M_r vicilin and a 60,000 M_r legumin (Domoney and Casey, 1983, 1985; see Chapter 6) were hybridized to mRNA in sections from young pea embryos. Although the presence of vicilin mRNA could not be detected any earlier than the corresponding protein (i.e. 10 mg fresh weight in JI 181), a distinct region of expression was observed close to the embryonic axis and the adaxial surface of the cotyledon (Fig. 5.9). This zone increased with increasing fresh weight down and across the embryo, so that, by 60 mg fresh weight, the whole of the cotyledons and that part of the axis adjoining the cotyledons showed the presence of vicilin mRNA (Fig. 5.10). A similar, but delayed, 'tear-drop' wave of expression was observed for legumin mRNA. No expression of vicilin and legumin genes was found in the epidermis or the provascular cells of the cotyledon as was reported in earlier studies by Harris *et al.* (1989). These authors, however, only detected a 'rapid rise in [silver] staining' but not a wave of expression. This discrepancy may have been due to the different genotype used by Harris *et al.* (1989) or to a lack of resolution afforded by growth at the high temperature used. The 'wave-like' pattern for vicilin and legumin was similar to that found in soyabean for the Kunitz trypsin inhibitor (KTi3) and a storage protein related to vicilin, β-conglycinin (Perez-Grau and Goldberg, 1989), except for one significant difference – it was in the opposite direction. Hauxwell *et al.* (1990), however, also correlated their pattern with another parameter in the cotyledons, that of cell division. DAPI staining of the nuclei allowed regions of mitotic activity to be discerned in wax sections equivalent to those showing storage protein gene expression. These regions were the provascular tissue, the axes and the distal and abaxial regions of the cotyledons. Hence storage protein gene expression occurred only in regions lacking mitotic activity.

Manipulation of cellular development

One can gain some insights into the relationship between cellular parameters, the expression of storage protein genes and the deposition of their corresponding proteins by manipulating the development of the embryo physically

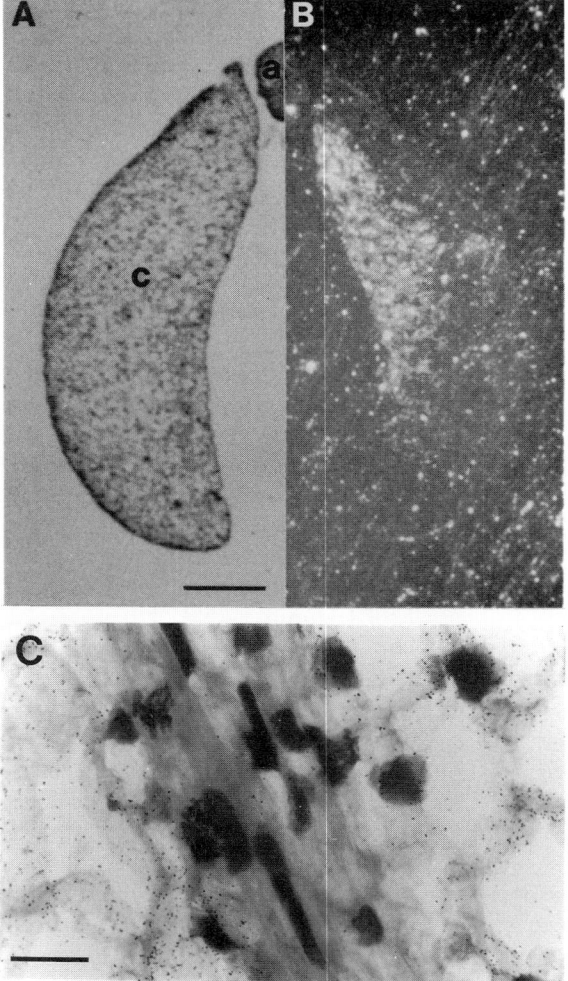

Fig. 5.9. *In situ* mRNA hybridization for storage protein in a 10-mg JI 181 embryo using ^{3}H-labelled antisense RNA corresponding to a 47,000 M_r vicilin cDNA. (A and B) Mirror-images of a cotyledon half (c) with axis (a). (A) Bright-field illumination. (B) Dark-field illumination with silver grains (and hence mRNA) showing as a white area towards inner surface of cotyledon. Bar = 350 μm.
(C) Bright-field illumination of an area of cotyledon across a provascular strand (denoted by broken line). The strand shows several cells in division with metaphase chromosomes. The silver grains (black) only occur over the cytoplasm of the parenchyma cells on either side of the strand. Bar = 19 μm.

Genetic and Developmental Analysis of the Seed

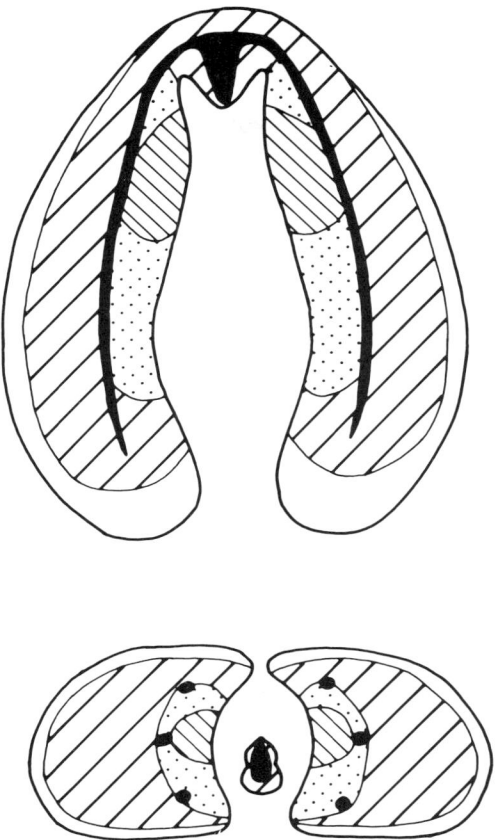

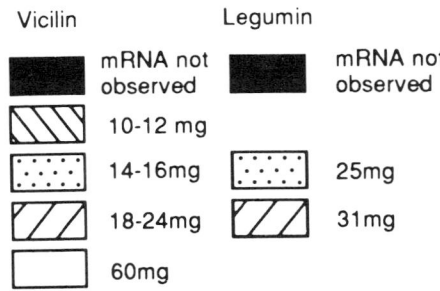

Fig. 5.10. Diagram illustrating the wave of mRNA accumulation in JI 181 corresponding to 47,000 M_r vicilin and 60,000 M_r legumin cDNAs.

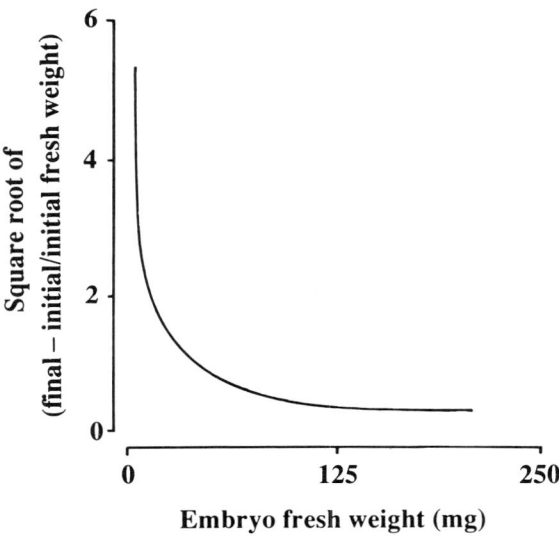

Fig. 5.11. The proportional increase in fresh weight of cultured embryos using a 10% sucrose, glutamine-containing medium (medium III; based on data from Cook *et al.*, 1988).

or genetically (see last section). A technique that has proved particularly useful for manipulating embryo development in the former manner is tissue culture. Whole embryos can be placed in a defined medium where they grow apparently normally, as measured by their morphology and fresh weight increases (Millerd *et al.*, 1975; Stafford and Davies, 1979). In the initial studies of Millerd *et al.* (1975), it was proposed that only vicilin could be accumulated in embryos under *in vitro* conditions, but it was later shown, using more sensitive immunoassay techniques, that legumin also could be produced in culture (Domoney *et al.*, 1980). In this latter investigation, tissue culture appeared to increase the protein content of the embryos.

Following optimization of the culture conditions to produce maximum growth (Cook *et al.*, 1988), it was found that young embryos showed increases in fresh weight far greater than those of more mature embryos (Fig. 5.11). The young embryos could also be readily germinated following culture under these conditions simply by reducing the osmotic pressure of the medium. When embryos from two specific stages in development were grown *in vitro* and *in vivo*, however, it was clear that those *in vitro* did not develop 'normally', since their cell population structures were very different; those from the cultured embryos showed a higher average cell volume with a much larger maximum cell area than those grown on the plant (Table 5.2; Ambrose *et al.*, 1987).

When such tissue culture studies were coupled to a cellular analysis using the small-seeded line, JI 181, it was shown that there was a steady decline in

Table 5.2. Effect of embryo culture on the cell populations of cotyledons.

	Embryo fresh weight (mg)			
	30		100	
Statistic	In vivo	In vitro	In vivo	In vitro
Mean ($\log_e \mu m^2$)	6.92	7.32	7.25	8.06
Standard deviation ($\log_e \mu m^2$)	0.65	0.76	0.87	0.69
Min. cell area (μm^2)	180	190	130	270
Max. cell area (μm^2)	5220	9200	6470	13,470

Data are taken from Ambrose et al. (1987).

the number of mitotic cells following culture and cell division ceased in the cotyledon cells within 60 hours (Corke et al., 1990a). In these cultured embryos, several parameters changed when compared with embryos of equivalent fresh weight *in vivo*: the proportion of cells containing vicilin increased up to threefold in 10 mg embryos; average distances between nuclei of cotyledon cells in sections from cultured embryos increased and the maximum DNA content of the cotyledon cells increased. The minimum level at which vicilin was detected in cells was unchanged, however, at 5C. Furthermore, enlarged cells containing vicilin could also be observed in the embryonic axis – a phenomenon not seen at the same stage *in vivo*. Linking these findings with the initial observation of Ambrose et al. (1987), that average cell areas increase, and of Cook et al. (1988), that proportional growth is greater in young embryos, it appears that the tissue culture conditions were favouring cell expansion over cell division. Thus the increased number of cells depositing protein may be associated with increased cell expansion. This also explains the earlier findings of Domoney et al. (1980), who demonstrated increased amounts of legumin in cultured embryos. The link between cell expansion and endoreduplication, as suggested by Davies (1976), however, still needs to be examined using the DAPI techniques described earlier.

In a detailed analysis of the expression of storage protein genes in embryos grown *in vivo* and *in vitro*, Yang et al. (1990) found that for all three classes of storage protein genes, vicilin, legumin and convicilin (see Chapter 6), the amounts of mRNA were altered on a temporal basis in cultured embryos. Generally, the earlier the embryo was placed in culture, the more storage protein mRNA it contained on a fresh weight basis, which corresponds to data obtained earlier on legumin protein (Domoney et al., 1980). This increased mRNA content in younger embryos was especially marked for legumin and convicilin genes whose mRNAs accumulated at a slower rate than those of vicilin in *in vivo* embryos. Hence, tissue culture favours an increase in the accumulation of storage proteins and their mRNAs. The magnitude of this increase in mRNA would depend on the potential of the embryo to undergo cell expansion. Young embryos with fewer fully expanded cells have a greater potential for cell expansion in culture and thus can accumulate more protein and mRNA. More mature embryos, in contrast, have more cells fully expanded and thus less potential for increases in storage protein or mRNA when compared with embryos from the plant (Yang *et al.*, 1990). On the basis of this hypothesis, culturing can be seen as an ageing process since the changes that occur during normal development are accelerated. This becomes very apparent when one examines the embryonic axis – the accumulation of storage protein in parenchyma cells of the axis occurs at a relatively low fresh weight in culture (Corke *et al.*, 1990a) but only occurs very late in normal development (Craig, 1986).

Although it is clear that the normal development of an embryo is changed when cultured, further modification to its growth and development can be achieved using xenobiotic compounds. In an attempt to break the relationship between DNA content and the deposition of storage protein, taxol (which affects microtubule stability), the auxin 2,4-D, and aphidicolin (a DNA replication inhibitor) were added to the embryo culture medium. Though taxol caused spindle abnormalities and chromosome condensation, and 2,4-D maintained the population of cells with diploid DNA levels, both as anticipated, neither was able to alter the minimum DNA level at which storage protein deposition could be detected. In contrast, endoreduplication was significantly blocked in embryos treated with aphidicolin, and storage protein was observed in cells with DNA contents less than 5C (Corke *et al.*, 1990b). This indicates that endoreduplication and storage protein deposition are concomitant rather than dependent events.

In summary, therefore, the accumulation of storage proteins and their respective mRNAs is linked to cell expansion in pea cotyledon cells, an event which is promoted in tissue culture. The concept of cell size regulating metabolism and development is not new and has been suggested as a general control mechanism for both plants and animals (Fantes and Nurse, 1981). Endoreduplication, however, also a characteristic of expanding cells, occurs concomitantly with protein accumulation and is likely to be a consequence of the expansion (see Davies, 1976). The work of the different authors (Millerd *et al.*, 1975; Domoney *et al.*, 1980; Corke *et al.*, 1990a; Yang *et al.*, 1990) has been carried out on different genotypes, and, as suggested by Davies (1976), no parallel measurements of cell area and DNA content have been made yet in cultured embryos. The universality of the cell expansion hypothesis, therefore, remains to be established.

Genetic Analysis of Development

The above sections describe the use of genetic variants to examine various cellular processes in the embryo such as testa/embryo interactions and the development of cell populations. Genetic analyses can also be used in a more direct approach to understanding morphogenesis, cellular development and the deposition of storage products, as described below.

Rugosus mutants

In his seminal studies, Gregor Mendel (1866) used the wrinkled appearance of some pea seeds as one of the characters he examined. This character, or locus, was later named *rugosus* (gene symbol *r*) by White (1917) from the Latin for wrinkled or shrivelled (see Fig. 5.12). Much later, a second locus, *rb*, causing wrinkled seeds was identified (Kooistra, 1962). These two loci were, until recently, the only ones known to affect the development of the embryo.

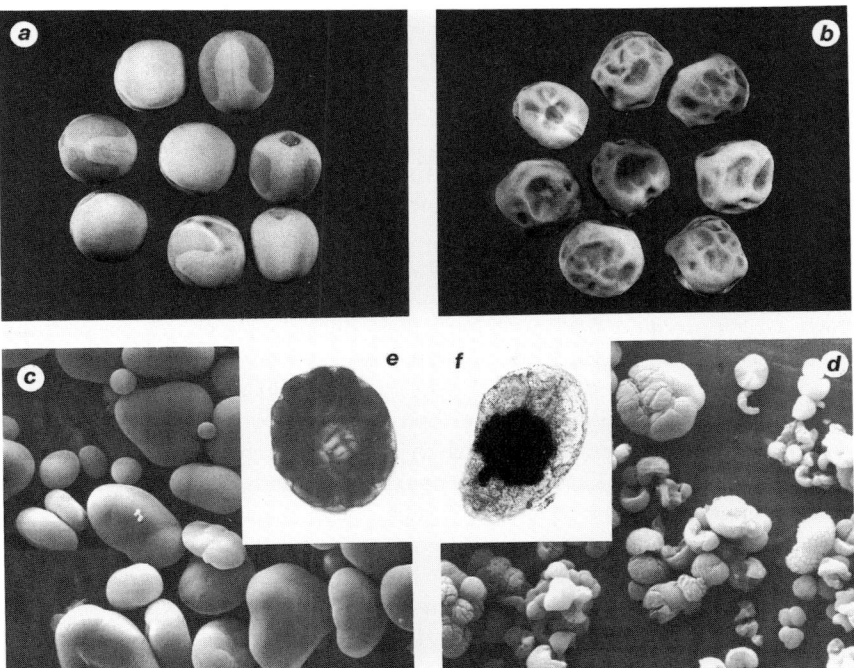

Fig. 5.12. A comparison between round-seeded (*RR*) and wrinkled-seeded (*rr*) near-isolines. Representative mature seeds (**a**, *RR*; **b**, *rr*), scanning electron micrographs of starch grains (**c**, *RR*; **d**, *rr*) and light micrographs of typical cotyledonary storage cells (**e**, *RR*; **f**, *rr*). (From Hedley *et al.* (1986) with permission of Academic Press.)

Mutants at both loci behave as single gene recessives. Other loci that cause wrinkled seeds are known but these affect the development of the testa (Hedley and Wang, 1987). The *rugosus* loci are highly pleiotropic and affect a number of physiological, molecular and developmental processes in the embryo, culminating in a change to the storage product content of the seed. Their importance rests, therefore, not only in their significance to research on seed development but also to commerce, the wrinkled-seeded pea being used for the frozen pea market (see below).

Early studies on round-seeded and wrinkled-seeded pea lines showed the mutant alleles to affect characteristics such as growth and sugar content, the latter proving so desirable to the food industry. These characteristics have been summarized elsewhere (Wang and Hedley, 1991). More recently, however, most investigations have relied on a series of near-isolines (see Chapter 2) that have made it possible to examine the pleiotropy of the loci and determine the true effects of the alleles. The simplest growth character that can be used to distinguish the mutants and wild type is the ratio between the fresh and dry weights of the embryo (Hedley *et al.*, 1986; Wang and Hedley, 1991). Both mutants show a greater water content in their developing embryos than the wild type, the *rr* mutant more so than the *rbrb*. Ambrose *et al.* (1987) showed that the *rr* embryo, once it had grown to 100 mg, possessed a cell population with a higher mean and maximum cell area than that of the *RR*. This difference was maintained until the embryo started to desiccate. Embryos of both *rugosus* mutants also possess a higher osmotic pressure than those of the wild type, which is probably the driving force for the additional uptake of water. It is the loss of this additional water during desiccation that accounts for the wrinkling of the seed.

The storage product content of the lines differs markedly (Table 5.3); the most dramatic effect is on the starch and amylose content. Round-seeded peas have *c*. 22% protein, 55% starch (with 30% of the starch as amylose) and 2% lipid, all on a dry weight basis. In both *rugosus* lines, starch is decreased to *c*. 35% of the dry weight from *c*. 55% in the wild type. The *rr* and *rbrb* mutants have opposite effects on the proportion of amylose in the starch, *rr* increasing and *rbrb* decreasing it, with the double mutants (*rrrbrb*) almost midway between the two. The *rbrb*, but not the *rr*, mutant decreases the absolute amount of amylose. The shape of the starch grain is also affected. The *rr* lines contain compound starch grains (Gregory, 1903; see Fig. 5.12) whereas those of *rbrb* are like the wild type in that they are classed as smooth or simple. However, in the *rbrb* mutant the grains are smaller than the wild type isoline (J. Lloyd, personal communication). The *rr* alleles also override the effect of (i.e. are epistatic to) *rbrb* alleles with respect to grain shape. Thus the double mutant has compound starch grains. Furthermore, there is an effect of these mutations at the *rugosus* loci on the storage proteins, with legumin being much decreased (Davies, 1980; Domoney and Casey, 1985; Perez *et al.*, 1993). It has been suggested, in the case of *rr* lines, that the latter effect is due to a

Table 5.3. Storage product content of near-isogenic lines for the *rugosus* loci.

Genotype	Starch	Amylose	Protein	Lipid
RRRbRb	55	30	22	1.9
rrRbRb	36	65	26	3.5
RRrbrb	36	20	27	4.5
rrrbrb	25	49	28	5.6

Starch data are in percent cotyledon dry weight from the analysis of mature dry seed and amylose data are as a percentage of the starch. Starch was analysed by enzymic degradation followed by measurement of the glucose using Glucose GOD-Perid reagent (Boehringer Mannheim GmbH), amylose by iodine binding, protein by measuring total nitrogen and lipid by fatty acid methyl ester analysis.

lower stability of legumin mRNAs compared with those for vicilins in the high sucrose environment of the mutant embryos (Turner *et al.*, 1990). With increasing dosage of recessive alleles, the lipid content also increases from a basal level of 1.9% to 5.6% in the double mutant (Table 5.3). Most recently, the *R* gene has been shown to encode a starch branching enzyme (SBE I; Bhattacharrya *et al.*, 1990) and there is much evidence to indicate that *Rb* encodes a subunit of ADPG pyrophosphorylase (see Chapter 6).

Hence, the primary lesions of the two *rugosus* loci lie in starch metabolism. The effect on starch was linked through a series of arguments to the final wrinkled seed by Wang and Hedley (1991) (see Fig. 5.13). This 'Jigsaw' was developed to explain the effects of the *r* locus but the principles hold equally well for *rb* on current data. A reduction in the accumulation of starch alters the pool of precursors, increases the embryo sucrose content and leads to an increase in the osmotic pressure in the cotyledons. More water is driven in and the cells expand. The fresh weight to dry weight ratio is thus increased. During desiccation the water is lost and the seed becomes wrinkled. This is, of course, a very simplified scheme, since we have no knowledge of the mechanisms operating between the pieces of the 'Jigsaw'. On the basis of the results of Turner *et al.* (1990) either the increased sucrose content or the increased osmotic pressure of the embryo in some way leads to a decreased stability of specific storage protein mRNA species. This, in turn, brings about a decrease in the synthesis of legumin in *rr* compared with *RR* embryos. The alteration in the pool of starch precursors may also contribute to the increase in lipid (see Table 5.3) in *rr* and *rb* mutants through glucose-6-phosphate, which is common to both starch and lipid synthesis in the plastid (see Chapter 6). This increase is on both a percentage and an absolute basis, with a linear progression from c. 4 mg per seed in *RR* to c. 12 mg per seed for the double mutant, *rrrbrb*.

The 'Jigsaw' illustrates a clear example of the interaction between metabolism and development; a change in primary metabolism bringing about indirect developmental and physiological changes. If such a mechanism is

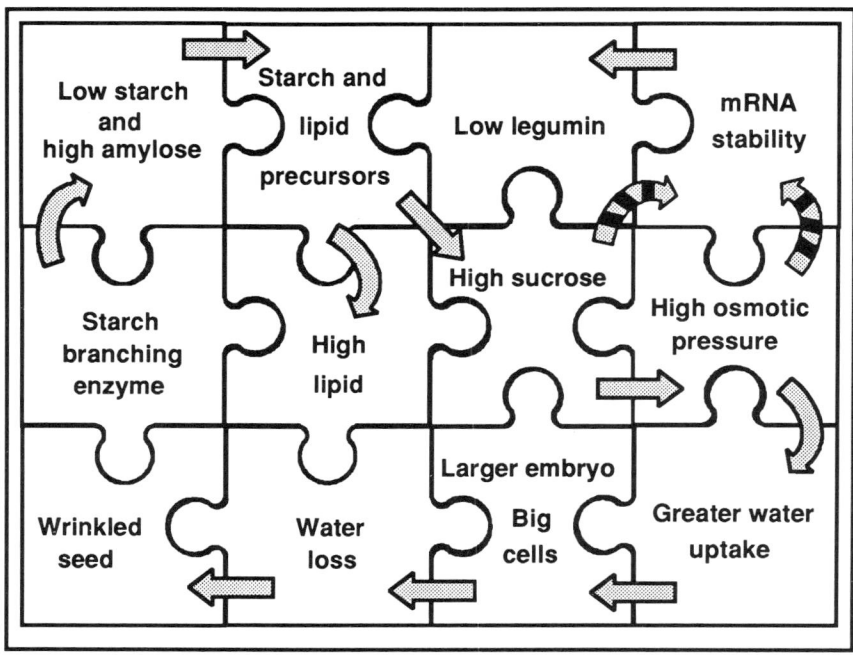

Fig. 5.13. The *r* locus 'Jigsaw' (see text for further details). From Wang and Hedley (1991) with permission of CAB International.

widespread, then the osmotic environment of the embryo will be crucial not only to the development of the embryo, as suggested by Walbot (1978), but also to gene expression and could represent a more universal process for the regulation of gene expression. Certainly, other such interactions have been identified, e.g. the regulation of the patatin promoter by sucrose (Rocha-Sosa *et al.*, 1989). We believe such mechanisms have been largely ignored in developmental biology. There are several corollaries to this model that have been explored before (Wang and Hedley, 1991). The first is that any process that brings about an increase in the osmotic pressure within the cotyledons should induce a wrinkled seed. Thus, it is likely that any decrease in starch production will create a wrinkled seed. It does not follow, however, that all wrinkled seeds will be affected in starch metabolism. Nevertheless, as a simple screen for storage product mutants, wrinkling provides a unique opportunity. Hence in 1986 a mutation programme was initiated to search for additional wrinkled-seeded mutants, to expand both the present range of storage product variation and to obtain mutants altered in development (Wang *et al.*, 1990).

Twenty thousand *RRRbRb* seed were chemically mutagenized and M2

Table 5.4. Complementation groups for the SIM lines in relation to rr and rbrb.

1	2	3	4	5
rr	rbrb			
53	15	1	91	51
54	101	32	201A	52
55	102	41		
56		42		
57		43		
58				
61				

The following SIM lines are unassigned as yet: 11, 14, 16, 59, 71, 81, 103, 103W, 201. Groups 3-5 have been assigned the gene symbols rug-3, rug-4 and rug-5 respectively.

plants screened for the wrinkled-seeded character. Round-seeded sister segregants were then resown to recover wrinkled seeds and determine if the trait was inherited. Segregating populations were chemically analysed to identify mutants affected in storage product content. In initial experiments (Wang et al., 1990), the total mutant variation was classified into a series of groups to indicate major differences between them. Some of the groups clearly showed similarities to the rr and rbrb mutants by virtue of their starch and amylose contents (see below); others were unique. In particular, one group contained very low starch levels and appeared to be amylose-free. In a more complete analysis (Wang and Hedley, 1991), 27 new wrinkled-seeded mutants were identified (termed SIM lines, for 'seed: induced mutant') and this list has been extended slightly since then. Allelism tests between the new mutants and the rr and rbrb isolines have shown that at least three new loci can be distinguished (Table 5.4). These have been assigned the gene symbols rug-3, rug-4 and rug-5 in agreement with the *Pisum* Genetics Association and in keeping with current genetic nomenclature. Furthermore, as is apparent from Table 5.4, additional alleles at both the r and rb loci have been isolated. Examples of each group are shown in Fig. 5.14.

A preliminary analysis of the new alleles at the r locus has revealed new variation for this gene. The rr isoline, derived from Mendel's original mutant, has been shown to have been caused by an insertion into the R gene (Bhattacharrya et al., 1990). The new alleles will probably differ from this line, therefore, since they are likely to be point or deletion mutants based on the predicted mode of action of the chemical mutagens used. The new alleles can be distinguished at the level of storage product content. In material which has been purified through selfing of heterozygotes for six generations, a statistical analysis of starch content showed that three phenotypes could be resolved: the round-seeded line (53.5% starch on a dry weight basis); a low-starch group

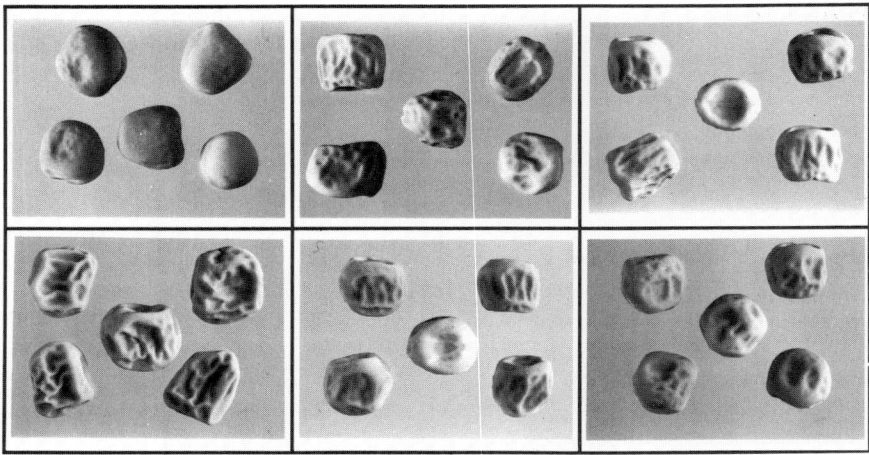

Fig. 5.14. Mature seeds of the new *rugosus* mutants compared with the round-seeded line. Top, left to right - the round-seeded line, *r* locus, *rb* locus; bottom, left to right - *rug-3* locus, *rug-4* locus, *rug-5* locus. (Approx. twice normal size.)

(35.7–37.2%; SIMs 53, 55 and 57) not significantly different from *rr*; and a lower-starch group (26.6–28.0%; SIMs 54, 56, 58 and 61). An analysis of transcripts and proteins produced by the different alleles, by Northern and Western blotting respectively, also revealed that, within the group with the lowest starch, SIM 56 was a null mutant producing neither SBE I mRNA nor protein (MacLeod *et al.*, 1991). Thus it is likely that the group containing SIM 56 all lack SBE I activity.

The *rug-3* alleles represent a very interesting series of mutants. Alleles at this locus have the widest range of starch contents of any group, from virtually no starch in SIMs 32, 42 and 43, through 10% of the wild type level in SIM 41, to 40% of the wild type in SIM 1. It has also been found that there is no iodine-positive response in leaves, demonstrating that the effect of the lesion in the mutants is not confined to the seed. Originally considered starchless and amylose-free (Wang *et al.*, 1990), some of these *rug-3rug-3* mutants have been found, in later generations, to possess low levels of amylose. Using near-isolines, the presence of amylose (c. 15% of the starch) has been confirmed in SIM 1, which bears the weakest allele (C. Harrison, personal communication). These lines may represent mutations in the gene encoding an enzyme in the starch pathway other than SBE I or ADPG pyrophosphorylase and, because of the severity of the mutation, the gene may be one by which to regulate starch synthesis. Starchless mutants are also known in other species and they were all selected on the basis of the iodine leaf test. In tobacco mutants, the enzyme plastidial phosphoglucomutase is deficient

(Hanson, 1990), whereas in *Arabidopsis* there are both plastidial phosphoglucomutase-deficient (Caspar et al., 1985) and ADPG pyrophosphorylase-deficient (Lin et al., 1988) mutants. Both enzymes are currently under investigation with respect to the *rug-3* locus. The effect of the mutants on the level of starch in the seed of these other species has not been reported.

Embryo morphology mutants

The *rugosus* mutants show an effect on embryo and seed development, but embryo morphology is virtually unchanged. The mechanisms underlying the morphogenesis of the embryo cannot be examined in such mutants. In other species such as maize (Clark and Sheridan, 1991) and *Arabidopsis* (Meinke, 1985, 1986, 1991; Mayer et al., 1991), it has proved possible to isolate embryo developmental mutants which may help us gain insights into the basic mechanisms underlying organogenesis. Although many of these are lethal mutants, it is possible to maintain them as heterozygotes. During the mutation programme mentioned above (Wang et al., 1990), therefore, material was screened for abnormal embryos by opening immature pods and seeds. Mutants were chosen if they showed abnormal cotyledon morphology or expansion. Lines segregating for both abnormal and normal embryos were then maintained through their heterozygotes. More recently, non-lethal morphology mutants in *Arabidopsis* have been isolated by screening seedling characteristics (Mayer et al., 1991). In pea, this may prove difficult since, as mentioned earlier, the meristems are relatively well developed in the seed, unlike brassicas.

Of the nine mutants that have been isolated, two broad categories can be distinguished by simple visual observation. The first are those with cotyledon abnormalities that appear as fused or single cotyledons; the second lack full expansion of the cotyledons. In some of the latter, the axis is unaffected but in others, the whole embryo shows abnormal growth. Examples of each are shown in Fig. 5.15. Preliminary analysis of serial sections has shown that cotyledonary cell expansion is affected in at least two mutants. In one of these, the cotyledon and hypocotyl cells are so grossly abnormal that cell walls are difficult to distinguish (Fig. 5.16A and 5.16C in comparison with 5.16D) and polyploid mitotic figures can often be observed (Fig. 5.16B). Such mutants should prove invaluable for examining the initiation of organogenesis, the development of cotyledons and the relationship between cellular development and the deposition of storage products.

Redesigning the Seed

Although many elements of the model for seed development proposed by Pate (1975) are still valid, our comprehension of this process is now much wider

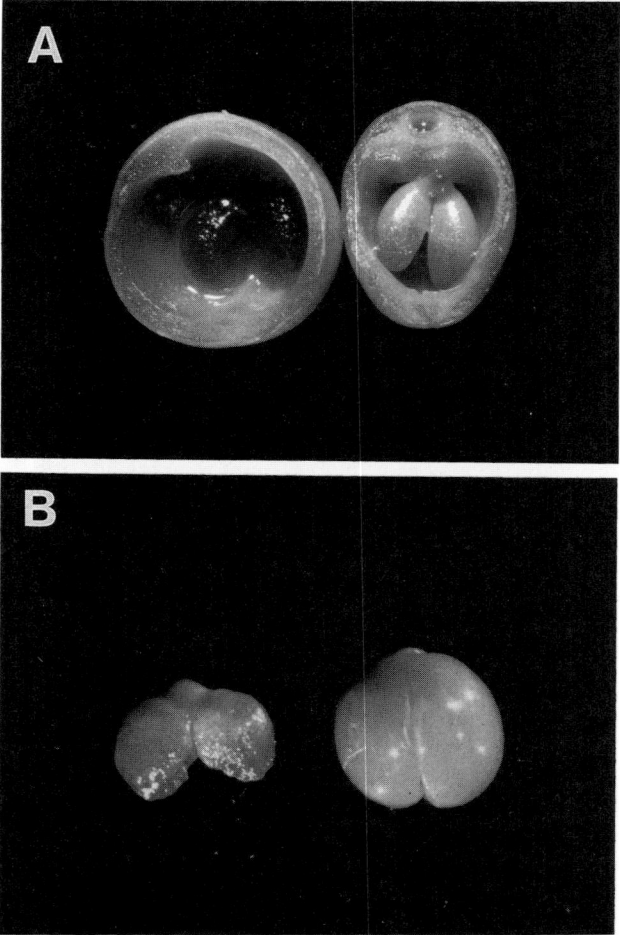

Fig. 5.15. Two embryo morphology mutants of pea. (A) E2748 mutant (left) and a normal seed embryo similar size (right) from a heterozygous plant. The seeds have been cut open to reveal the embryos inside the seed coat. Note the difference is size of the seeds for the same size of embryo. (B) E1735 mutant (left) and normal (right) embryos removed from seeds in the same pod on a heterozygous plant.

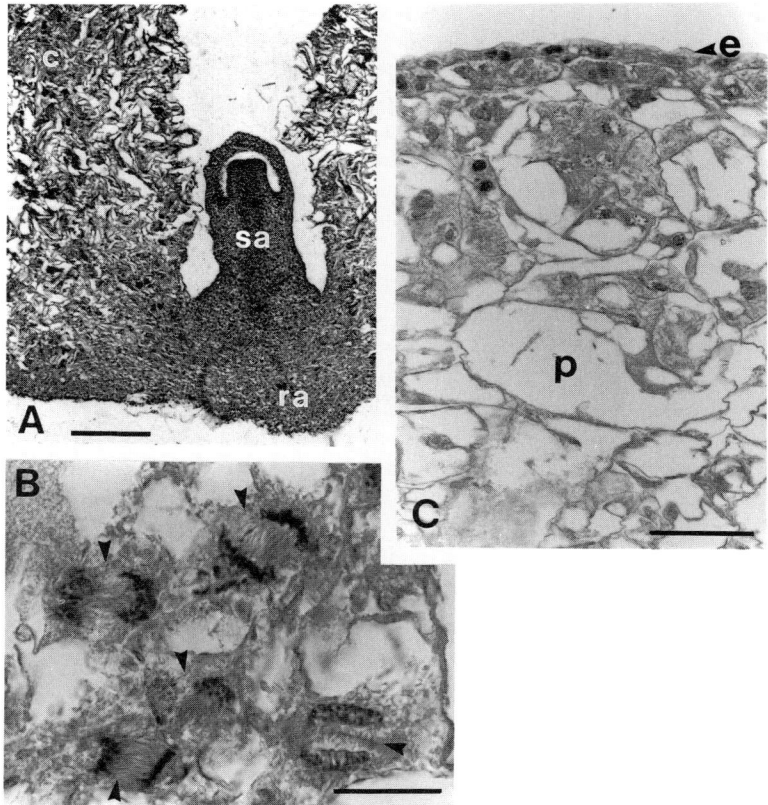

Fig. 5.16. Detail of embryo mutant E1735. (A) Section showing root and shoot axes (ra and sa respectively) and cotyledons (c); bar = 300 μm. (B) A cotyledonary parenchyma cell showing multiple nuclear divisions (arrowed); bar = 20 μm. (C) Section of cotyledon with epidermal layer (e) and underlying parenchyma tissue (p) to demonstrate irregular cell structure (see D) in both layers; bar = 50 μm. (D) Line diagram to illustrate the normal structure of the cotyledon, with epidermal layer (e) and underlying parenchyma tissue (p) (to compare with C above).

and more detailed. We are aware of genetic variation for most of the characters defining embryo growth as well as for the interaction between the embryonic and the maternal tissues comprising the seed, the effect of the latter not being covered by Pate's model. Armed with this information one can develop a new understanding of how the pea seed develops and an appreciation of the possibilities for modifying the composition of the seed to suit a wider range of uses.

The final size and often the final shape of the seed are greatly influenced by the genotype of the testa, a maternal tissue. This can be seen best from the difference between reciprocal crosses mentioned earlier. How the testa exerts this influence is little understood but it is known that the testa affects the growth and development of the embryo and, in particular, its final cell population structure. It is well established also that the testa can regulate all the nutrients entering the developing embryo because there is no continuous symplastic route from the mother plant to the embryo. It is not surprising, therefore, that the interaction between the maternal and filial tissues is a major contributory factor to the complex growth curves observed for developing peas.

The other major contributory factor to seed growth is the genotype of the embryo. It is apparent that the separation of embryo development into discrete division and expansion phases for cotyledon cells is simplistic. For a given size of embryo, there is genetic variation for the proportion of the overall cell population that is mitotic. There is also genetic variation for the final number and size distribution of cells in the mature seed. The recently identified mutants, mentioned earlier, that are affected in their cellular development will hopefully allow us to probe the mechanisms controlling these cellular changes.

During development, cotyledon cells appear to progress through a preprogrammed sequence from division to expansion and then maturation, which includes the synthesis and accumulation of storage products and the endoreduplication of the genome. An understanding of the underlying mechanisms in the relationship between cellular development, endoreduplication and storage product deposition will provide information, in the long term, for us to improve the capacity of cotyledons to store materials. At present, the extent to which storage product accumulation is influenced by differences in cell populations is not known. Nevertheless, it is clear that models that consider cotyledon development in pea as a series of distinct phases where the cells are synchronized or entrained are not tenable; there is much variation and each cell acts as a storage unit, behaving more like an individual rather than part of a synchronized population.

Although genetic variation has been described for many of the organ or cellular components that determine the development of the seed, very little is known of the inheritance of such characters. On the other hand, major genes have been identified for individual seed components, in particular those encoding specific storage proteins and those affecting starch synthesis (Chapter 6). In many cases, the role of these genes in controlling steps in biosynthetic pathways has been determined and the genes have been cloned and sequenced. Some genes, such as those at the *r* and *rb* loci, control specific steps in a pathway, while, in addition, affecting the overall development of the seed. The pleiotropic effects of these loci extend to modifying the synthesis of legumins, one of the major groups of storage proteins (see Chapter 6). The combination of genetic variation for seed development and known major genes affecting storage product synthesis provides us with a wealth of opportunities for manipulating the size, shape and chemical composition of the pea seed. There are possibilities for improving existing uses of the seed as well as for opening up new markets by radically changing the seed storage products.

At present, a major niche for the dried pea crop in Western industrialized countries is as a protein source in compounded animal feed. The crop would be more acceptable for this purpose if the protein content of the seed could be increased and the protein quality altered (see Chapter 6). A link between the cessation of cell division and the initiation of storage protein synthesis has been demonstrated. There is also genetic variation for the cellular composition of cotyledons during embryo development. It may be possible, therefore, to increase protein content by manipulating the proportion of dividing to expanding cells in the cotyledon cell population.

The variation for protein content in peas has been examined by many authors but, in the more extensive surveys (e.g. Jermyn and Slinkard, 1977; Matthews and Arthur, 1985), the conclusion was reached that there was little genetic variation for percent protein and most of the variation observed could be accounted for by an environmental effect. More recently, Arthur *et al.* (1991) have shown that there is a marked genotype-environment effect for percent protein and seed weight which is particularly genotype-specific; different genotypes respond differently to environments and treatments affecting the two characters. Any assessment of variation for percent protein, therefore, must take account of seed size, with only large-seeded genotypes likely to show a good correlation between seed weight and percent protein. Hence, improvement of percent protein in peas will not be easy (Jermyn and Slinkard, 1977; Arthur *et al.*, 1991).

Despite major remodelling of the pea plant to suit agricultural practices in the Western industrialized countries, the seed has been changed little. The widespread introduction of the *r* mutant to extend the use of peas to the freezer market represents the only exploited modification of the seed's growth and development. As stated earlier, one of the effects of the mutation at the *r* locus is to increase the sucrose level of the seed during development. This characteristic is the basis for its adoption by that market, for which the pea seeds are harvested and eaten when immature. The primary effect of this mutation, on starch composition, has not been exploited, however, even though the high amylose starch produced is a commodity much sought after for industrial uses.

Observation of the seeds of many cultivated peas, especially those grown in northern Europe for dry seeds, leads one to the conclusion that man has selected for larger seeds with more cells and that such seeds accumulate considerable amounts of starch, the driving force (to man) being increased yields. Nevertheless, evidence from the new *rug-3* starchless mutants indicates that the starch is not needed either for the seed to develop or for it to germinate, since neither process is deficient in such mutants. The selection of mutants that have very low starch contents, however, gives rise to the exciting prospect of designing a pea seed with a storage product composition similar to that of soyabean. The major storage products of such a pea seed would be protein and

Table 5.5. Variation for fatty acid composition (% FAME) in pea seeds.

	Highest		Lowest	
	Genotype	%	Genotype	%
Palmitic 16:0	JI 1388	18.6	JI 114	11.2
Stearic 18:0	JI 569	5.6	JI 716	1.3
Oleic 18:1	JI 425	46.2	JI 1098	10.7
Linoleic 18:2	JI 1307	60.8	JI 569	32.1
Linolenic 18:3	JI 85	14.6	JI 1219	2.5

Genotypes (from the John Innes Germplasm Collection) which contain the highest or lowest amount (as % of total lipid) of the individual fatty acid (measured as fatty acid methyl esters).

oil but, unlike soya, the seed would be produced on a plant that will grow well in cool temperate climates. Soya contains *c.* 40% protein and 20% lipid, whereas round-seeded peas contain only *c.* 22% protein and 2% lipid. Pea embryo protein contents of over 35% and lipid contents of 8% have been noted, however, in some of the new mutants (Wang *et al.*, 1990; Perez *et al.*, 1993). Alleles at the *r* and *rb* loci have a marked positive effect on lipid content (Jones *et al.*, 1990) and there is much variation for individual fatty acids within existing germplasm (Table 5.5). There are possibilities, therefore, for both qualitative and quantitative improvements in the lipid content of pea. The observation that low starch is not detrimental to seed development or germination supports the possibility that the carbohydrate that would normally enter the seed and be made into starch could be used for other compounds. This presents opportunities for engineering seeds in which this carbohydrate is deliberately redirected into products other than starch, both characteristic of (e.g. lipid) and foreign to pea. It also requires, however, that we gain much more information concerning source–sink relationships in seed development and concerning the interactions between embryo and testa. In both instances, the new genetic variation outlined above should be a considerable aid.

The *rugosus* mutants also provide opportunities for widening the uses of the seed directly by dramatically altering its composition, in particular the content and composition of the starch. Using combinations of the mutant genes it may be possible to replace some of the chemically modified starches used in processed food with 'naturally modified' starch. There are also numerous potential and existing uses for natural starches as raw material in paper, cellulose, chemical, pharmaceutical and synthetic polymer industries. By exploiting natural genetic variation in the pea seed, our understanding of its growth and development has improved dramatically over the past few years; the identification of new mutants will help us to build on this knowledge. The storage product compositions that accompany changes in development are perhaps the features of the seed that can be most readily capitalized on. We can anticipate, therefore, a major redesigning of the pea seed over the next few years just as we have seen for the plant over the last decade.

Acknowledgements

We should like to thank our colleagues for their help and the provision of unpublished data: Lorraine Barber, Chris Harrison, Sue Johnson, Alan Jones, Chun-Ming Liu, James Lloyd and Calum MacLeod.

References

Ambrose, M.J., Wang, T.L., Cook, S.K. and Hedley, C.L. (1987) An analysis of seed development in *Pisum sativum*. IV. Cotyledon cell populations *in vivo* and *in vitro*. *Journal of Experimental Botany* 38, 1909–1920.

Arthur, A.E., Adams, H., Strouts, K., Jones, D.A., Wang, T.L. and Hedley, C.L. (1991) An analysis of seed development in *Pisum sativum*. XV. The influence of seed size on protein content. *Seed Science Research* 1, 203–208.

Bain, J.M. and Mercer, F.V. (1966) Subcellular organisation of the developing cotyledons of *Pisum sativum* L. *Australian Journal of Biological Sciences* 19, 49–67.

Bhattacharrya, M.K., Smith, A.M., Ellis, T.H.N., Hedley, C.L. and Martin, C.R. (1990) The wrinkled-seed character of pea described by Mendel is caused by a transposon-like insertion in a gene encoding starch-branching enzyme. *Cell* 60, 115–122.

Bisson, C.S. and Jones, H.A. (1932) Changes accompanying fruit development in the garden pea. *Plant Physiology* 7, 91–105.

Burrows, W.J. and Carr, D.J. (1970) Cytokinin content of pea seeds during their growth and development. *Physiologia Plantarum* 23, 1064–1070.

Carr, D.J. and Skene, K.G.M. (1961) Diauxic growth curves of seeds, with special reference to french beans (*Phaseolus vulgaris* L.). *Australian Journal of Biological Sciences* 14, 1–12.

Caspar, T., Huber, S.C. and Somerville, C. (1985) Alterations in growth, photosynthesis, and respiration in a starchless mutant of *Arabidopsis thaliana* (L.) deficient in chloroplast phosphoglucomutase activity. *Plant Physiology* 79, 11–17.

Clark, J.K. and Sheridan, W.F. (1991) Isolation and characterization of 51 embryo-specific mutations of maize. *Plant Cell* 3, 935–951.

Cook, S.K., Adams, H., Hedley, C.L., Ambrose, M.J. and Wang, T.L. (1988) An analysis of seed development in *Pisum sativum*. VII. Embryo development and precocious germination *in vitro*. *Plant Cell, Tissue and Organ Culture* 14, 89–101.

Cooper, D.C. (1938) Embryology of *Pisum sativum*. *Botanical Gazette* 100, 123–132.

Cooper, G.O. (1938) Cytological Investigations of *Pisum sativum*. *Botanical Gazette* 99, 584–591.
Corke, F.M.K., Hedley, C.L., Shaw, P.J. and Wang, T.L. (1987) An analysis of seed development in *Pisum sativum*. V. Fluorescence triple staining for investigating cotyledon cell development. *Protoplasma* 140, 164–172.
Corke, F.M.K., Hedley, C.L. and Wang, T.L. (1990a) An analysis of seed development in *Pisum sativum*. Cellular development and the deposition of storage protein in immature embryos grown *in vivo* and *in vitro*. *Protoplasma* 155, 127–135.
Corke, F.M.K., Hedley, C.L. and Wang, T.L. (1990b). An analysis of seed development in *Pisum sativum*. *In vitro* manipulation of embryo development using xenobiotic compounds. *Protoplasma* 155, 136–143.
Craig, S. (1986) Compartmentation of albumin and globulin storage proteins in protein deposits of pea embryo axis cells. *Protoplasma* 32, 107–109.
Cullis, C.A. and Davies, D.R. (1975) Ribosomal DNA amounts in *Pisum sativum*. *Genetics* 81, 485–492.
Davies, D.R. (1975) Studies of seed development in *Pisum sativum*. I. Seed size in reciprocal crosses. *Planta* 124, 297–302.
Davies, D.R. (1976) DNA and RNA contents in relation to cell and seed weight in *Pisum sativum*. *Plant Science Letters* 7, 17–25.
Davies, D.R. (1980) The r_a locus and legumin synthesis in *Pisum sativum*. *Biochemical Genetics* 18, 1207–1219.
Davies, D.R. and Brewster, V. (1975) Studies of seed development in *Pisum sativum*.II. Ribosomal RNA in reciprocal crosses. *Planta* 124, 303–309.
Dhillon, S.S. and Miksche, J.P. (1983) DNA, RNA, protein and heterochromatin changes during embryo development and germination of soybean (*Glycine max* L.). *Histochemical Journal* 15, 21–37.
Domoney, C. and Casey, R. (1983) Cloning and characterization of complementary DNA for convicilin, a major seed storage protein in *Pisum sativum* L. *Planta* 159, 446–453.
Domoney, C. and Casey, R. (1985) Measurement of gene number for seed storage proteins in *Pisum*. *Nucleic Acids Research* 13, 687–699.
Domoney, C., Davies, D.R. and Casey, R. (1980) The initiation of legumin synthesis in immature embryos of *Pisum sativum* L. grown *in vivo* and *in vitro*. *Planta* 149, 454–460.
Dure, L.S., III. (1975) Seed formation. *Annual Review of Plant Physiology* 26, 259–278.
Dure, L.S., III. (1985) Embryogenesis and gene expression during seed formation. *Oxford Surveys of Plant Molecular and Cell Biology* 2, 179–197.
Eeuwens, C.J. and Schwabe, W.W. (1975) Seed and pod wall development in *Pisum sativum* L. in relation to extracted and applied hormones. *Journal of Experimental Botany* 26, 1–14.
Fantes, P.A. and Nurse, P. (1981) Division timing: controls, models and mechanisms. In: John, P.C.L. (ed.), *The Cell Cycle*. Cambridge University Press, Cambridge pp. 11–33,
Flinn, A.M. and Pate, J.S. (1968) Biochemical and physiological changes during maturation of fruit of the field pea (*Pisum arvense* L.). *Annals of Botany* 32, 479–495.
Frydman, V.M., Gaskin, P. and MacMillan, J. (1974) Qualitative and quantitative

analyses of gibberellins throughout seed maturation in *Pisum sativum* cv. Progress No. 9. *Planta* 118, 123-132.

Gregory, P.P. (1903) The seed characteristics of *Pisum sativum*. *New Phytologist* 2, 226-228.

Gunning, B.E.S. and Pate, J.S. (1969) 'Transfer cells'. Plant cells with wall ingrowths, specialized in relation to short distance transport of solutes – their occurrence, structure, and development. *Protoplasma* 68, 107-133.

Hanson, K.R. (1990) Regulation of starch and sucrose synthesis in tobacco species. Studies of *Nicotiana tabacum* and a mutant of *N. sylvestris* deficient in plastid phosphoglucomutase. In: Zelitch, I. (ed.), *Perspectives in Biochemical and Genetic Regulation of Photosynthesis*. Alan R. Liss, Inc., New York, pp. 69-84.

Harris, N., Grindley, H., Mulchrone, J. and Croy, R.R.D. (1989) Correlated *in situ* hybridization and immunochemical studies of legumin storage protein deposition in pea (*Pisum sativum* L.). *Cell Biology International Reports* 13, 23-35.

Hauxwell, A.J., Corke, F.M.K., Hedley, C.L. and Wang, T.L (1990) Storage protein gene expression is localised to regions lacking mitotic activity in developing pea embryos. An analysis of seed development in *Pisum sativum* L. XIV. *Development* 110, 283-289.

Hedley, C.L. and Ambrose, M.J. (1980) An analysis of seed development in *Pisum sativum* L. *Annals of Botany* 46, 89-105.

Hedley, C.L. and Smith, C.M. (1985) Genetic variation for pea seed development. In: Hebblethwaite, P.D., Heath, M.C. and Dawkins, T.C.K. (eds), *The Pea Crop*. Butterworths, London, pp. 329-337.

Hedley, C.L. and Wang, T.L. (1987) Seed and foliar mutations in *Pisum*. In: Thomas, H. and Grierson, D. (eds), *Developmental Mutants in Higher Plants*. Cambridge University Press, Cambridge pp. 219-244.

Hedley, C.L., Smith, C.M., Ambrose, M.J., Cook, S.K. and Wang, T.L. (1986) An analysis of seed development in *Pisum sativum*. II. The effect of the *r*-locus on the growth and development of the seed. *Annals of Botany* 58, 371-379.

Hughes, D.W. and Galau, G.A. (1989) Temporally modular gene expression during cotyledon development. *Genes and Development* 3, 358-369.

Jermyn, W.A. and Slinkard, A.E. (1977). Variability of percent protein and its relationship to seed yield and seed shape in peas. *Legume Research* 1, 33-37.

Johri, B.M. (1984) *Embryology of Angiosperms*. Springer-Verlag, Berlin.

Johri, B.M., Ambegaokar, K.B. and Srivastava, P.S. (1992) *Comparative Embryology of Angiosperms*. Springer-Verlag, Berlin, Heidelberg, New York.

Jones, D.A., Arthur, A.E., Adams, H.M., Coxon, D.T., Wang,T.L. and Hedley, C.L. (1990) An analysis of seed development in *Pisum sativum* L. IX. Genetic analysis of lipid content. *Plant Breeding* 104, 144-151.

Kooistra, E. (1962) On the differences between smooth and three types of wrinkled peas. *Euphytica* 11, 357-373.

Le Gal, M.F., Hallet, J.N. and Rey, L. (1984) DNA and protein content relationship in the cells of mature cotyledons of *Pisum sativum*. *Protoplasma* 120, 113-122.

Lin, T,P., Caspar, T., Somerville, C. and Preiss, J. (1988) Isolation and characterisation of a starchless mutant of *Arabidopsis thaliana* (L.) Heynh. lacking ADP glucose pyrophosphorylase activity. *Plant Physiology* 81, 642-645.

MacLeod, M.R., Hedley, C.L., Martin, C.R. and Wang, T.L. (1991). Genetic analysis of new wrinkled-seeded pea mutants at the r locus. In: Froud-Williams R.J.,

Gladders, P., Heath, M.C., Jenkyn, J.F., Knott, C. M., Lane, A and Pink, D. (eds), *Production and Protection of Legumes, Aspects of Applied Biology*, Vol. 27. Association of Applied Biologists, Wellesbourne, Warwick, pp. 263–265.

Marinos, N.G. (1970) Embryogenesis of the pea (*Pisum sativum*).I. The cytological environment of the developing embryo. *Protoplasma* 70, 261–279.

Matthews, P. and Arthur, A.E. (1985). Genetic and environmental components of variation in protein content of peas. In: Hebblethwaite P.D., Heath M.C., and Dawkins T.C.K. eds), *The Pea Crop*. Butterworths, London, pp. 369–381.

Mayer, U., Ruiz, R.A.T., Berleth, T., Misera, S. and Jurgens, G. (1991) Mutations affecting body organization in the *Arabidopsis* embryo. *Nature* 353, 402–407.

McKee, H.R., Robertson, R.N. and Lee, J.B. (1955) Physiology of the pea fruits. I. The developing fruit. *Australian Journal of Biological Sciences* 8, 137–162.

Meinke, D.W. (1985) Embryo-lethal mutants of *Arabidopsis thaliana*: analysis of mutants with a wide range of lethal phases. *Theoretical and Applied Genetics* 69, 543–552.

Meinke, D.W. (1986) Embryo-lethal mutants and the study of plant embryo development. *Oxford Surveys of Plant Molecular and Cell Biology* 3, 122–165.

Meinke, D.W. (1991) Perspectives on genetic analysis of plant embryogenesis. *Plant Cell* 3, 857–866.

Mendel, G. (1866) Versuche über Pflanzen-hybriden. *Verhandlungen des naturforschenden Vereins in Brünn* 4.

Millerd, A. and Whitfield, P.R. (1973) Deoxyribonucleic acid and ribonucleic acid synthesis during the cell expansion phase of cotyledon development in *Vicia faba* L. *Plant Physiology* 51, 1005–1010.

Millerd, A., Spencer, D., Dudman, W.F. and Stiller, M. (1975) Growth of immature pea cotyledons in culture. *Australian Journal of Plant Physiology* 2, 51–59.

Murfet, I.C. (1985) The influence of genes *ar* and *n* on senescence in *Pisum sativum* L. *Annals of Botany* 55, 675–683.

Murray, D.R. (1987) Nutritive role of seedcoats in developing legume seeds. *American Journal of Botany* 74, 1122–1137.

Natesh, S. and Rau, M.A. (1984) The embryo. In: Johri, B.M. (ed.), *Embryology of Angiosperms*. Springer-Verlag, Berlin, Heidelberg, New York, Tokyo, pp. 375–443.

Pate, J.S. (1975) Pea. In: Evans, L.T. (ed.), *Crop Physiology*. Cambridge University Press, Cambridge, pp. 191–224.

Pate, J.S. and Gunning, B.E.S. (1972) Transfer Cells. *Annual Review of Plant Physiology* 23, 173–196.

Perez M.D., Chambers, S.J., Bacon, J.R., Lambert, N., Hedley, C.L. and Wang, T.L. (1993) Seed protein content and composition of near-isogenic and induced mutant pea lines. *Seed Science Research* 3 (in press).

Perez-Grau, L. and Goldberg, R.B. (1989) Soybean seed protein genes are regulated spatially during embryogenesis. *Plant Cell* 1, 1095–1109.

Raghavan, V. (1986) *Embryogenesis in Angiosperms*. Cambridge University Press, Cambridge.

Reeve, R.M. (1948) Late embryogeny and histogenesis in *Pisum*. *American Journal of Botany* 35, 591–602.

Reid, J.B. (1989) Seed development in *Pisum*. The influence of genes *ar* and *n*. *Journal of Plant Physiology* 134, 751–755.

Rocha-Sosa, M., Sonnewald, U., Frommer, W., Stratmann, M., Schell, J. and Willmitzer, L. (1989) Both developmental and metabolic signals activate the promoter of a class I patatin gene. *EMBO Journal* 8, 23–29.

Rochat, C. and Boutin, J.-P. (1991) Metabolism of phloem-borne amino acids in maternal tissues of fruit of nodulated or nitrate-fed pea plants (*Pisum sativum* L.). *Journal of Experimental Botany* 42, 207–214.

Rochat, C. and Boutin, J.-P. (1992) Temporary storage compounds and sucrose-starch metabolism in seed coats during pea seed development (*Pisum sativum*). *Physiologia Plantarum* 85, 567–572.

Rowan, K.S. and Turner, D.H. (1957) Physiology of pea fruits. V. Phosphate components in the developing seed. *Australian Journal of Biological Sciences* 10, 414–425.

Scharpé, A. and Van Parijs, R. (1973) The formation of polyploid cells in ripening cotyledons of *Pisum sativum* L. in relation to ribosome and protein synthesis. *Journal of Experimental Botany* 24, 216–222.

Smith, D.L. (1973) Nucleic acid, protein, and starch synthesis in developing cotyledons of *Pisum arvense* L. *Annals of Botany* 37, 795–804.

Smith, D.L. and Flinn, A.M. (1967) Histology and histochemistry of the cotyledons of *Pisum arvense* L. during germination. *Planta* 74, 72–85.

Stafford, A. and Davies, D.R. (1979) The culture of immature pea embryos. *Annals of Botany* 44, 315–321.

Tedin, H. (1920) Inheritance of flower colour in *Pisum*. *Hereditas* 1, 68–97.

Thorne, J.H. (1980) Kinetics of ^{14}C-photosynthate uptake by developing soybean fruit. *Plant Physiology* 65, 975–979.

Turner, J.F., Turner, D.H. and Lee, J.B. (1957) Physiology of pea fruits. IV. Change in sugars in the developing seed. *Australian Journal of Biological Sciences* 10, 407–413.

Turner, S.R., Barratt, D.H.P. and Casey, R. (1990) The effect of different alleles at the *r* locus on the synthesis of seed storage proteins in *Pisum sativum*. *Plant Molecular Biology* 14, 793–803.

Tykarska, T. (1979) Rape embryogenesis. II. Development of the embryo proper. *Acta Societatis Botanicorum Poloniae* 48, 391–421.

Walbot, V. (1978) Control mechanisms for plant embryogeny. In: Clutter, M.E. (ed.), *Dormancy and Developmental Arrest*. Academic Press, New York, pp. 114–165.

Wang, T.L. and Hedley, C.L. (1991) Seed development in peas: knowing your three 'r's' (or four, or five). *Seed Science Research* 1, 3–14.

Wang, T.L., Hadavizideh, A., Harwood, A., Welham, T.J., Harwood, W.A., Faulks, R. and Hedley, C.L. (1990) An analysis of seed development in *Pisum sativum*. XIII. The chemical induction of storage product mutants. *Plant Breeding* 105, 311–320.

Wellensiek, S.J. (1925) Pisum crosses. *Genetica* 7, 1–25.

White, O.E. (1917) Studies of inheritance in Pisum. II. The present state of knowledge of heredity and variation in peas. *Proceedings of the American Philosophical Society* 56, 487–588.

Wolswinkel, P. (1985) Effect of inhibitors on solute efflux from seed-coat halves and cotyledons of *Pisum sativum* L., after uptake from a bathing medium. The difference between sucrose and amino acids. *Journal of Plant Physiology* 120, 419–429.

Wolswinkel, P. and Ammerlaan, A. (1984) Turgor-sensitive sucrose and amino acid transport into developing seeds of *Pisum sativum*. Effect of a high sucrose or mannitol concentration in experiments with empty ovules. *Physiologia Plantarum* 61, 172–182.

Wolswinkel, P. and Ammerlaan, A. (1985) Effect of potassium on sucrose and amino acid release from the seed coat of developing seeds of *Pisum sativum*. *Annals of Botany* 56, 35–43.

Yang, L.-J., Barratt, D.H.P., Domoney, C., Hedley, C.L. and Wang, T.L. (1990) An analysis of seed development in *Pisum sativum* L. Expression of storage protein genes in cultured embryos. *Journal of Experimental Botany* 41, 283–288.

Yeung, E.C. and Clutter, M.E. (1978) Embryogeny of *Phaseolus coccineus*: growth and microanatomy. *Protoplasma* 94, 19–40.

Biochemistry and Molecular Biology of Seed Products

R. CASEY, C. DOMONEY AND A.M. SMITH
John Innes Institute, Colney Lane, Norwich, NR4 7UH, UK

The previous chapter has described the development of the pea seed and the genetics of deposition of storage reserves within the developing embryo. This chapter considers in detail the structure and synthesis of starch and protein, the major storage components of pea seeds, and discusses their biotechnological implications and potential for directed modification.

Starch

The structure of starch

Starch occurs as water-insoluble, semicrystalline granules inside plastids (see Chapter 5 for a description of granule size and morphology in pea embryos). About 70% by weight of most starches consists of amylopectin, a large and highly branched polymer of glucose (Fig. 6.1). Linear chains of α 1,4-linked glucose units, typically 15–45 glucose units in length, are joined together by α 1,6 linkages to form a complex branching pattern. Clusters of branches are arranged at regular intervals of about 9 nm along the long axis of the amylopectin molecule (Fig. 6.1). Within the starch granule, amylopectin molecules lie with the free, non-reducing ends of the branches pointing outwards. Adjacent branches within a cluster associate to form double helices, and these pack together in ordered arrays which extend across the clusters of many adjacent amylopectin molecules. About 50% of the matrix of the granule may consist of these ordered arrays.

The other major component of starch is amylose, a much smaller and less branched glucose polymer than amylopectin. It consists of a few, usually quite

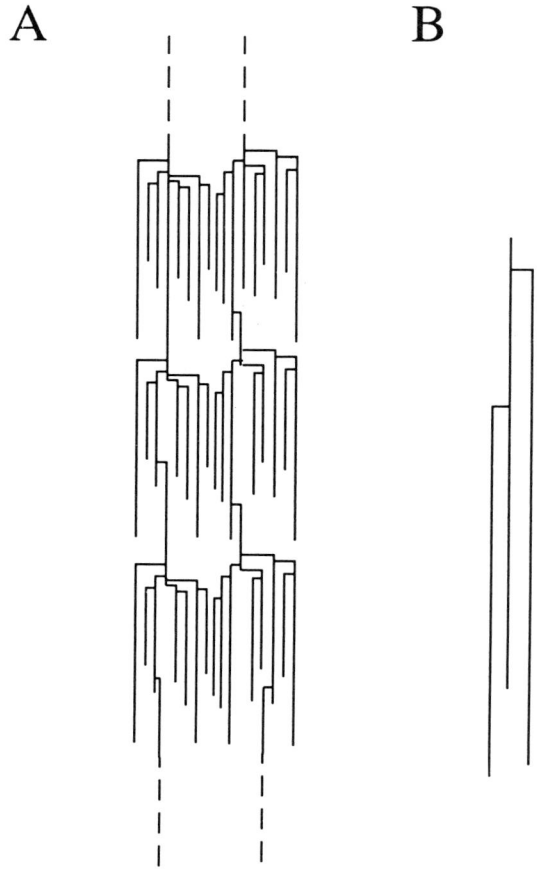

Fig. 6.1. Generalized structures of (A) three clusters of branches within an amylopectin molecule, and (B) an amylose molecule. Lines represent chains of 1,4-linked glucose units. (After Gidley and Bociek, 1985.)

long chains of α 1,4-linked glucose units, joined by α 1,6 linkages. Amylose is thought to exist in an amorphous phase within starch granules, distinct from the semicrystalline phase formed by the ordered arrays of the amylopectin molecules (French, 1984; Gidley and Bociek, 1985, 1988; Kainuma, 1988).

The starch of mature embryos of wild-type peas conforms to the general description given above. The M_r of its amylopectin is approximately 8×10^7 (Colonna and Mercier, 1984). The amylose of pea starch shows a wide distribution of M_r, with an average value of approximately 0.5×10^6. A typical amylose molecule from pea probably consists of two or three long chains of glucose units (Colonna and Mercier, 1984). Wild-type pea starch differs from the storage starches of tubers and cereals in the way in which the amylopectin molecules pack into ordered arrays within the granule. X-ray crystallography reveals that cereal starches have closely packed arrays (referred to as an A structure), while those of tuber starches are more loosely packed (referred to as a B structure). The difference in packing is thought to be due to the fact that the average length of branches within the clusters of amylopectin molecules is greater in tuber than in cereal starches. Wild-type pea starch displays a third type of X-ray crystallographic pattern (referred to as a C structure), which probably reflects a degree of packing, and hence an average branch length, intermediate between that of cereal and tuber starches (Hizukuri *et al.*, 1983; Guilbot and Mercier, 1985).

The percentage of amylose in starch from pea embryos increases during development. It rises from a value of about 15% during the early stages of starch synthesis to a final value of over 30% (Greenwood and Thompson, 1962; Chapter 5). This change during development is typical of the starches of storage organs (Banks and Muir, 1980).

The starch of pea embryos with recessive alleles at the *r* locus (*rr* embryos) differs considerably from that of wild-type embryos. The final amylopectin content of the starch is only about 30%, and both amylose and amylopectin molecules have lower average M_rs than their wild-type counterparts (Colonna and Mercier, 1984; Chapter 5). The *r* locus is in these respects similar to the *amylose-extender* locus of maize, which affects the starch of the endosperm (Banks *et al.*, 1974). The alteration in polymer composition and structure in *rr* embryos changes the packing of ordered arrays from a C to a B structure as defined by X-ray crystallography, and reduces considerably the proportion of the granule which consists of ordered arrays (Colonna *et al.*, 1982; Colonna and Mercier, 1985).

The synthesis of starch polymers

The supply of carbon to the seed

In common with other storage and structural components of the embryo, starch is synthesized from sucrose made in photosynthesis in the leaves, stipules and pods. Sucrose enters the testa via the phloem. It is unloaded from phloem-associated cells into the apoplast, from which it is taken up by the cells of the developing embryo (Smith and Denyer, 1992; Wolswinkel, 1992). In the parenchymatous cells of the cotyledons, where most of the storage starch is synthesized, sucrose is metabolized via sucrose synthase in the cytosol (Edwards and ap Rees, 1986b). The products of this reaction enter glycolysis, intermediates of which form the starting points for the pathway of respiration and synthesis of the storage and structural components of the cell (Edwards and ap Rees, 1986a; Fig. 6.2).

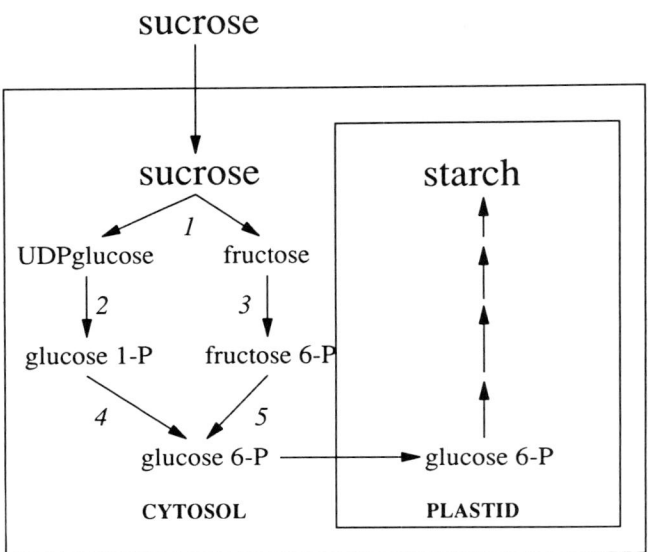

Fig. 6.2. The supply of carbon to the amylopast in a cell of a developing pea cotyledon: (1) sucrose synthase; (2) UDPglucose pyrophosphorylase; (3) hexokinase; (4) phosphoglucomutase; (5) phosphoglucose isomerase.

The synthesis of starch is quantitatively the most important fate for carbon from glycolysis in the developing embryo. In all higher plants, starch synthesis takes place exclusively inside plastids. During most of its development the embryo contains a range of plastids, from organelles which are heavily committed to the synthesis of storage starch to organelles which retain most of their chloroplast-like features and contain little starch (Smith *et al.*, 1990).

The form in which carbon destined for starch synthesis enters the plastid from the cytosol in storage organs such as the pea embryo is the subject of continuing controversy (Okita, 1992; Smith and Martin, 1993). Pea embryos have proved relatively easy to study in this respect because, unlike many storage organs, their plastids can be isolated rapidly in a metabolically active state (Denyer and Smith, 1988; Hill and Smith, 1991). Experiments with these isolated plastids indicate that glucose 6-phosphate enters as the substrate for starch synthesis. Glucose 6-phosphate is the only compound among hexose phosphates, hexoses and triose phosphate synthesized from sucrose in the cytosol (Fig. 6.2) which can support significant starch synthesis by the plastids. It enters the plastid via a translocator, which can exchange it for phosphate (and perhaps triose phosphate) and which cannot transport other hexose phosphates (L. Hill, personal communication). The rate of starch synthesis supported by glucose 6-phosphate in isolated plastids is sufficient to account for the rate of starch synthesis in an intact embryo. These characteristics of

Biochemistry and Molecular Biology of Seed Products   125

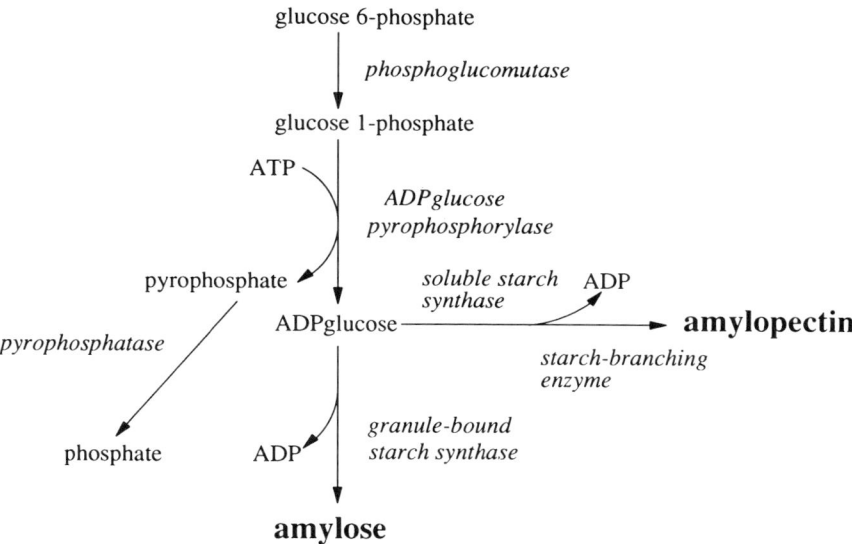

Fig. 6.3. The pathway of synthesis of starch polymers in a plastid of a developing pea embryo.

glucose 6-phosphate uptake and its metabolism by isolated plastids indicate strongly that this is the route by which carbon is supplied to starch synthesis *in vivo* (Fig. 6.2).

Inside the plastid, the glucose 6-phosphate is converted to glucose 1-phosphate via phosphoglucomutase (Fig. 6.3); high levels of activity of this enzyme are detected in starch-synthesizing plastids (Foster and Smith, 1993). Glucose 1-phosphate is the substrate for the reactions which are exclusive to the pathway of starch synthesis, and which lead to the formation of the starch granule.

ADPglucose pyrophosphorylase

ADPglucose, the substrate for the synthesis of the starch polymers, is synthesized from glucose 1-phosphate via ADPglucose pyrophosphorylase. The reaction is close to equilibrium *in vivo*, but the hydrolysis of pyrophosphate drives it in the direction of ADPglucose synthesis (Gross and ap Rees, 1986; Fig. 6.3). In most of the plants from which it has been purified, the enzyme consists of two different polypeptides, referred to as the large and small subunits (Smith and Martin, 1993). The enzyme from pea embryos, however, comprises four different polypeptides (Hylton and Smith, 1992). The relationship between these polypeptides and the large and small subunits of the enzyme from other species is not fully understood.

cDNA clones encoding the large and small subunits of the enzyme from potato tubers have been used as probes to isolate clones from cDNA libraries of pea embryos. The relationship of these pea cDNA clones to the polypeptides of the pea enzyme and to the potato cDNA clones is being established (R. Burton, personal communication). The recessive mutation at the *rb* locus (Chapter 5) provides a valuable tool for studying the enzyme because it affects its specific activity, regulatory properties and polypeptide composition. The mutation affects the overall morphology and composition of the seed, and has a particularly marked effect on starch synthesis (Kooistra, 1962; Wang and Hedley, 1991; Chapter 5). Measurements of the maximum catalytic activities of enzymes on the pathway from sucrose to starch in the developing *rbrb* embryo revealed that the activity of ADPglucose pyrophosphorylase is reduced 10-to 40-fold relative to the wild-type, but other enzymes are not affected (Smith *et al.*, 1989). The reduction in maximum catalytic activity of ADPglucose pyrophosphorylase results from a large reduction in specific activity, which is associated with the loss of one of the four polypeptides. The enzyme from *rbrb* embryos is also more sensitive to regulation by metabolites than the wild-type enzyme (Hylton and Smith, 1992).

Starch synthase

Pea embryos contain three different isoforms of starch synthase. Two of these are soluble proteins of M_r 60,000 and 77,000, with the latter isoform accounting for most of the soluble activity (Denyer and Smith, 1992; K. Denyer, personal communication). The third is a protein of M_r 59,000 located exclusively on the starch granule (Smith, 1990; Denyer *et al.*, 1993). This granule-bound isoform is very similar in amino-acid sequence to the products of the *waxy* genes of cereals, which encode the granule-bound starch synthases of the endosperm (Dry *et al.*, 1992). Mutations in the *waxy* genes result in the loss of amylose from starch, indicating that the granule-bound class of starch synthases is responsible for amylose synthesis (Preiss and Levi, 1982). The effects of the *waxy* mutations have allowed a model for the synthesis of amylose and amylopectin to be proposed (Robyt, 1984), as follows. The soluble starch synthases and starch-branching enzyme (see below) act together at the extreme periphery of the granule to elongate and branch the polymers that become the amylopectin component of the granule. The newly synthesized amylopectin molecules associate in ordered arrays to form the matrix of the granule, where the granule-bound starch synthase is located. The product of this starch synthase is formed at or within the matrix surface where it is physically or chemically unavailable to starch-branching enzyme. It thus becomes the amylose component of the starch. Differences between the major, soluble starch synthase of M_r 77,000 and the exclusively granule-bound starch synthase of pea embryos have been revealed by sequencing cDNA clones and have interesting implications for this model. The soluble enzyme is very similar to the granule-bound enzyme over most of its length, particularly in a region around a Lys–Thr–Gly–Gly motif at which UDP glucose or ADPglucose is thought to bind in starch and glycogen synthases generally (Furukawa et al., 1990). The major difference between the two enzymes is an N-terminal stretch of 203 amino acids, rich in serine, which is unique to the soluble enzyme (Dry et al., 1992). It is tempting to speculate that this sequence determines the difference between the two enzymes in their affinity for the granule matrix, or allows an interaction between the soluble starch synthase and starch-branching enzyme.

In addition to their presence in the soluble fraction of the embryo, both the 60,000 and the 77,000 M_r starch synthases are bound to the starch granule. Immunogold labelling using an antibody to the 77,000 M_r protein reveals that this isoform is not simply confined to the periphery of the granule, but is present throughout the matrix. Both the 60,000 and the 77,000 M_r proteins can be solubilized in an active form from purified starch by partial digestion with amylase (Smith, 1990; Denyer et al., 1993). It seems probable that these enzymes are active in the soluble fraction, but become incorporated into the granule as their products associate to form the matrix. It is highly unlikely that the proteins contribute to starch synthesis when trapped inside the matrix, both because much of the matrix is an ordered, crystalline structure and because ADPglucose is unlikely to enter the matrix. The 59,000 M_r granule-bound starch synthase is also present throughout the matrix (Denyer et al., 1993), so this too may be incorporated as the granule grows.

The presence of all of the starch synthases within the granule does not necessarily invalidate the model for amylose and amylopectin synthesis. The active fractions of the soluble and exclusively granule-bound isoforms of the enzyme may still be separated. so that the products of one are accessible to branching enzyme whereas the products of the other are not.

Starch-branching enzyme

Pea embryos have two isoforms of starch-branching enzyme, one of M_r 110,000 (SBE I) and the other of M_r 100,000 (SBE II). The activities of the two isoforms, in assays in which the branching of either preformed amylose or a growing glucan polymer is measured, indicate that they have distinctly different kinetic properties (Smith, 1988). Like the soluble starch synthases, these proteins are present both in the soluble fraction and within the matrix of the granule (Denyer et al., 1993). It is likely that they become trapped in the matrix in the same way as the starch synthases, and are active only as soluble proteins in association with the starch synthases at the extreme periphery of the granule.

Further information about the roles of the two isoforms is being sought through a study of the effects of the mutation at the *r* locus. Like the *rb* mutation, this affects both the morphology and composition of the seed, and has its major effect on starch (Chapter 5; see above). The fact that the starch content is reduced and the apparent amylose content of the starch is considerably increased suggests that the mutation affects starch-branching enzyme, and studies of the enzymes of starch synthesis showed that mutant embryos have very low activities of starch-branching enzyme (Matters and Boyer, 1981; Edwards *et al.*, 1988; Smith, 1988). Immunoblotting experiments with an antibody raised against SBE I from wild-type embryos revealed that both the activity and the protein of this isoform are absent from mutant embryos, which accounts for their low total SBE activity (Smith, 1988; Bhattacharyya *et al.*, 1990). cDNA clones corresponding to wild-type SBE I RNA were used to show that the SBE I gene in mutant embryos is disrupted by an insertion of about 0.8 kb into an exon, towards the 3' end of the coding sequence; the result of this would be the loss of the last 61 amino acids of SBE I. The insertion has an inverted repeat of 12 bp at its ends, and is generally similar in structure to the Ac class of transposable elements of maize. The gene with the insertion is transcribed in the mutant embryo into an aberrantly large mRNA, the amounts of which are very low compared with those of the wild-type SBE I gene and no stable protein is produced (Bhattacharyya *et al.*, 1990). The RFLP created by this insertion maps to the *r* locus, indicating that the insertion is responsible for all of the effects of the *r* mutation.

The regulation of the pathway

A description of the proteins that catalyse the pathway of starch synthesis, however detailed, does not reveal how the flux of carbon through the pathway is regulated. This information is of both applied and academic interest. A knowledge of the importance of individual proteins in regulating the amount and the structure of starch in embryos permits a rational approach to the genetic manipulation of starch to improve the value of the crop. Little is known yet about the regulation of the pathway of starch synthesis in non-photosynthetic organs generally, and the main purpose of this section is to indicate methods by which it can be studied in pea embryos.

THE USE OF MUTANTS

The most direct way to study regulation is to make changes in the amounts of each of the proteins on the pathway, and measure the effect of those alterations on the rate of starch synthesis and the structure of the starch. The mutations at the *r* and *rb* loci provide a means of altering the activities of starch-branching enzyme and ADPglucose pyrophosphorylase, respectively. For both of the mutations, the particular enzyme affected has an activity in heterozygote embryos that is intermediate between those of the parents

(K. Denyer, personal communication). If the mutant and wild-type alleles are in near-isogenic lines, this allows the effects of three different levels of the activity of a single enzyme – in the homozygous dominants and recessives and the heterozygotes – to be studied in otherwise genetically identical backgrounds.

The full potential of this approach has not yet been exploited, but preliminary studies of the control of the flux of carbon into starch indicate that, during most of development, neither starch-branching enzyme nor ADPglucose pyrophosphorylase is of major importance. For both mutations, the rate of starch synthesis in heterozygote embryos is the same as that in wild-type embryos (K. Denyer, personal communication). This result indicates that either an enzyme other than ADPglucose pyrophosphorylase or starch-branching enzyme is of major importance in controlling flux, or the control is partitioned evenly between several steps so that no one enzyme is of great importance. This situation is very different from that in leaves, where ADPglucose pyrophosphorylase is of overriding importance in regulating the flux of carbon into starch (Neuhaus and Stitt, 1990). The rate of starch synthesis in leaves must respond to short-term changes in the rate of carbon assimilation and the demands for carbon by the sink organs of the plant. The sensitivity of ADPglucose pyrophosphorylase in leaves to modulation by metabolites allows rapid changes in activity, and hence in the flux of carbon into starch. The rate of starch synthesis in embryos, in contrast, is quite constant over long periods of development. It is not surprising, therefore, that the partitioning of control between steps on the pathway is different from that in leaves, or that the ADPglucose pyrophosphorylase of embryos lacks the sensitivity to metabolite modulation displayed by the enzyme from leaves (Hylton and Smith, 1992).

This approach also gives information about the regulation of the size and structure of starch polymers. The branched polymers of rr and wild-type embryos differ in these respects (Colonna and Mercier, 1984). The polymers in wild-type embryos result from the actions of SBE I and II, whereas those in rr embryos result from the action of SBE II alone (see above). The r mutation thus allows the roles of the two branching enzymes in determining starch structure to be studied. This information must, however, be treated with caution. Although the mutation does not change the maximum catalytic activities of enzymes of starch synthesis other than starch-branching enzyme, it causes changes in other proteins involved in starch metabolism. First, the 60,000 M_r soluble starch synthase contributes a much greater proportion of the total soluble starch synthase activity in mutant than in wild-type embryos (K. Denyer, personal communication). Second, the starch of mutant embryos contains at least two proteins which are absent from the starch of wild-type embryos (A. Smith, unpublished data). These proteins are not antigenically related to starch synthases or to starch-branching enzyme, but their presence in the starch granule suggests strongly that they are involved in starch metabolism. The precise significance of these secondary changes is not known, but their existence may mean that changes in starch structure caused by the mutation at the *r* locus are not attributable solely to the loss of one isoform of starch-branching enzyme.

The mutation at the *rb* locus also affects the structure of starch. The starch of mutant embryos has a lower amylose to amylopectin ratio than that of wild-type embryos (Chapter 5). It is not immediately clear how a mutation that affects the rate of synthesis of ADPglucose rather than the synthesis of starch polymers can affect starch structure, but there are at least two possibilities. First, the mutation may have subtle secondary effects on proteins directly involved in starch metabolism, of the type described above for the mutation at the *r* locus. Second, a decreased supply of ADPglucose may alter the balance between the mechanisms responsible for amylose and amylopectin synthesis. If this latter proposal is correct, it suggests that any modification of the rate of starch synthesis will alter the structure of the starch as well.

The information that can be gained about the regulation of starch synthesis through a study of mutants is at present severely restricted by the lack of suitable mutant material. The three new mutations that alter the amount and structure of starch in the embryo (Chapter 5) may well be in genes encoding enzymes on the pathway of starch synthesis. These will provide valuable new material for studying the regulation of the pathway. The development of a transformation system for peas (Chapter 10) will expand this approach still further. Transformation with sense and antisense constructs containing genes encoding enzymes on the pathway will enable the activities of these enzymes to be increased or decreased relative to the wild-type embryo. The use of embryo-specific promoters – for example those for the genes encoding the storage proteins (see later) – will enable these changes to be targeted specifically to the embryo, and to particular periods during development. Since cDNAs encoding all of the enzymes specifically involved in the synthesis of starch polymers (see above) have now been cloned, transformation will allow much wider and more sensitive manipulation of the pathway of starch synthesis than is currently possible, and hence the development of a comprehensive picture of regulation.

CHANGES DURING DEVELOPMENT

Analysis of the regulation of the pathway of starch synthesis is complicated by the fact that the nature of the pathway itself changes during the development of the embryo. The genes encoding isoforms of starch synthase and starch-branching enzyme appear to have different patterns of expression during development. Amounts of mRNA for the 77,000 M_r starch synthase and SBE I are maximal at an early stage of development and then fall, whereas those for the exclusively granule-bound, 59,000 M_r starch synthase and SBE II reach maximal levels only at an advanced stage of development (Dry *et al.*, 1992; C. Martin, personal communication). The patterns of expression of the genes encoding starch synthases indicate that activity of the granule-bound isoform, which is thought to synthesize amylose (see above), may increase relative to that of the soluble isoform during development. This is consistent with the observation that the amylose to amylopectin ratio increases during embryo development (Greenwood and Thompson, 1962). The fact that the gene encoding SBE I is expressed at an early stage of development is consistent with the effect of the mutation at the *r* locus on branching enzyme activity during development. The reduction in activity caused by the mutation is much greater during early development than at later stages when SBE II is also being produced at high levels (Smith, 1988).

These examples indicate that differential expression of genes encoding isoforms of enzymes of starch synthesis gives rise to continuous change in the enzymic composition of the pathway, and hence to the nature of its products, during embryo development. This probably means that the partitioning of control between steps on the pathway also changes continuously. Measurement of the partitioning of control must be made at several, defined developmental stages if the regulation of the final amount and composition of starch in the mature seed is to be understood.

Potential for the modification of pea starch

The recent development of methods for transforming peas will allow the alteration of activities and isoform composition of enzymes on the pathway of starch synthesis (see above), and hence the modification of the structure and amount of starch in the seed. Rational modification, to produce starches with defined and commercially useful properties, will not be possible until a better understanding of the regulation of starch synthesis is achieved. It is already possible, however, by exploiting the mutations at the *r*, *rb* and *rug* loci (Chapter 5) to make potentially useful changes in the amylose to amylopectin ratio, in the structure of amylopectin and in the amount of starch in the seed.

Production of pea amylose-free starch or a series of high-amylose starches, comparable with the high-value waxy and amylose-extender starches of maize, may be achieved through a reduction in the activities of the granule-bound starch synthase and the starch-branching enzymes, respectively.

Alterations in the structure of amylopectin, leading to changes in the physical properties of the starch, may be achieved in a variety of ways. For example, manipulation of SBE I may have a different effect on the structure of amylopectin from manipulation of SBE II, because the two forms have different kinetic properties and are maximally active at different stages of development. Restriction or extension of the period of maximal expression of one isoform, through the use of developmentally regulated promoters, introduction of starch-branching enzymes with different kinetic properties, and perhaps alteration of the activity of soluble starch synthase, can be expected to modify amylopectin structure.

Reductions in the activities of starch synthase and starch-branching enzyme to achieve desirable changes in starch structure may also reduce the rate of starch synthesis. It might be possible to compensate for this potential loss of yield by identifying and manipulating the activity of an enzyme catalysing a 'rate-limiting step' on the pathway. This will be difficult if control is partitioned evenly between several steps, or if it changes radically during development. Manipulating flux through the pathway might of itself change starch structure, and it may not be possible to alter the amount of starch in the seed independently of its structure.

Reduction of the amount of starch in the seed has long been shown to have commercial value. The *r* and *rb* mutations and the new mutations at the *rug* loci (Chapter 5) result in the accumulation of sucrose, and most cultivars of peas used for direct human consumption carry either the *r* or *rb* mutation. Reduction of starch content is associated with increased lipid content, and it may be possible to produce lines with commercially useful amounts of lipid from low-starch material (Chapter 5).

Low-starch lines of peas could also provide a suitable vehicle for the introduction of new biosynthetic pathways via genetic manipulation. Introduction of enzymes to utilize the sucrose, hexoses or hexose phosphates which accumulate in embryos of such lines should result in the synthesis of large amounts of novel products in place of these carbohydrates.

Proteins

Pea seeds, like those of many other grain legumes, contain a relatively high proportion, typically $\sim 25\%$, of protein (see Chapter 5). Pea seed protein falls into two classical solubility fractions (Osborne, 1924), the majority being globulin (i.e. protein that is soluble in dilute salt, but insoluble in water) and the remainder being water-soluble albumin. The globulins are synthesized during seed development, stored in the dry seed in organelles termed protein bodies and hydrolysed on germination to provide carbon skeletons and nitrogen for the developing seedling; consequently the two terms 'globulin' and 'storage protein' have been used interchangeably. Pea seed proteins characteristically contain a relatively high proportion of nitrogen (being rich in lysine, arginine, glutamine and asparagine), resulting in a relatively low factor (about 5.3) for the conversion of total nitrogen content to protein content for pea seeds (Holt and Sosulski, 1979). The globulins also have low contents of the sulphur-containing amino acids cysteine and methionine. The albumin fraction is relatively sulphur-rich and heterogeneous, containing a large number of minor proteins, some of which have important biotechnological implications.

Globulins

Pea seed globulins can be divided into two distinct classes, termed 7S and 11–12S, on the basis of their sedimentation coefficient (for references see Derbyshire *et al.*, 1976; Casey *et al.*, 1986; Wright, 1987). The 11–12S component is also termed legumin; the 7S proteins comprise vicilin and convicilin. The ratio of legumin (11S) : vicilin/convicilin (7S) varies from one pea genotype to another; 11 : 7S values as high as 4 have been reported (Croy and Gatehouse, 1985), although the range tends generally to be from ~0.5 to ~1 (Casey *et al.*, 1982; Gueguen, 1983).

Structure and synthesis

LEGUMIN

Pea legumin is an oligomeric 12S protein of M_r ~360,000–400,000, the majority of which comprises acidic (α) and basic (β) polypeptides of M_r ~40,000 and 20,000, respectively, arranged as disulphide-bonded $\alpha\beta$ subunit pairs of M_r ~60,000, six of which form an individual legumin molecule. Although there are no clear physical data concerning the overall shape and dimensions of pea legumin, a number of lines of evidence suggest that it exists as a trigonal antiprism [two trimeric rings superimposed and twisted about 60° (Fig. 6.4; see also Lambert and Yarwood, 1992)]. It seems likely that at least part of each α polypeptide occupies a position to the exterior of the

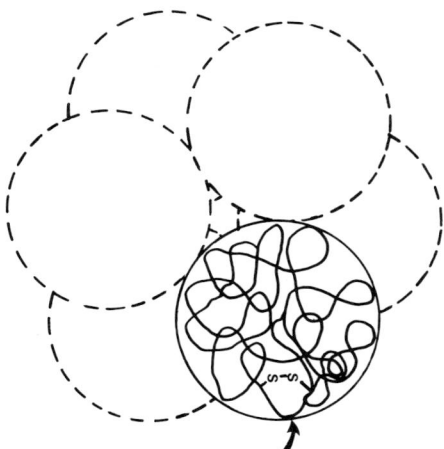

Fig. 6.4. Diagrammatic representation of legumin as a trigonal antiprism in which each morphological subunit is a disulphide-linked $\alpha\beta$ dimer; the $\alpha\beta$ processing site (arrow) is depicted as being at the exterior of the assembled hexamer.

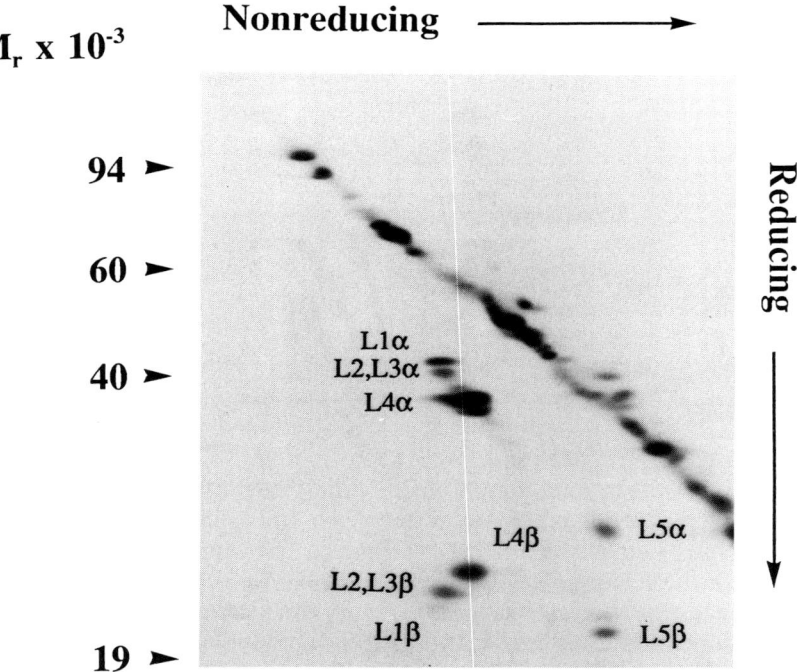

Fig. 6.5. Two-dimensional gel electrophoretic separation of a total pea seed extract stained with Coomassie blue to visualize proteins. Both dimensions are in polyacrylamide gel in the presence of the denaturant sodium dodecyl sulphate (SDS). After separation in the absence of reducing agent, the gel strip containing the separated bands is reduced to break disulphide bonds and the products separated in the second dimension. Those polypeptides that were not disulphide-bonded remain on the diagonal, whereas the disulphide-bonded subunit pairs of legumin (L1–L5) separate into constituent α and β polypeptides which fall beneath the diagonal.

molecule and that the β polypeptides are buried in the interior (see Lambert and Yarwood, 1992). Examination of the subunit pairs and polypeptides of pea legumin by gel electrophoresis under a variety of conditions suggests that there are at least five subunit pairs [designated L1–L5 (Matta *et al.*, 1981)] (see Fig. 6.5); collectively the various subunit pairs are thought to contain as many as 22 α and 11 β polypeptides (Matta *et al.*, 1981). Legumin is, therefore, a heterogeneous collection of molecules with different polypeptide structures.

Three different classes of cDNA and two different classes of gene (*leg*A and *leg*J) corresponding to pea legumin polypeptides have been isolated and sequenced (Lycett *et al.*, 1984; Domoney *et al.*, 1986a; Gatehouse *et al.*, 1988; Rerie *et al.*, 1990, 1991; Thompson *et al.*, 1991). The *leg*J- and *leg*A-type gene sequences are about 50% identical and contain different numbers of introns (see Casey *et al.*, 1986). The N-terminal sequences of the polypeptides from the major (L4) subunit pairs (Casey *et al.*, 1981a,b) correspond to the protein sequence predicted by the *leg*A genes. It is possible that the L2 subunit pairs are derived from the *leg*J genes (Gatehouse *et al.*, 1988). Sequencing of the polypeptides derived from the minor L1 subunit pair (see Fig. 6.5) shows them to be about 50–60% identical to the sequences predicted by either the *leg*A or the *leg*J genes (March *et al.*, 1988). The L5 subunit pairs differ from all the others in containing α polypeptides of $M_r \sim 25,000$ rather than 40,000 (Matta *et al.*, 1981) and have also been described as the subunit pairs of 'small legumin' (Matta and Gatehouse, 1982). From cDNA, gene and protein sequencing at least four classes of *Pisum* legumin polypeptide can be distinguished on the basis of primary structure.

Legumin is synthesized as precursors of $M_r \sim 60,000-65,000$, with a minor fraction of $M_r \sim 80,000$ also being produced (Croy *et al.*, 1980a; Spencer and Higgins, 1980; Domoney and Casey, 1984). The 60,000 M_r precursors contain contiguous α and β polypeptide sequences in the order NH_2-α-β-COOH (Croy *et al.*, 1982); they undergo folding and oligomerization to form a 7–9S half-molecule in which the disulphide-bonded α and β polypeptides are still covalently associated by a peptide bond (Spencer and Higgins, 1980). One consequence of this synthesis of legumin polypeptides as precursor molecules is specificity in the pairing between α and β polypeptides (Matta *et al.*, 1981). Processing of legumin precursors of $M_r \sim 60,000$ and $\sim 63,000-65,000$ probably gives rise to the L1–L4 subunit pairs; the L5 subunit pair appears to be the product of the 80,000 M_r precursors (Yang *et al.*, 1990). The 'nicking' of the peptide bond between the α and β polypeptides [usually Asn–Gly but occasionally Asn–Phe (Domoney *et al.*, 1986a; March *et al.*, 1988)] is a late event, which probably takes place in the protein body 2–3 h after initial synthesis (see Higgins, 1984). Such observations are compatible with the suggestion that the αβ processing site, which is close to a highly polar region in the C-terminal part of the α polypeptide, is situated on the outside of the molecule (see Casey *et al.*, 1986). The nature and length of the α polypeptide polar region varies considerably between different classes of legumin polypeptide [from 27 to 100 amino acids (Domoney *et al.*, 1986a)], making such regions, with their inherent variability, good targets for directed sequence modification aimed at improving protein quality.

VICILIN

Pea vicilin is extremely heterogeneous in terms of polypeptide composition, consisting of a range of polypeptides of M_r from $\sim 12,000$ to $\sim 70,000$ (Thomson *et al.*, 1978). Such heterogeneity derives from the combination of several factors, including the production of vicilin polypeptides from several small gene families encoding different primary sequences, differential proteolytic processing and differential glycosylation. Most polypeptides of vicilin seem to derive from precursor molecules of $M_r \sim 47,000-50,000$ or $\sim 68,000$. Some of these precursors are proteolytically processed, others are not (see below),

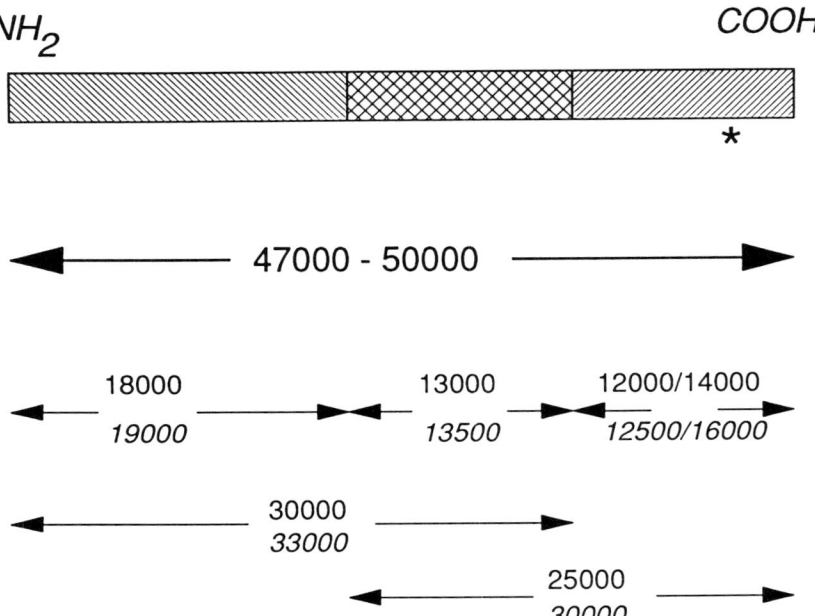

Fig. 6.6. Diagrammatic representation of the post-translational processing of vicilin precursors of M_r 47,000-50,000; *represents the glycosylation site. The apparent M_r of the various processed polypeptides are as determined by Gatehouse *et al.* (1983) (italics) or Spencer *et al.* (1983). The apparent M_r of the C-terminal fragment depends on whether or not it is glycosylated.

and the range of polypeptides that can be derived by such processing and differential glycosylation of vicilin precursors of M_r ~47,000–50,000 is shown diagrammatically in Fig. 6.6. The putative processing regions are hydrophilic (Lycett *et al.*, 1983) and may be at the exterior of the molecule. The polypeptide fragments that derive from processing remain non-covalently associated in the native molecule. The heterogeneity of pea vicilin in terms of size, electrophoretic mobility and polypeptide composition has long been established (Thomson *et al.*, 1978, 1980), but it is still not clear how the many vicilin polypeptides relate to each other in the native proteins. The quaternary structure of vicilins has been established as a trimer of morphological subunits of M_r ~50,000, with two structural domains within each subunit (I'Anson *et al.*, 1988) giving a pseudohexagonal appearance; this is confirmed by scanning tunnelling electron microscopy (Welland *et al.*, 1989).

There are few sequence data directly determined from pea vicilin or convicilin polypeptides (but see Hirano *et al.*, 1982); small amounts of N-terminal and internal protein sequence have been used to identify clones and processing sites, but the majority of sequence data are from the analysis of cDNA and genomic clones. Complete gene sequences have been determined for two convicilin and one vicilin polypeptide; several part-length and complete cDNA sequences are known for vicilin polypeptides of M_r ~50,000 and ~68,000 and for convicilin (Gatehouse et al., 1983; Lycett et al., 1983; Spencer et al., 1983; Casey et al., 1984; Casey and Domoney, 1985; Bown et al., 1988; Higgins et al., 1988; Domoney and Casey, 1990; Newbigin et al., 1990; see also Higgins, 1984). There are at least five classes of vicilin/convicilin sequence in pea; homologies among the three types of vicilin polypeptide of M_r 47,000–50,000 are ~85% (Gatehouse et al., 1983; Lycett et al., 1983; Spencer et al., 1983; Higgins, 1984), and between the convicilin and the vicilin M_r ~68,000 polypeptides they are ~33% (Domoney and Casey, 1990).

Convicilin is ~80% identical to the ~50,000 M_r vicilin polypeptides over a great deal of its length, but contains an additional highly hydrophilic sequence of 120–166 amino acids at, or near, the N-terminus (Bown et al., 1988; Newbigin et al., 1990). The composition of this 'insert', which is similar to that of the polar region of the legumin α polypeptides (see above), may permit directed modification of amino acid composition without deleterious effects on structure and synthesis. Crosslinking data indicate that convicilin is also a trimer (Casey and Sanger, 1980; Newbigin et al., 1990), rather than a tetramer (Croy et al., 1980b), of subunits of M_r 70,000–75,000.

Vicilins are synthesized as precursors of M_r ~47,000–50,000 or ~68,000 (see Spencer, 1984; Casey and Domoney, 1985, for references). The precursors of M_r ~47,000–50,000 can undergo cleavage after translation at one, both or neither of two internal processing sites (Gatehouse et al., 1983; Spencer et al., 1983). Some members of this class are also glycosylated (Badenoch-Jones et al., 1981; Chrispeels et al., 1982a,b). At least three classes of precursor of M_r ~47,000–50,000 can be recognized on the basis of cDNA sequences. There is insufficient information to suggest the exact nature of the sequence that does or does not get processed in vicilin precursors, but it appears that the sequences Gly–Leu–Arg and Ser–Leu–Lys do not permit cleavage, whereas Lys–Glu–Asn is hydrolysed at the carboxyl side of Asn (Gatehouse et al., 1983). More sequences of processed and non-processed molecules are required to clarify the processing sites. A minor vicilin-related polypeptide of M_r ~68,000 has also been observed in newly synthesized vicilin and appears to be processed to a ~50,000 M_r minor polypeptide (Chrispeels et al., 1982b). The convicilin polypeptides do not appear to undergo any further processing beyond signal peptide removal and are not glycosylated.

Pea legumin and vicilin polypeptides are synthesized as precursors, containing N-terminal signal peptides (Spencer and Higgins, 1980; Higgins and Spencer, 1981) which are removed co-translationally on the rough endoplasmic reticulum, and then targeted into vacuolar protein bodies (Craig et al., 1980; Chrispeels et al., 1982a,b). The nature of targeting signals to the vacuole is just beginning to be defined through the use of broad-bean (*Vicia faba*) legumin mutants (Saalbach et al., 1991). Although the enzymes involved in the proteolytic 'nicking' of legumin and vicilin precursor polypeptides have not been identified, such processing can be carried out *in vitro* (Yang et al., 1990). Some preliminary identification of the corresponding enzymes in soyabean has been reported (Scott et al., 1992) and it is clear from *in vitro* processing studies of soyabean glycinin that 'nicking' is a prerequisite to correct assembly (Dickinson et al., 1989). Some sort of 'chaperone' function is required for folding and interpolypeptide contacts; chaperone proteins in seeds are just beginning to attract attention (Domoney et al., 1991b; Fontes et al., 1991; Marocco et al., 1991; D'Amico et al., 1992).

Genetics

Variation among *Pisum* genotypes has been observed in the net charge of the native proteins, the numbers of polypeptides corresponding to a given protein, and the charge and apparent M_r of individual polypeptides (Thomson and Schroeder, 1978; Casey, 1979a,b; Casey and Sanger, 1980; Davies, 1980; Matta and Gatehouse, 1982; Mahmoud and Gatehouse, 1984; see Fig. 6.7). Differences in polypeptide composition between genotypes is unlikely to be great enough to lead to significant alterations in protein chemical composition (Casey and Short, 1981), but qualitative differences can be used as heritable characteristics for mapping storage protein genes in relation to classical morphological markers (see Chapters 2 and 7). Thus, the genes for major legumin polypeptides have been mapped to linkage group 7, near *r* (Davies, 1980; Matta and Gatehouse, 1982); a convicilin locus has been mapped close to *k*, on linkage group 2 (Matta and Gatehouse, 1982); and a vicilin locus has been assigned to linkage group 7, at a position indistinguishable from *r* (Mahmoud and Gatehouse, 1984). RFLP analyses have extended these data.

Using three legumin cDNAs, variant hybridization patterns were used to map the *leg*A genes (for 60,000 M_r precursors) to linkage group 7, ~10 map units from *r*, the *leg*J genes (for the 63,000–65,000 M_r precursors) to linkage group 1, near *a* and the genes for the 80,000 M_r precursors to within 1–2 map units of *a* in one, but not another, cross (Domoney et al., 1986b).

The number of genes in the *leg*A gene class appears to be five. One of the four *Eco*RI fragments that contain *leg*A genes (see Domoney and Casey, 1985; Casey et al., 1986) has two genes, *leg*A and *leg*D, of which *leg*A is potentially functional. *leg*D is a pseudogene and the two genes are ~8 kb apart (Lycett et al., 1984; Bown et al., 1985). Other cloned *Eco*RI fragments contain only one legumin gene (R.R.D. Croy, personal communication). There are two or three copies of the *leg*J gene per haploid genome (Domoney and Casey, 1985) clustered at a single locus (Domoney et al., 1986b). Two of the genes (*leg*J and *leg*K) are 6–7 kb apart in pea genomic DNA (Thompson et al., 1991). Recent data (Turner et al., 1993) suggest that there are four members of the *leg*J gene class in the cv. 'Dark Skinned Perfection', and also that legumin

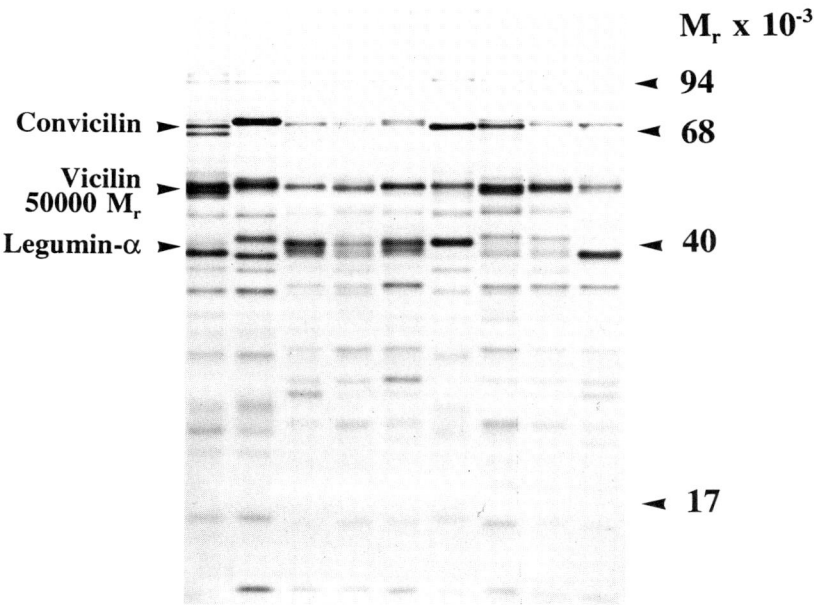

Fig. 6.7. Separation of total pea seed extracts from nine different pea genotypes by SDS-polyacrylamide gel electrophoresis in one dimension; Coomassie blue stained. The pattern varies from one genotype to another; the bands corresponding to convicilin, the ~50,000 M_r polypeptides of vicilin and the legumin α polypeptides are indicated.

gene numbers can vary between genotypes as a consequence of homologous recombination. This important point, which applies to all genes present in multiple copies, is emphasized in Chapter 2. Sequence data (Domoney et al., 1986a; R. Casey, J.F. March and C. Forster, unpublished) have shown there to be at least two different mRNA sequences corresponding to the 80,000 M_r precursors in the cv. 'Birte'. No DNA probes, and therefore no gene copy number data, are available for the L1 subunit pair.

Hybridization analyses (Domoney and Casey, 1985; Domoney et al., 1986b) combined with the isolation and sequencing of genomic clones (Bown et al., 1985; Lycett et al., 1985; Gatehouse et al., 1988; Rerie et al., 1990, 1991; Thompson et al., 1991) and the sequencing of polypeptides (March et al., 1988) indicate that there are more than ten legumin genes in the haploid pea genome that map to at least three genetic loci.

The only definitive correlation that can be made between legumin genes and mature polypeptides is that between legA genes and major (L4 subunit pairs) legumin α and β polypeptides.

Hybridization studies exploiting four vicilin/convicilin genomic clones and four cDNA clones have been used to define seven different genetic loci for pea vicilins/convicilins. Three of these loci have been mapped to positions on linkage group 7 (Ellis et al., 1986); vicilin genes corresponding to the ~68,000 M_r precursors have been mapped to a locus near a (Ellis et al., 1992). A further genetic locus has been recognized through hybridizations to a cDNA corresponding to a 50,000 M_r precursor. The vicilin locus mapped on the basis of polypeptide variation (Mahmoud and Gatehouse, 1984 – see earlier) has yet to be correlated with any loci identified by hybridizations (Ellis et al., 1986).

There are between two (Gatehouse et al., 1983) and seven (Domoney and Casey, 1985) genes corresponding to the 47,000 M_r vicilin precursor. These genes are located on *Eco*RI fragments of ~7.5 kb and the evidence favours the possibility that each fragment of ~7.5 kb contains only one copy of the gene (Ellis et al., 1986). There are two classes of gene corresponding to vicilin precursors of 50,000 M_r, each apparently comprising approximately five copies (Gatehouse et al., 1983; Domoney and Casey, 1985).

Hybridization (Domoney and Casey, 1985) and sequence (Casey et al., 1984; Bown et al., 1988; Newbigin et al., 1990) data indicate the existence of two convicilin genes per haploid genome, which have slightly different sequences but which are genetically linked (Ellis et al., 1986); it is not clear how the two types of sequence relate to mature convicilin polypeptides.

Hybridization at high stringency of a cDNA corresponding to the vicilin precursors of M_r ~68,000 (see earlier) indicates the presence of three copies of this sequence per haploid genome; lower-stringency analyses reveal further copies of related genes that are distinct from the genes for convicilin or the M_r 47,000–50,000 vicilin polypeptides (Domoney and Casey, 1990). There is, therefore, a minimum of 24 vicilin genes in the pea genome at a minimum of seven loci. In most cases it is not possible to correlate specific vicilin/convicilin genes with specific polypeptides.

Gene expression

The genes encoding pea legumins and vicilins are highly regulated during seed development; they are not expressed at any other stage of plant development and each has a particular temporal pattern of expression (Gatehouse et al., 1982, 1986; Chandler et al., 1984; Higgins, 1984; Boulter et al., 1987). The synthesis of vicilin precedes that of legumin and convicilin in the few genotypes examined so far (Guldager, 1978; Millerd et al., 1978; Croy et al., 1980b; Ersland et al., 1983; Chandler et al., 1984; Gatehouse et al., 1986). Analysis of the relative levels of mRNAs corresponding to legumin 60,000 M_r, vicilin 47,000–50,000 M_r and convicilin precursors shows the vicilin mRNAs reach their maximum level early in seed development, whereas those for legumin 60,000 M_r precursor and convicilin are most abundant midway through development (Domoney and Casey, 1983; Chandler et al., 1984; see Fig. 6.8). RNA corresponding to the vicilin precursor of M_r ~68,000 also appears relatively

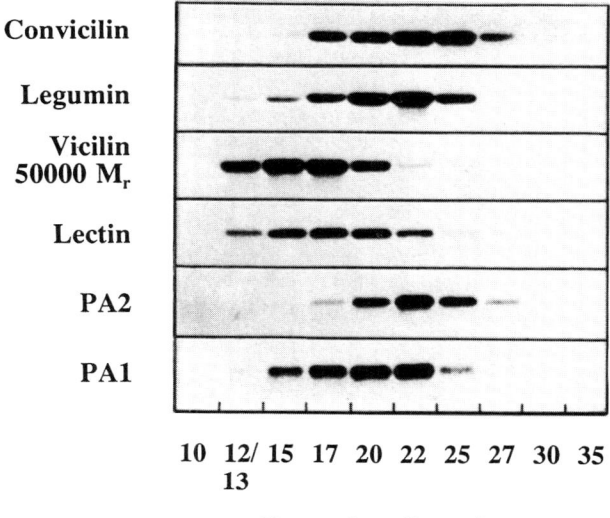

Fig. 6.8. Changes in the relative amounts of mRNA for convicilin, legumin, vicilin M_r 50,000, lectin, albumin PA2 and albumin PA1 precursors. Equal amounts of RNA from different stages of cotyledon development were fractionated by agarose gel electrophoresis, transferred to filters and hybridized to ^{32}P-labelled cDNA plasmids corresponding to the specific mRNAs. From Chandler *et al.* (1984), with permission.

late in seed development (Domoney and Casey, 1990). The mRNAs corresponding to legumin precursors of 63,000–65,000 and 80,000 M_r have different patterns of accumulation from those of 60,000 M_r (Domoney and Casey, 1987; Thompson *et al.*, 1991). The relative timing of the appearance of cytoplasmic mRNA corresponding to pea seed storage proteins seems to reflect transcriptional activity of the legumin, vicilin and convicilin genes (Evans *et al.*, 1984; Beach *et al.*, 1985; Thompson *et al.*, 1989) but the relative amounts of mRNA from the different gene families are subject to appreciable post-transcriptional modulation (Thompson *et al.*, 1989; Turner *et al.*, 1990).

Details of the regulation of pea seed protein genes vary between genotypes and growing conditions. Gene copy number may vary as a consequence of homologous recombination (see Chapter 2); genes may be active in one, but not another, genotype as a consequence of specific mutations (see Thompson *et al.*, 1991, for example); genes such as the *r* locus can have pleiotropic effects on seed protein synthesis (Turner *et al.*, 1990); and nutrient, especially sulphur, deficiency can profoundly affect gene expression (Chandler *et al.*, 1984; Beach *et al.*, 1985; Evans *et al.*, 1985). The two last (the *r* locus and sulphur deficiency) are of interest because each has a differential effect on mRNA stability. The nature of the allele at the *r* locus seems to have a specific effect on legumin mRNA stability, with the high sucrose/osmotic pressure of the *rr* (wrinkled-seeded) mutant (see Chapter 5) leading to decreased stability (Turner *et al.*, 1990). Under conditions of sulphur deficiency, the synthesis of legumin and of the albumin PA1 (see later) is also severely impaired as a consequence of decreased mRNA stability (Chandler *et al.*, 1984; Beach *et al.*, 1985; Spencer *et al.*, 1990), whereas the regulation of vicilin mRNA levels under the same conditions is transcriptional. In neither case is the mechanism of decreased legumin mRNA stability understood. The expression of PA1 in the seeds and/or leaves of transgenic tobacco is still subject to profound modulation by sulphur deficiency (Spencer *et al.*, 1990), suggesting that the sequences responsible for recognizing sulphur status have been included in the PA1 transgene and that the mechanism for regulating the expression of 'sulphur-sensitive' genes is present in tobacco as well as peas (Spencer *et al.*, 1990).

Studies of seed protein genes from legumes, especially soyabean, *Phaseolus vulgaris*, pea and *Vicia faba*, have contributed to our understanding of gene regulation during seed development. The coding regions (plus introns) and adjacent 5' and 3' regions of pea legumin (Shirsat *et al.*, 1989; Rerie *et al.*, 1991), vicilin (Higgins *et al.*, 1988) and convicilin (Newbigin *et al.*, 1990) genes have been introduced into transgenic tobacco. In each case expression was confined to seeds and temporal regulation was similar to that in peas, although the pattern of post-translational processing was not always the same in tobacco as in pea (Newbigin *et al.*, 1990).

Much attention has been paid to the role of sequences located 5' to the coding region of pea legumin, vicilin and convicilin genes in peas (Lycett *et al.*, 1984, 1985; Bown *et al.*, 1985, 1988; Higgins *et al.*, 1988; Shirsat, 1988; Newbigin *et al.*, 1990; Rerie *et al.*, 1990; Thompson *et al.*, 1991). Conserved elements in the 5'-flanking regions of vicilin and legumin genes, referred to as 'vicilin and legumin boxes' (Bäumlein *et al.*, 1986; Gatehouse *et al.*, 1986), were suggested to be involved in, but not solely responsible for, transcriptional control (Gatehouse *et al.*, 1986). The 'legumin box' is a 28 bp sequence of which 22 bp are identical between soyabean, pea and *Vicia faba* 'legumin' genes; its core 'RY-repeat' sequence, CATGCATG (Dickinson *et al.*, 1988; Bäumlein *et al.*, 1992), is found in slightly modified form in a vicilin-like gene in soyabean (Chamberland *et al.*, 1992). 'Vicilin' genes from pea and *Phaseolus vulgaris* have a conserved 'vicilin box' in which 32 out of 42 bp are identical (the pea *leg*A gene also has the first 8 bp of the vicilin box (GCCACCTC) in its 5'-flanking region). Analyses of deletions that remove the 'legumin box' (Bäumlein *et al.*, 1991), destruction of the central CATGCATG motif by a 6 bp deletion (Bäumlein *et al.*, 1992), and progressive deletions of the upstream sequence that leave the CATGCATG motif intact (Shirsat *et al.*, 1990; Bäumlein *et al.*, 1992) all suggest that the 'legumin box' core element is necessary for optimal legumin gene promoter function but can operate optimally only through interaction with other upstream elements. Shirsat et al. (1990) have demonstrated that nuclear proteins from developing pea cotyledons do not bind to a pea legumin gene fragment containing the 'legumin box', but bind instead to regions further upstream (see also Meakin and Gatehouse, 1991). The presence of the CATGCATG motif in other plant genes suggests a possible role in the formation of a functional transcription complex; mutagenesis of CATGCATG sequences in soyabean vicilin-related genes show that two such motifs act together to increase transcription by an order of magnitude (Chamberland et al., 1992).

Albumins

There is some uncertainty over the relative proportions of albumin and globulin in pea seeds (Schroeder, 1984; Chambers et al., 1992), and values for the proportion of albumin in total seed protein range from 15–50%. Such variation may result from differences between the environments in which the plants are grown, genetic effects on albumin amounts, and differences in the methods for measuring albumins. The albumin fraction contains a large number of proteins including protease inhibitors, lipoxygenases, lectins and two specific albumins (PA1 and PA2), some of which are abundant enough to be visible on a stained gel of total pea seed albumin (see Fig. 6.9).

Albumins PA1 and PA2

PA1 contributes less than 10% of the total seed protein and yet provides over 50% of the seed protein sulphur-containing amino acids (Higgins et al., 1986). PA1 from mature pea seeds contains polypeptides of M_r ~6000 and ~4000 that are soluble in 60% aqueous methanol. It is synthesized as a precursor of M_r ~13,000 from which a signal sequence is removed co-translationally. The resultant protein (M_r ~11,000) is subsequently cleaved post-translationally to yield the mature albumins PA1a [M_r ~6000, also known as PsaLA (Gatehouse et al., 1985)] and PA1b (M_r ~4000); both PA1a and PA1b appear to be post-translationally processed by the removal of six to eight C-terminal residues (Higgins et al., 1986). mRNA corresponding to PA1 appears at about the same time as that for legumin (Chandler et al., 1984; see Fig. 6.8). Nucleotide and amino acid sequence data indicate that there are at least four, slightly diverged, PA1 genes expressed in developing pea cotyledons. No biological function has been clearly assigned to PA1, although PA1b has a low level of homology to the active sites of protease inhibitors and to other low-molecular-weight seed albumins (Higgins et al., 1986). The homologous protein in *Phaseolus vulgaris* may be involved in vacuolar sorting (Tanchak and Chrispeels, 1989). The high cysteine content (11%) of PA1 makes it an important source of protein sulphur; significantly increasing the amounts of PA1 in pea seeds would improve the content of sulphur-containing amino acids in pea seed protein.

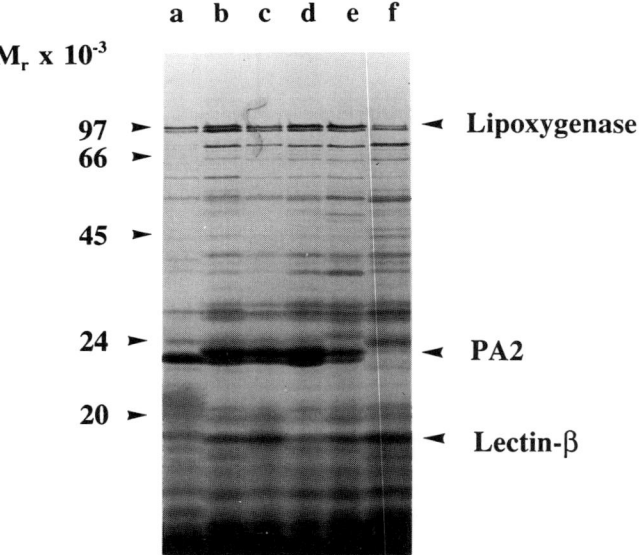

Fig. 6.9. SDS-PAGE analysis of albumins from six *Pisum* genotypes. Variants with one (a), two (b-e) or none (f) of two PA2 polypeptides are shown.

PA2 (Higgins *et al.*, 1987), also known as PsaMA, PMA, MA, PMA-L and PMA-S (Croy *et al.*, 1984; Harris and Croy, 1985; Rao *et al.*, 1989), consists of two distinct polypeptides (PA2a and PA2b) of M_r ~24,000–26,000, the relative proportions of which can vary significantly between genotypes (Schroeder, 1984; Rao *et al.*, 1989; see Fig. 6.9). mRNA corresponding to PA2 appears relatively late in seed development (Chandler *et al.*, 1984; see Fig. 6.8). The amino acid sequence of PA2 predicted from a cDNA clone is 230 amino acids in length, containing four imperfect repeat sequences of approximately 57 amino acids (Higgins *et al.*, 1987). PA2 is unlike most other major pea seed proteins in being synthesized without a signal sequence, undergoing no major post-translational modification, being localized in the cytosol rather than protein bodies (Croy *et al.*, 1984; Harris and Croy, 1985; Higgins *et al.*, 1987) and being relatively resistant to breakdown during germination (Croy *et al.*, 1984), which may reflect its tightly folded, non-helical, globular structure (Gruen *et al.*, 1987).

Variation in the apparent M_r of PA2a and PA2b has been used to show they are the products of single Mendelian genes at a single locus (Rao *et al.*, 1989). Mutant lines have been identified that lack PA2a and/or PA2b (Rao *et al.*, 1989; see Fig. 6.9); a *Pisum* accession that lacks both PA2 polypeptides appears to have a structural deletion of the PA2 genes (Rao *et al.*, 1989).

The physiological function of the PA2 proteins is unclear; they are not nitrogen-rich, are relatively resistant to breakdown during germination and are not sufficiently rich in sulphur-containing amino acids (~2.5 residues percent) to represent a significant sulphur store.

Lipoxygenases

Lipoxygenases are commercially important enzymes that catalyse the oxygenation of fatty acids and other compounds that contain a *cis,cis*,1,4 pentadiene system. In soyabean seeds, three isoforms comprise 1–2% of the total seed protein; in pea seeds the amounts of lipoxygenases are considerably lower. One of the soyabean isoforms (lipoxygenase-2) is primarily responsible for the generation of off-flavours and the so-called 'grassy-beany' flavour of soyabean protein products (Davies *et al.*, 1987). Pea seeds contain two major lipoxygenase polypeptides that correspond in sequence to lipoxygenases-2 and -3 from soyabean (Casey *et al.*, 1985; Ealing and Casey, 1988, 1989). There are also three relatively minor pea seed lipoxygenases (Domoney *et al.*, 1990). Genetic analyses of major lipoxygenase polypeptide variants (M_r ~97,000 on SDS gels – see Fig. 6.9) have mapped the genes encoding these major polypeptides to a locus (*lox*) on linkage group four, close to the *le* locus (North *et al.*, 1989). RFLP analyses have also identified a lipoxygenase locus on linkage group four but have identified additional, unlinked lipoxygenase gene loci (Domoney *et al.*, 1991a; see also Chapter 2). Each of the classes of gene encoding the two major seed polypeptides contains two to three members (Domoney *et al.*, 1991a). Pea seed lipoxygenase genes are developmentally regulated, with the mRNAs corresponding to the two major seed lipoxygenases appearing at a relatively late stage of seed development (Domoney *et al.*, 1990; see also Chapter 5).

Most interest has centred on the seed lipoxygenases because of their relative abundance, their role in the generation of undesirable tastes and their implications for the long-term storage of frozen foods; if lipoxygenase(s) are not inactivated by heating (blanching) before freezing, unpleasant (rancid) tastes will develop and limit shelf-life. Selective removal of specific lipoxygenases is therefore of potential importance to the frozen (vining) pea industry. Pea genotypes in which one or other of the major seed lipoxygenase polypeptides is either absent or greatly reduced in amount (North, 1990) will prove significant in this context. Lipoxygenases that are capable of co-oxidation reactions, such as those from soyabean, *Vicia faba* and pea seeds, have a beneficial effect on the rheological properties of dough in breadmaking (Frazier, 1979), possibly through the co-oxidation of thiol groups in wheat gluten. As part of a project to construct a genetically engineered baker's yeast that secretes active lipoxygenase during breadmaking, a full-length cDNA clone for one of the major pea seed lipoxygenases has been transformed into baker's yeast which subsequently produces active lipoxygenase (Knust and von Wettstein, 1992).

Additional lipoxygenase polypeptides, distinct from those in seed, have been identified in stems and roots (Domoney et al., 1990), and attention has recently focused on such lipoxygenases in vegetative tissues because they have been implicated in plants' responses to wounding, pathogen and pest attack and water stress, and in the synthesis of the plant growth regulators methyl jasmonate and abscisic acid (Hildebrand et al., 1988; Bell and Mullet, 1991; Tranbarger et al., 1991; Creelman et al., 1992); there is, however, no definitive evidence that lipoxygenases play any physiological role in many stress situations, and the evidence that they are involved in the synthesis of abscisic acid (Creelman et al., 1992) or jasmonate (Vick and Zimmerman, 1983) is still tenuous.

Protease inhibitors

A heterogeneous group of compounds from legume seeds is currently receiving considerable attention, primarily as a result of their negative influence on the nutritional quality of seed. These antinutritional factors include diverse seed constituents, such as tannins, lectins, alkaloids, phytates, saponins and the protein inhibitors of digestive enzymes, all of which are relatively minor in amount; more major protein constituents of seeds, however, can also be regarded as antinutritional where they invoke important antigenic responses in animals.

The protein inhibitors of animal digestive enzymes are of interest for a variety of reasons. From a nutritional viewpoint, these inhibitors bind more or less irreversibly to their target digestive enzyme, causing reduced protein digestibility and absorption and, in many animals, reduced weight gain. Similar effects on animal digestion can also be elicited, however, through the action of other seed compounds (e.g. tannins or phytates) on protein; the results of many nutritional studies and assays of crude extracts should therefore be treated with caution, as the extent of interaction between various antinutritional factors is largely unknown. Many reports exist which document variation in the content of trypsin inhibitor in peas. The results of different workers are difficult to correlate for many reasons (Savage, 1989), but a survey of 63 pea genotypes has revealed a tenfold range in apparent trypsin inhibitor content (Domoney and Welham, 1992). Although an environmental influence on trypsin inhibitor content is indicated by many studies, there is a strong genotype component to the variation observed and selected genotypes have been shown to be consistently at the upper or lower ends of the range, irrespective of growing conditions (Domoney and Welham, 1992). The accumulation pattern of trypsin inhibitors during pea seed development suggests that a significant proportion is synthesized during the very late or desiccation phase of seed development, a feature that is shared with other classes of seed proteins, such as the pea seed abscisic-acid-responsive proteins (Barratt et al., 1989;

Domoney and Welham, 1992; see below). Detailed analyses of the genes encoding these diverse proteins will be necessary to decide whether the controls influencing late expression are similar for these gene classes.

Up to five trypsin inhibitors have been distinguished in selected pea genotypes; purification and sequence analysis have revealed that two pea inhibitors have identical N-terminal sequences and a third is related to these but lacks the two terminal amino acids (Domoney et al., 1993). The inhibitors with identical N-terminal sequences must differ elsewhere in the molecule to account for their different behaviour on gels and columns; these differences may directly reflect differences in gene structure. Alternatively, post-translational modifications (such as differential C-terminal processing, glycosylation or phosphorylation, for example) could be responsible firstly for generating heterogeneous inhibitors and secondly for generating inhibitors which are more or less active than the parent molecule. The pea inhibitors are homologous to the Bowman-Birk class of inhibitors (Domoney et al., 1993), which are double-headed inhibitors, being capable of the simultaneous inhibition of two enzymes, most usually trypsin and chymotrypsin (for review, see Richardson, 1991). Although it has been shown that a number of pea inhibitors are related antigenically (Gaborit et al., 1989), it remains to be seen whether all are members of the Bowman-Birk class.

Manipulation of the content of pea seed trypsin inhibitors, as with any other seed or plant constituent, requires an understanding of their biology and an assessment of the consequences of their removal or modification. The physiological role of inhibitors is unclear, but enzyme inhibitors have been implicated in the protection of plants from predation; this role is clearest in the case of cowpea and potato inhibitors, where protection of tobacco plants from predation was achieved through gene transfer (Hilder et al., 1987, 1990; Johnson et al., 1989). Furthermore, a Bowman-Birk-type inhibitor is induced by wounding in alfalfa leaves (Brown et al., 1985; Graham et al., 1985a,b). Other inhibitors have been implicated in nodule development (Manen et al., 1991). Clearly it is advisable to be aware of the beneficial roles played by particular inhibitors in seed or plant development and protection before attempting their removal. Another possible general role for inhibitors lies in the protection of seed reserves from premature hydrolysis (Poerio et al., 1989), or, alternatively, in the protection of labile proteases which are required for germination (Gennis and Cantor, 1976). Although an endogenous protease with trypsin-like activity has been identified in pea seeds, the *in vitro* activity of this protease is unaffected by addition of any of the pea trypsin inhibitors so far identified (Domoney and Welham, 1992; Domoney et al., 1993); these data do not support, therefore, a role for pea trypsin inhibitors in the inhibition of at least this endogenous enzyme. Obviously if inhibitors play crucial roles in seed longevity and/or seedling vigour, their complete removal from plants could have disastrous consequences. It is, however, conceivable that the seed inhibitors that are the most problematical from a nutritional viewpoint may not be the ones that play critical roles in plant protection and/or development. Therefore a thorough examination of individual inhibitors is necessitated by these possibilities. Finally, it is worth noting that the pea trypsin inhibitors so far characterized (Domoney et al., 1993), like other Bowman-Birk inhibitors, are proteins that are rich in sulphur-containing amino acids, specifically cysteine, and, as such, are of positive nutritional significance. As a result, modification of their active sites, while retaining their beneficial features, may be the preferred route by which pea seed trypsin inhibitors should be manipulated.

Abscisic acid-responsive (ABR) proteins

Two albumins which are produced late in pea seed development *in vivo* have been characterized and shown to be synthesized in embryos cultured *in vitro* in response to exogenous abscisic acid (ABA) (Barratt et al., 1989). The two proteins, which are related immunologically, have apparent M_rs of 17,000 and 18,000, appear to lack signal peptides, are glycosylated and are not confined to seed (Barratt et al., 1989; Barratt and Clark, 1991). ABR18 and its mRNA have been detected in the testa during early seed development, although their synthesis is confined to the very late stages of embryo (cotyledon) development; ABR17 protein is present in non-stressed leaves and roots at a low level, although ABR17 mRNA increased during severe desiccation of pea leaves, where ABA content increased 30-fold (Barratt and Clark, 1991). The roles played by these proteins during desiccation is not clear, but their increased synthesis in response to increased ABA concentrations may be an indirect effect of alteration of water relations and the proteins may function as protectants. The N-terminal sequences of the ABR proteins show 56% identity to each other and show homology to two proteins that have been implicated in the resistance of peas to pathogens (see Chapter 8), as well as to a pollen allergen from birch (Barratt and Clark, 1991). The ABR proteins may thus play a defence role; their lack of homology with other late-embryogenesis proteins suggests that they are a distinct group of stress-induced proteins.

Lectins

Lectins are carbohydrate-binding proteins of non-immune origin which agglutinate cells or precipitate polysaccharides or glycoconjugates (Goldstein et al., 1980). The lectin from pea seeds, which represents ~2% of total dry seed protein (Gatehouse et al., 1986), is D-mannose/D-glucose-specific, consisting of two light (α) and two heavy (β) chains (Trowbridge, 1974). The primary sequences of each have been deduced from cDNA sequences (Higgins et al., 1983a) and biosynthetic data (Higgins et al., 1983b) show that the α and β polypeptides are synthesized as a covalently linked $\alpha\beta$ precursor, which is subsequently proteolytically split after removal of an N-terminal signal peptide. The final α and β chains are non-covalently attached in the quaternary structure (which has the composition $\alpha_2\beta_2$); the crystal structure of pea seed lectin has been determined to 3 Å resolution (Einspahr et al., 1986). It has been argued that lectins may act as defensive agents, carbohydrate transporters, recognition agents, storage proteins or growth regulators (see, for example, Pusztai, 1991), but experimental evidence in support of biological roles is scanty. Lectin mRNA appears early in seed development (Chandler et al., 1984; see Fig. 6.8). Although lectins are present in protein bodies within seeds, they are also found in roots and shoots/buds of peas (Gatehouse and Boulter, 1980; Buffard et al., 1988; Pak et al., 1992). Using a pea seed lectin cDNA probe, mRNA of the same size as that in seeds appears to be present in pea roots at a level ~4000 times less than that in seeds; the highest levels of root lectin are observed just before nodulation (Buffard et al., 1988). Lectins, including those from pea, have long been implicated in the specificity of the *Rhizobium*–legume symbiosis (see Pusztai, 1991, for a review). Diaz et al. (1989) have broken the host-specificity barrier by introducing a pea lectin gene into white clover roots, resulting in clover (*Trifolium repens*) that has roots which have become susceptible to infection by *Rhizobium leguminosarum* bv. *viciae*. There are, however, many questions that need to be answered (see Pusztai, 1991) before the essential nature of lectin involvement in *Rhizobium*–pea symbiosis can be accepted (see also Chapter 9). A recently described mutant, in which conversion of Asn^{125} to Asp in the lectin eliminates sugar binding, may prove useful in this respect (van Eijsden et al., 1992).

Although there are at least four genes encoding the pea seed lectin precursor, only one of them is functional (Kaminski et al., 1987) and this appears to be expressed in seed and roots (Buffard et al., 1988; Thompson et al., 1989). In contrast, at least four members of a highly conserved lectin multigene family, whose members are poorly related to the pea seed lectin, are expressed in pea shoots (Pak et al., 1992).

There are no reports of qualitative lectin variants in peas, but Rao et al. (1989) have shown that seed lectin amounts are reduced in a *Pisum sativum* line that lacks PA2 (see above). The significance of this is not clear.

One of the most important interactions between plant lectins and animals occurs in the latter's alimentary canals, and the biological responses to dietary lectins are dependent on their degree of resistance to gut proteolysis. Many lectins show appreciable resistance and are detrimental to the growth of animals (e.g. phytohaemagglutinin (PHA) from *Phaseolus vulgaris*). The nutritional utilization of peas is only slightly improved by heat treatments and pea lectin is considerably more susceptible to proteolysis than PHA, for instance (Pusztai et al., 1990). Pea lectin is therefore not generally recognized as an antinutritional factor (see Pusztai, 1991).

Seed protein biotechnology

There are a number of potential objectives in terms of improvement of seed proteins, including: increased cysteine/methionine content; increased digestibility; decreased nutrient/mineral binding; increased foaming and emulsification capacity; more varied gelling and texturizing properties; increased flavour; increased water and fat binding; lower heat stability – all of which would be of importance to the food industry for one reason or another. Since pea genotypes with widely different amounts and proportions of seed proteins can be found in germplasm collections, it may well be possible to identify, or to produce by conventional breeding methods, strains that are better suited to the food industry. For example, legumin and vicilin have different physical properties and their ratio within a protein preparation will influence protein quality parameters (Wright and Bumstead, 1984); pea protein isolates from genotypes with a low 11S : 7S ratio, for instance, have improved emulsification properties (Dagorn-Scaviner et al., 1986). Again, genotypes with reduced amounts of seed lipoxygenases or of protease inhibitors (see earlier) may be of significance in the context of flavour formation and digestibility, respectively.

In principle, recombinant protein technology ('protein engineering') can play a major role in seed protein improvement (see, for example, Shotwell and Larkins, 1991; Shewry and Gutteridge, 1992); this should be attainable eventually through the introduction into plants of 'designer' seed proteins with the appropriate desirable properties. There are, however, a number of factors that may constrain changes to protein structure and synthesis and these need to be considered before modifications are made. Alterations to solubility may affect aggregation properties and deposition in protein bodies; introduction of cysteine residues could lead to spurious disulphide bond formation and anomalous secondary structures; modification of (as yet largely undefined) critical residues could affect polypeptide folding and assembly, interfere with post-translational processing and assembly, decrease protein stability (leading to premature hydrolysis), or disrupt targeting signals within the coding sequence that are required for proper transport. For example, Hoffman et al. (1988) modified a gene for β-phaseolin (the equivalent of vicilin in *Phaseolus vulgaris*) by the introduction of an oligonucleotide containing six methionine codons and expressed it in transgenic tobacco seeds. The resultant high-methionine phaseolin was synthesized in the endoplasmic reticulum (ER), assembled into trimers and transported to the Golgi apparatus but was then degraded. Subsequent studies (Lawrence et al., 1990) showed that the site into which the additional methionine residues were introduced is important for trimer stability, emphasizing the need for detailed understanding of the physical chemistry of a protein prior to directed modification.

Considerable progress has been made by Nielsen and his colleagues (Dickinson et al., 1989; Scott et al., 1992; see also Shotwell and Larkins, 1991) in defining the types and degree of change to soyabean storage globulins that are still compatible with assembly and oligomerization, through *in vitro* processing and assembly experiments. Modifications to the basic polypeptide of glycinin (the soyabean equivalent of legumin) usually eliminated the ability of the protein to assemble *in vitro*. Alterations in the acidic polypeptide changed solubility characteristics, but did not eliminate assembly. Extensive deletions and insertions were made in the highly variable polar region of the acidic polypeptide (described earlier in this chapter under legumin structure) without perturbing subunit assembly; some of the insertions increased significantly the methionine content of the polypeptide (Dickinson *et al.*, 1990). Such experiments permit the rapid evaluation of the effects of alterations before introducing modified genes into transgenic plants. Little has been done towards the engineering of pea seed storage proteins, beyond the demonstration that their precursors can be processed *in vitro* (Yang *et al.*, 1990) and the introduction of disulphide bridges into vicilin (Dale *et al.*, 1991).

A better understanding of the limitations to, and likely consequences of, directed modifications will be achieved through a knowledge of three-dimensional structures; the production of crystals for such studies will almost certainly come through the *in vitro* expression of large amounts of specific polypeptides, since the natural heterogeneity of the storage proteins is likely to preclude the production of highly diffractible crystals. Such an *in vitro* product will, however, not reflect the *in vivo* situation, being composed of only one type of monomer; further, some specific polypeptides appear to be incapable of oligomerization/assembly in the absence of others (see Shotwell and Larkins, 1991).

Given a successfully modified gene, it is still necessary to introduce it into peas (see Chapter 10), where it must be expressed in a temporally and spatially controlled way; experiments using transgenic plants have indicated that pea seed storage protein genes contain elements that regulate both the timing and the organ-specificity of expression (Higgins *et al.*, 1988; Shirsat *et al.*, 1989; Newbigin *et al.*, 1990; Rerie *et al.*, 1991). The amount of modified product must also be significantly above that from the unmodified genes, which will require the identification of strong, seed-specific promoters. In principle it should also be possible to reduce expression from genes encoding less desirable products by transformation with constructs that produce antisense RNA. In practice this approach may founder if both the normal and the modified gene products are recognized by the antisense RNA.

A number of attempts to express pea storage protein genes in leaves of transgenic plants have led to low-level accumulation (see, for example, Higgins and Spencer, 1991). Provision of the ER targeting signal Lys–Asp–Glu–Leu at the C-terminus of a pea vicilin polypeptide results in high-level accumulation in transgenic tobacco and lucerne leaves (Wandelt *et al.*, 1992); it is believed that retention in the ER leads to increased stability by protecting the foreign protein from vacuolar proteases. Such steps have, however, proved to be unnecessary in the expression of pea lectin in potato leaves (Edwards *et al.*,

1991), and the factors that govern stability of foreign seed proteins in leaves are not clear at present.

If the overall objective is to increase the cysteine/methionine content of pea seed protein, it may be better to express a chemically synthesized gene for a high-quality polypeptide or a heterologous high-sulphur protein from another source (e.g. the 2S albumins from Brazil nut or sunflower), in transgenic pea seeds, as has been done in tobacco and rapeseed (Altenbach *et al.*, 1989, 1992). Expression of high-quality proteins in seeds and their protection by sequestration into protein bodies undoubtedly has great potential, exemplified by the production of the peptide pharmaceutical Leu-enkaphalin in transgenic rapeseed (Vandekerckhove *et al.*, 1989).

At present directed modification of legume seed protein is in its infancy (see Shewry and Gutteridge, 1992). A combination of protein chemistry, protein engineering, plant transformation and the use of genetic variants of peas will eventually lead to improvements in the quality of seed protein, either through directed alterations in amino acid composition, changes in protein physical properties or the elimination of undesirable activities.

References

Altenbach, S.B., Pearson, K.W., Meeker, G., Staraci, L.C. and Sun, S.S.M. (1989) Enhancement of the methionine content of seed proteins by the expression of a chimeric gene encoding a methionine-rich protein in transgenic plants. *Plant Molecular Biology* 13, 513–522.

Altenbach, S.B., Kuo, C.-C., Staraci, L.C., Pearson, K.W., Wainwright, C., Georgescu, A. and Townsend, J. (1992) Accumulation of a Brazil nut albumin in seeds of transgenic canola results in enhanced levels of seed protein methionine. *Plant Molecular Biology* 18, 235–245.

Badenoch-Jones, J., Spencer, D., Higgins, T.J.V. and Millerd, A. (1981) The role of glycosylation in storage-protein synthesis in developing pea seeds. *Planta* 153, 201–209.

Banks, W. and Muir, D.D. (1980) Structure and chemistry of the starch granule. In: Preiss, J. (ed.), *The Biochemistry of Plants, Vol. 3: Carbohydrates: Structure and Function*, Academic Press, New York, pp. 321–369.

Banks, W., Greenwood, C.T. and Muir, D.D. (1974) Studies on starches of high amylose content. *Stärke* 23, 289–300.

Barratt, D.H.P. and Clark, J.A. (1991) Proteins arising during the late stages of embryogenesis in *Pisum sativum* L. *Planta* 184, 14–23.

Barratt, D.H.P., Domoney, C. and Wang, T.L. (1989) Purification and partial characterisation of two abscisic acid-responsive proteins induced in cultured embryos of *Pisum sativum* L. *Planta* 180, 16–23.

Bäumlein, H., Wobus, U., Pustell, J. and Kafatos, F.C. (1986) The legumin gene family: structure of a B-type gene of *Vicia faba* and a possible legumin gene specific regulatory element. *Nucleic Acids Research* 14, 2707–2720.

Bäumlein, H., Boerjan, W., Nagy, I., Panitz, R., Inzé, D. and Wobus, U. (1991)

Upstream sequences regulating legumin gene expression in heterologous transgenic plants. *Molecular and General Genetics* 225, 121–128.

Bäumlein, H., Nagy, I., Villarroel, R., Inzé, D. and Wobus, U. (1992) *Cis*-analysis of a seed protein gene promoter: the conservative RY repeat CATGCATG within the legumin box is essential for tissue-specific expression of a legumin gene. *Plant Journal* 2, 233–239.

Beach, L.R., Spencer, D., Randall, P.J. and Higgins, T.J.V. (1985) Transcriptional and post-transcriptional regulation of storage protein gene expression in sulfur-deficient pea seeds. *Nucleic Acids Research* 13, 999–1013.

Bell, E. and Mullet, J.E. (1991) Lipoxygenase gene expression is modulated in plants by water deficit, wounding, and methyl jasmonate. *Molecular and General Genetics* 230, 456–462.

Bhattacharyya, M.K., Smith, A.M., Ellis, T.H.N., Hedley, C. and Martin, C. (1990) The wrinkled-seed character of pea described by Mendel is caused by a transposon-like insertion in a gene encoding starch-branching enzyme. *Cell* 60, 115–122.

Boulter, D., Evans, I.M., Ellis, J.R., Shirsat, A., Gatehouse, J.A. and Croy, R.R.D. (1987) Differential gene expression in the development of *Pisum sativum*. *Plant Physiology and Biochemistry* 25, 283–289.

Bown, D., Levasseur, M., Croy, R.R.D., Boulter, D. and Gatehouse, J.A. (1985) Sequence of a pseudogene in the legumin gene family of pea (*Pisum sativum* L.). *Nucleic Acids Research* 13, 4527–4538.

Bown, D., Ellis, T.H.N. and Gatehouse, J.A. (1988) The sequence of a gene encoding convicilin from pea (*Pisum sativum* L.) shows that convicilin differs from vicilin by an insertion near the N-terminus. *Biochemical Journal* 251, 717–726.

Brown, W.E., Takio, K., Titani, K. and Ryan, C.A. (1985) Wound-induced trypsin inhibitor in alfalfa leaves: identity as a member of the Bowman-Birk inhibitor family. *Biochemistry* 24, 2105–2108.

Buffard, D., Kaminski, P.A. and Strosberg, A.D. (1988) Lectin-gene expression in pea (*Pisum sativum* L.) roots. *Planta* 173, 367–372.

Casey, R. (1979a) Immunoaffinity chromatography as a means of purifying legumin from *Pisum* (pea) seeds. *Biochemical Journal* 177, 509–520.

Casey, R. (1979b) Genetic variability in the structure of the α-subunits of legumin from *Pisum*-a two-dimensional gel electrophoresis study. *Heredity* 43, 265–272.

Casey, R. and Domoney C. (1985) Variation in pea-seed storage proteins. In: Hebblethwaite, P.D., Dawkins, T.C.K. and Heath, M.C. (eds), *The Pea Crop*. Butterworths, London, pp. 359–367.

Casey, R. and Sanger, E. (1980) Purification and some properties of a 7S seed storage protein from *Pisum* (pea). *Biochemical Society Transactions* 8, 658.

Casey, R. and Short, M.N. (1981) Variation in amino acid composition of legumin from *Pisum*. *Phytochemistry* 20, 21–23.

Casey, R., March, J.F. and Sanger, E. (1981a) N-terminal amino acid sequence of β-subunits of legumin from *Pisum sativum*. *Phytochemistry* 20, 161–163.

Casey, R., March, J.F., Sharman, J.E. and Short, M.N. (1981b) The purification, N-terminal amino acid sequence and some other properties of an α^M-subunit of legumin from the pea (*Pisum sativum* L.). *Biochimica et Biophysica Acta* 670, 428–432.

Casey, R., Sharman, J.E., Wright, D.J., Bacon, J.R. and Guldager, P. (1982) Quantitative variability in *Pisum* seed globulins: its assessment and significance. *Qualitas*

Plantarum, *Plant Foods for Human Nutrition* 31, 333-346.

Casey, R., Domoney, C. and Stanley, J. (1984) Convicilin mRNA from pea (*Pisum sativum* L.) has sequence homology with other legume 7S storage protein mRNA species. *Biochemical Journal* 224, 661-666.

Casey, R., Domoney, C. and Nielsen, N.C. (1985) Isolation of a cDNA clone for pea seed lipoxygenase. *Biochemical Journal* 232, 79-85.

Casey, R., Domoney, C. and Ellis, T.H.N. (1986) Legume storage proteins and their genes. *Oxford Surveys of Plant Molecular and Cell Biology* 3, 1-95.

Chamberland, S., Daigle, N. and Bernier, F. (1992) The legumin boxes and the 3' part of a soybean β-conglycinin promoter are involved in seed expression in transgenic tobacco plants. *Plant Molecular Biology* 19, 937-949.

Chambers, S.J., Bacon, J.R. and Lambert, N. (1992) The quantitative analysis of seed proteins from peas using high performance liquid chromatography. *Phytochemical Analysis* 3, 49-54.

Chandler, P.M., Spencer, D., Randall, P.J. and Higgins, T.J.V. (1984) Influence of sulfur nutrition on developmental patterns of some major pea seed proteins and their mRNAs. *Plant Physiology* 75, 651-657.

Chrispeels, M.J., Higgins, T.J.V., Craig, S. and Spencer, D. (1982a) The role of the endoplasmic reticulum in the synthesis of reserve proteins and the kinetics of their transport to protein bodies in developing pea cotyledons. *Journal of Cell Biology* 93, 5-14.

Chrispeels, M.J., Higgins, T.J.V. and Spencer, D. (1982b) Assembly of storage protein oligomers in the endoplasmic reticulum and processing of the polypeptides in the protein bodies of developing pea cotyledons. *Journal of Cell Biology* 93, 306-313.

Colonna, P. and Mercier, C. (1984) Macromolecular structure of wrinkled-and smooth-pea starch components. *Carbohydrate Research* 126, 233-247.

Colonna, P. and Mercier, C. (1985) Gelatinization and melting of maize and pea starches with normal and high-amylose genotypes. *Phytochemistry* 24, 1667-1674.

Colonna, P., Buleon, A., Lemaguer, M. and Mercier, C. (1982) *Pisum sativum* and *Vicia faba* carbohydrates. IV. Granular structure of wrinkled pea starch. *Carbohydrate Polymers* 2, 43-59.

Craig, S., Millerd, A. and Goodchild, D.J. (1980) Structural aspects of protein accumulation in developing pea cotyledons. III. Immunocytochemical localization of legumin and vicilin using antibodies shown to be specific by the enzyme-linked immunosorbent assay (ELISA). *Australian Journal of Plant Physiology* 7, 339-351.

Creelman, R.A., Bell, E. and Mullet, J.E. (1992) Involvement of a lipoxygenase-like enzyme in abscisic acid biosynthesis. *Plant Physiology* 99, 1258-1260.

Croy, R.R.D. and Gatehouse, J.A. (1985) Genetic engineering of seed proteins: current and potential applications. In: Dodds, J.H. (ed.), *Plant Genetic Engineering*. Cambridge University Press, Cambridge, pp. 143-268.

Croy, R.R.D., Gatehouse, J.A., Evans, I.M. and Boulter, D. (1980a) Characterisation of the storage protein subunits synthesised *in vitro* by polyribosomes and RNA from developing pea (*Pisum sativum* L.) I. Legumin. *Planta* 148, 49-56.

Croy, R.R.D., Gatehouse, J.A., Tyler, M. and Boulter, D. (1980b) The purification and characterization of a third storage protein (convicilin) from the seeds of pea (*Pisum sativum* L.). *Biochemical Journal* 191, 509-516.

Croy, R.R.D., Lycett, G.W., Gatehouse, J.A., Yarwood, J.N. and Boulter, D. (1982)

Cloning and analysis of cDNAs encoding plant storage protein precursors. *Nature, London* 295, 76–79.
Croy, R.R.D., Hoque, M.S., Gatehouse, J.A. and Boulter, D. (1984) The major albumin proteins from pea (*Pisum sativum* L.). *Biochemical Journal* 218, 795–803.
Dagorn-Scaviner, C., Gueguen, J. and Lefebvre, J. (1986) A comparison of interfacial behaviours of pea legumin and vicilin at air/water interfaces. *Die Nahrung* 30, 337–347.
Dale, A.S., Elvin, P., Yarwood, J.N. and Gatehouse, J.A. (1991) Introduction of sulphydryl groups into pea vicilin: formation of intra-and inter-polypeptide disulphide bonds. *Journal of the Science of Food and Agriculture* 55, 551–562.
D'Amico, L., Valsasina, B., Daminati, M.G., Fabbrini, M.S., Nitti, G., Bollini, R., Ceriotti, A. and Vitale, A. (1992) Bean homologs of the mammalian glucose-regulated proteins: induction by tunicamycin and interaction with newly synthesized seed storage proteins in the endoplasmic reticulum. *Plant Journal* 2, 443–455.
Davies, C.S., Nielsen, S.S. and Nielsen, N.C. (1987) Flavor improvement of soybean preparations by genetic removal of lipoxygenase-2. *Journal of the American Oil Chemists' Society* 64, 1428–1433.
Davies, D.R. (1980) The r_a locus and legumin synthesis in *Pisum sativum*. *Biochemical Genetics* 18, 1207–1219.
Denyer, K. and Smith, A.M. (1988) The capacity of plastids from developing pea cotyledons to synthesize acetyl CoA. *Planta* 173, 172–182.
Denyer, K. and Smith, A.M. (1992) The purification and characterisation of two isoforms of soluble starch synthase from developing pea embryos. *Planta* 186, 609–617.
Denyer, K., Sidebottom, C., Hylton, C.M. and Smith, A.M. (1993) Soluble isoforms of starch synthase and starch-branching enzyme also occur within starch granules in developing pea embryos. *Plant Journal* 4, 191–198.
Derbyshire, E., Wright, D.J. and Boulter, D. (1976) Legumin and vicilin, storage proteins of legume seeds. *Phytochemistry* 15, 3–24.
Diaz, L.C., Melchers, L.S., Hooykaas, P.J.J., Lugtenberg, B.J.J. and Kijne, J.W. (1989) Root lectin as a determinant of host-plant specificity in the *Rhizobium*–legume symbiosis. *Nature, London* 338, 579–581.
Dickinson, C.D., Evans, R.P. and Nielsen, N.C. (1988) RY repeats are conserved in the 5'-flanking region of legume seed protein genes. *Nucleic Acids Research* 16, 371.
Dickinson, C.D., Hussein, E.H.A. and Nielsen, N.C. (1989) Role of posttranslational cleavage in glycinin assembly. *Plant Cell* 1, 459–469.
Dickinson, C.D., Scott, M.P., Hussein, E.H.A., Argos, P. and Nielsen, N.C. (1990) Effect of structural modifications on the assembly of a glycinin subunit. *Plant Cell* 2, 403–413.
Domoney, C. and Casey, R. (1983) Cloning and characterization of complementary DNA for convicilin, a major seed storage protein in *Pisum sativum* L. *Planta* 159, 446–453.
Domoney, C. and Casey, R. (1984) Storage protein precursor polypeptides in cotyledons of *Pisum sativum* L. Identification of, and isolation of a cDNA clone for, an 80,000 M_r legumin-related polypeptide. *European Journal of Biochemistry* 139, 321–327.

Domoney, C. and Casey, R. (1985) Measurement of gene number for seed storage proteins in *Pisum*. *Nucleic Acids Research* 13, 687–699.
Domoney, C. and Casey, R. (1987) Changes in legumin messenger RNAs throughout seed development in *Pisum sativum* L. *Planta* 170, 562–566.
Domoney, C. and Casey, R. (1990) Another class of vicilin gene in *Pisum*. *Planta* 182, 39–42.
Domoney, C. and Welham, T. (1992) Trypsin inhibitors in *Pisum*: variation in amount and pattern of accumulation in developing seed. *Seed Science Research* 2, 147–154.
Domoney, C., Barker, D. and Casey, R. (1986a) The complete deduced amino acid sequences of legumin β-polypeptides from different genetic loci in *Pisum*. *Plant Molecular Biology* 7, 467–474.
Domoney, C., Ellis, T.H.N. and Davies, D.R. (1986b) Organization and mapping of legumin genes in *Pisum*. *Molecular and General Genetics* 202, 280–285.
Domoney, C., Firmin, J.L., Sidebottom, C., Ealing, P.M., Slabas, A. and Casey, R. (1990) Lipoxygenase heterogeneity in *Pisum sativum*. *Planta* 181, 35–43.
Domoney, C., Casey, R., Turner, L. and Ellis, N. (1991a) *Pisum* lipoxygenase genes. *Theoretical and Applied Genetics* 81, 800–805.
Domoney, C., Ellis, N., Turner, L. and Casey, R. (1991b) A developmentally regulated early-embryogenesis protein in pea (*Pisum sativum* L.) is related to the heat-shock protein (HSP70) gene family. *Planta* 184, 350–355.
Domoney, C., Welham, T. and Sidebottom, C. (1993) Purification and characterization of *Pisum* seed trypsin inhibitors. *Journal of Experimental Botany* 44, 701–709.
Dry, I., Smith, A., Edwards, E.A., Bhattacharyya, M., Dunn, P. and Martin, C. (1992). Characterisation of cDNAs encoding two isoforms of granule-bound starch synthase which show differential expression in developing storage organs of pea and potato. *Plant Journal* 2, 193–202.
Ealing, P.M. and Casey, R. (1988) The complete amino acid sequence of a pea (*Pisum sativum*) seed lipoxygenase predicted from a near full-length cDNA. *Biochemical Journal* 253, 915–918.
Ealing, P.M. and Casey, R. (1989) The cDNA cloning of a pea (*Pisum sativum*) seed lipoxygenase. Sequence comparisons of the two major pea seed lipoxygenase isoforms. *Biochemical Journal* 264, 929–932.
Edwards, G.A., Hepher, A., Clerk, S.P. and Boulter, D. (1991) Pea lectin is correctly processed, stable and active in leaves of transgenic potato plants. *Plant Molecular Biology* 17, 89–100.
Edwards, J. and ap Rees, T. (1986a). Sucrose partitioning in developing embryos of round and wrinkled varieties of *Pisum sativum*. *Phytochemistry* 25, 2027–2032.
Edwards, J. and ap Rees, T. (1986b) Metabolism of UDPglucose by developing embryos of round and wrinkled varieties of *Pisum sativum*. *Phytochemistry* 25, 2033–2039.
Edwards, J., Green, J.H., and ap Rees, T. (1988). Activity of branching enzyme as a cardinal feature of the Ra locus in *Pisum sativum*. *Phytochemistry* 27, 1615–1620.
Einspahr, H., Parks, E.H, Suguna, K., Subramanian, E. and Suddath, F.L. (1986) The structure of pea lectin at 3Å resolution. *Journal of Biological Chemistry* 261, 16518–16527.
Ellis, T.H.N., Domoney, C., Castleton, J., Cleary, W. and Davies, D.R. (1986) Vicilin genes of *Pisum*. *Molecular and General Genetics* 205, 164–169.
Ellis, T.H.N., Turner, L., Hellens, R.P., Lee, D., Harker, C.L., Enard, C., Domoney,

C. and Davies, D.R. (1992) Linkage maps in pea. *Genetics* 130, 649–663.
Ersland, D.R., Brown, J.W.S., Casey, R. and Hall, T.C. (1983) The storage proteins of *Phaseolus vulgaris* L., *Vicia faba* L. and *Pisum sativum* L. In: Gottschalk, W. and Muller, H.P. (eds), *Seed Proteins. Biochemistry, Genetics, Nutritive Value*. Martinus Nijhoff/Dr W. Junk, The Hague, pp. 355–375.
Evans, I.M., Gatehouse, J.A., Croy, R.R.D. and Boulter, D. (1984) Regulation of the transcription of storage-protein mRNA in nuclei isolated from developing pea (*Pisum sativum* L.) cotyledons. *Planta* 160, 559–568.
Evans, I.M., Gatehouse, J.A. and Boulter, D. (1985) Regulation of storage protein synthesis in pea (*Pisum sativum* L.) cotyledons under conditions of sulphur deficiency. *Biochemical Journal* 232, 261–265.
Fontes, E.B.P., Shank, B.B., Wrobel, R.L., Moose, S.P., O'Brian, G.R., Wurtzel, E.T. and Boston, R.S. (1991) Characterization of an immunoglobulin binding protein homolog in the maize *floury-2* endosperm mutant. *Plant Cell* 3, 483–496.
Foster, J.M. and Smith, A.M. (1993) Metabolism of glucose 6-phosphate by plastids from developing pea embryos. *Planta* 190, 17–24.
Frazier, P. (1979) Lipoxygenase action and lipid binding in breadmaking. *Bakers Digest* 53, 8–29.
French, D. (1984) Organisation of starch granules. In: Whistler, R.L., BeMiller, J.N. and Paschall, E.F. (eds), *Starch Chemistry and Technology*. Academic Press, Orlando, pp. 183–247.
Furukawa, K., Tagaya, M., Inoye, M., Preiss, J. and Fukui, T. (1990) Identification of lysine 15 at the active site in *Escherichia coli* glycogen synthase. *Journal of Biological Chemistry* 265, 2086–2090.
Gaborit, T., Delort-Laval, J., Thanh, L.P. and Paraf, A. (1989) Trypsin inhibitors from *Pisum sativum* L. exhibit identical epitopes. *Journal of the Science of Food and Agriculture* 48, 15–27.
Gatehouse, J.A. and Boulter, D. (1980) Isolation and properties of a lectin from the roots of *Pisum sativum* (garden pea). *Physiologia Plantarum* 49, 437–442.
Gatehouse, J.A., Evans, I.M., Bown, D., Croy, R.R.D. and Boulter, D. (1982) Control of storage-protein synthesis during seed development in pea (*Pisum sativum* L.). *Biochemical Journal* 208, 119–127.
Gatehouse, J.A., Lycett, G.W., Delauney, A.J., Croy, R.R.D. and Boulter, D. (1983) Sequence specificity of the post-translational proteolytic cleavage of vicilin, a seed storage protein of pea (*Pisum sativum* L.). *Biochemical Journal* 212, 427–432.
Gatehouse, J.A., Gilroy, J., Hoque, M.S. and Croy, R.R.D. (1985) Purification, properties and amino acid sequence of a low-M_r abundant seed protein from pea (*Pisum sativum* L.). *Biochemical Journal* 225, 239–247.
Gatehouse, J.A., Evans, I.M., Croy, R.R.D. and Boulter, D. (1986) Differential expression of genes during legume seed development. *Philosophical Transactions of the Royal Society of London Series B* 314, 367–384.
Gatehouse, J.A., Bown, D., Gilroy, J., Levasseur, M., Castleton, J. and Ellis, T.H.N. (1988) Two genes encoding 'minor' legumin polypeptides in pea (*Pisum sativum* L.). *Biochemical Journal* 250, 15–24.
Gennis, L.S. and Cantor, C.R. (1976) Double-headed protease inhibitors from black-eyed peas. I. Purification of two new protease inhibitors and the endogenous protease by affinity chromatography. *Journal of Biological Chemistry* 251, 734–740.
Gidley, M.J. and Bociek, S. (1985) Molecular organisation in starches: a ^{13}C CP/MAS

NMR study. *Journal of the American Chemical Society* 107, 7040–7044.
Gidley, M.J. and Bociek, S.M. (1988) ^{13}C CP/MAS NMR studies of amylose inclusion complexes, cyclodextrins, and the amorphous phase of starch granules. *Journal of the American Chemical Society* 110, 3820–3829.
Goldstein, I.J., Hughes, R.C., Monsigny, M., Osawa, T. and Sharon, N. (1980) What should be called a lectin. *Nature, London* 285, 66.
Graham, J.S., Pearce, G., Merryweather, J., Titani, K., Ericsson, L. and Ryan, C.A. (1985a) Wound-induced proteinase inhibitors from tomato leaves. I. The cDNA-deduced primary structure of pre-inhibitor I and its post-translational processing. *Journal of Biological Chemistry* 260, 6555–6560.
Graham, J.S., Pearce, G., Merryweather, J., Titani, K., Ericsson, L.H. and Ryan, C.A. (1985b) Wound-induced proteinase inhibitors from tomato leaves. II. The cDNA-deduced structure of pre-inhibitor II. *Journal of Biological Chemistry* 260, 6561–6564.
Greenwood, C.T. and Thompson, J. (1962) Studies on the biosynthesis of starch granules. 2. The properties of components of starches from smooth and wrinkled seeded peas during growth. *Biochemical Journal* 82, 156–164.
Gross, P. and ap Rees, T. (1986) Alkaline inorganic pyrophosphatase and starch synthesis in amyloplasts. *Planta* 167, 140–145.
Gruen, L.C., Guthrie, R.E. and Blagrove, R.J. (1987) Structure of a major pea seed albumin: implication of a free sulphydryl group. *Journal of the Science of Food and Agriculture* 41, 167–178.
Gueguen, J. (1983) Legume seed protein extraction, processing and end-product characteristics. *Qualitas Plantarum, Plant Foods for Human Nutrition* 32, 267–303.
Guilbot, A. and Mercier, C. (1985) Starch. In: Aspinall, G.O. (ed.), *The Polysaccharides*, vol. 3, Academic Press, London, pp. 209–282.
Guldager, P. (1978) Immunoelectrophoretic analysis of seed proteins from *Pisum sativum* L. *Theoretical and Applied Genetics* 53, 241–250.
Harris, N. and Croy, R.R.D. (1985) The major albumin protein from pea (*Pisum sativum* L.). Localisation by immunocytochemistry. *Planta* 165, 522–526.
Higgins, T.J.V. (1984) Synthesis and regulation of major proteins in seeds. *Annual Review of Plant Physiology* 35, 191–221.
Higgins, T.J.V. and Spencer, D. (1981) Precursor forms of pea vicilin subunits. Modification by microsomal membranes during cell-free translation. *Plant Physiology* 67, 205–211.
Higgins, T.J.V. and Spencer, D. (1991) The expression of a chimeric cauliflower mosaic virus (CaMV-35S)-pea vicilin gene in tobacco. *Plant Science* 74, 89–98.
Higgins, T.J.V., Chandler, P.M., Zurawski, G., Button, S.C. and Spencer, D. (1983a) The biosynthesis and primary structure of pea seed lectin. *Journal of Biological Chemistry* 258, 9544–9549.
Higgins, T.J.V., Chrispeels, M.J., Chandler, P.M. and Spencer, D. (1983b) Intracellular sites of synthesis and processing of lectin in developing pea cotyledons. *Journal of Biological Chemistry* 258, 9550–9553.
Higgins, T.J.V., Chandler, P.M., Randall, P.J., Spencer, D., Beach, L.R., Blagrove, R.J., Kortt, A.A. and Inglis, A.S. (1986) Gene structure, protein structure, and regulation of the synthesis of a sulfur-rich protein in pea seeds. *Journal of Biological Chemistry* 261, 11124–11130.

Higgins, T.J.V., Beach, L.R., Spencer, D., Chandler, P.M., Randall, P.J., Blagrove, R.J., Kortt, A.A. and Guthrie, R.E. (1987) cDNA and protein sequence of a major pea seed albumin (PA 2: M_r ~26 000). *Plant Molecular Biology* 8, 37–45.

Higgins, T.J.V., Newbigin, E.J., Spencer, D., Llewellyn, D.J. and Craig, S. (1988) The sequence of a pea vicilin gene and its expression in transgenic tobacco plants. *Plant Molecular Biology* 11, 683–695.

Hildebrand, D.F., Hamilton-Kemp, T.R., Legg, C.S. and Bookjans, G. (1988) Plant lipoxygenases: occurrence, properties and possible functions. *Current Topics in Plant Biochemistry and Physiology* 7, 201–219.

Hilder, V.A., Gatehouse, A.M.R., Sheerman, S.E., Barker, R.F. and Boulter, D. (1987) A novel mechanism of insect resistance engineered into tobacco. *Nature, London* 330, 160–163.

Hilder, V.A., Gatehouse, A.M.R. and Boulter, D. (1990) Genetic engineering of crops for insect resistance using genes of plant origin. In: Grierson, D. and Lycett, G. (eds), *Genetic engineering of crop plants*, Butterworths, London, pp. 51–66.

Hill, L.M. and Smith, A.M. (1991). Evidence that glucose 6-phosphate is imported as the substrate for starch synthesis by the plastids of developing pea embryos. *Planta* 185, 91–96.

Hirano, H., Gatehouse, J.A. and Boulter, D. (1982) The complete amino acid sequence of a subunit of the vicilin seed storage protein of pea (*Pisum sativum* L.). *FEBS Letters* 145, 99–102.

Hizukuri, S., Kaneko, T. and Takeda, Y. (1983) Measurement of the chain length of amylopectin and its relevance to the origin of crystalline polymorphism of starch granules. *Biochimica et Biophysica Acta* 670, 188–191.

Hoffman, L.M., Donaldson, D.D. and Herman, E.M. (1988) A modified storage protein is synthesized, processed, and degraded in the seeds of transgenic plants. *Plant Molecular Biology* 11, 717–729.

Holt, N.W. and Sosulski, F.W. (1979) Amino acid composition and protein quality of field peas. *Canadian Journal of Plant Science* 59, 653–660.

Hylton, C.M. and Smith, A.M. (1992). The *rb* mutation of peas causes structural and regulatory changes in ADPglucose pyrophosphorylase from developing embryos. *Plant Physiology* 99, 1626–1634.

I'Anson, K.J., Miles, M.J., Bacon, J.R., Carr, H.J., Lambert, N., Morris, V.J. and Wright, D.J. (1988) Structure of the 7S globulin from pea. *International Journal of Biological Macromolecules* 10, 311–317.

Johnson, R., Narvaez, J., An, G. and Ryan, C.A. (1989) Expression of proteinase inhibitors I and II in transgenic tobacco plants: effects on natural defence against *Manduca sexta* larvae. *Proceedings of the National Academy of Sciences USA* 86, 9871–9875.

Kainuma, K. (1988) Structure and chemistry of the starch granule. In: Preiss, J. (ed.), *The Biochemistry of Plants, Vol. 14: Carbohydrates*. Academic Press, San Diego, pp. 141–180.

Kaminski, P.A., Buffard, D. and Strosberg, A.D. (1987) The pea lectin gene family contains only one functional gene. *Plant Molecular Biology* 9, 497–507.

Knust, B. and von Wettstein, D. (1992) Expression and secretion of pea-seed lipoxygenase isoenzymes in *Saccharomyces cerevisiae*. *Applied Microbiology and Biotechnology* 37, 342–351.

Kooistra, E. (1962) On the differences between smooth and three types of wrinkled peas. *Euphytica* 11, 357-373.
Lambert, N. and Yarwood, J.N. (1992) Engineering legume seed storage proteins. In: Shewry, P.R. and Gutteridge, S. (eds), *Plant Protein Engineering*. Cambridge University Press, Cambridge, pp. 167-187.
Lawrence, M.C., Suzuki, E., Varghese, J.N., Davis, P.C., Van Donkelaer, A., Tulloch, P.A. and Colman, P.M. (1990) The three-dimensional structure of the seed storage protein phaseolin at 3 Å resolution. *EMBO Journal* 9, 9-15.
Lycett, G.W., Delauney, A.J., Gatehouse, J.A., Gilroy, J., Croy, R.R.D. and Boulter, D. (1983) The vicilin gene family of pea (*Pisum sativum* L.): a complete cDNA coding sequence for preprovicilin. *Nucleic Acids Research* 11, 2367-2380.
Lycett, G.W., Croy, R.R.D., Shirsat, A.H. and Boulter, D. (1984) The complete nucleotide sequence of a legumin gene from pea (*Pisum sativum* L.). *Nucleic Acids Research* 12, 4493-4506.
Lycett, G.W., Croy, R.R.D., Shirsat, A.H., Richards, D.M. and Boulter, D. (1985) The 5'-flanking regions of three pea legumin genes: comparison of the DNA sequences. *Nucleic Acids Research* 13, 6733-6743.
Mahmoud, S.H. and Gatehouse, J.A. (1984) Inheritance and mapping of vicilin storage protein genes in *Pisum sativum* L. *Heredity* 53, 185-191.
Manen, J.-F., Simon, P., VanSlooten, J.-C., Østerås, M., Frutiger, S. and Hughes, G.J. (1991) A nodulin specifically expressed in senescent nodules of winged bean is a protease inhibitor. *Plant Cell* 3, 259-270.
March, J.F., Pappin, D.J.C. and Casey, R. (1988) Isolation and characterization of a minor legumin and its constituent polypeptides from *Pisum sativum* (pea). *Biochemical Journal* 250, 911-915.
Marocco, A., Santucci, A., Cerioli, S., Motto, M., Di Fonzo, N., Thompson, R. and Salamini, F. (1991) Three high-lysine mutations control the level of ATP-binding HSP70-like proteins in the maize endosperm. *Plant Cell* 3, 507-515.
Matta, N.K. and Gatehouse, J.A. (1982) Inheritance and mapping of storage protein genes in *Pisum sativum* L. *Heredity* 48, 383-392.
Matta, N.K., Gatehouse, J.A. and Boulter, D. (1981) Molecular and subunit heterogeneity of legumin of *Pisum sativum* L. (garden pea) – a multidimensional gel electrophoretic study. *Journal of Experimental Botany* 32, 1295-1307.
Matters, G.L. and Boyer, C.D. (1981). Starch synthases and starch-branching enzymes from *Pisum sativum*. *Phytochemistry* 20, 1805-1809.
Meakin, P.J. and Gatehouse, J.A. (1991) Interaction of seed nuclear proteins with transcriptionally-enhancing regions of the pea (*Pisum sativum* L.) *leg*A gene promoter. *Planta* 183, 471-477.
Millerd, A., Thomson, J.A. and Schroeder, H.E. (1978) Cotyledonary storage proteins in *Pisum sativum* III. Patterns of accumulation during development. *Australian Journal of Plant Physiology* 5, 519-534.
Neuhaus, H.E. and Stitt, M. (1990) Control analysis in photosynthetic partitioning. Impact of reduced activity of ADP-glucose pyrophosphorylase or plastid phosphoglucomutase on the fluxes to starch and sucrose in *Arabidopsis thaliana* (L.) Heynh. *Planta* 182, 445-454.
Newbigin, E.J., deLumen, B.O., Chandler, P.M., Gould, A., Blagrove, R.J., March, J.F., Kortt, A.A. and Higgins, T.J.V. (1990) Pea convicilin: structure and primary

sequence of the protein and expression of a gene in the seeds of transgenic tobacco. *Planta* 180, 461–470.

North, H. (1990) Pea seed lipoxygenase variants. PhD Thesis, University of East Anglia.

North, H., Casey, R. and Domoney, C. (1989) Inheritance and mapping of seed lipoxygenase polypeptides in *Pisum*. *Theoretical and Applied Genetics* 77, 805–808.

Okita, T.W. (1992) Is there an alternative pathway for starch synthesis? *Plant Physiology* 100, 560–564.

Osborne, T.B. (1924) *The Vegetable Proteins*. Longman, Green and Co., London.

Pak, J.H., Hendrickson, T. and Dobres, M.S. (1992) Predicted sequence and structure of a vegetative lectin in *Pisum sativum*. *Plant Molecular Biology* 18, 857–863.

Poerio, E., Carrano, L., Garzillo, A.M. and Buonocore, V. (1989) A trypsin inhibitor from the water-soluble protein fraction of wheat kernel. *Phytochemistry* 28, 1307–1311.

Preiss, J. and Levi, C. (1982) Starch biosynthesis and degradation. In: Preiss, J. (ed.), *The Biochemistry of Plants, Vol. 3: Carbohydrates: Structure and Function*. Academic Press, San Diego, pp. 371–423.

Pusztai, A. (ed.) (1991) *Plant Lectins*. Cambridge University Press, Cambridge.

Pusztai, A., Ewen, S.W.B., Grant, G., Peumans, W.J., van Damme, E.J.M., Rubio, L. and Bardocz, S. (1990) The relationship between survival and binding of plant lectins during small intestinal passage and their effectiveness as growth factors. *Digestion* 46 (suppl. 2) 308–316.

Rao, R., Costa, A., Croy, R.R.D., Boulter, D. and Gatehouse, J.A. (1989) Variation in polypeptides of the major albumin protein of pea (*Pisum sativum* L.): inheritance and molecular analysis. *Molecular and General Genetics* 219, 277–281.

Rerie, W.G., Whitecross, M.I. and Higgins, T.J.V. (1990) Complete nucleotide sequence of an A-type legumin gene in pea. *Nucleic Acids Research* 18, 55.

Rerie, W.G., Whitecross, M. and Higgins, T.J.V. (1991) Developmental and environmental regulation of pea legumin genes in transgenic tobacco. *Molecular and General Genetics* 225, 148–157.

Richardson, M. (1991) Seed storage proteins: the enzyme inhibitors. *Methods in Plant Biochemistry* 5, 259–305.

Robyt, J. (1984) Enzymes in the synthesis and hydrolysis of starch. In: Whistler, R.L., BeMiller, J.N. and Paschall, E.F. (eds), *Starch: Chemistry and Technology*. Academic Press, Orlando, pp. 87–123.

Saalbach, G., Jung, R., Kunze, G., Saalbach, I., Adler, K. and Müntz, K. (1991) Different legumin protein domains act as vacuolar targeting signals. *Plant Cell* 3, 695–708.

Savage, G.P. (1989) Antinutritive factors in peas. In: Huisman, J., van der Poel, T.F.B. and Liener, I.E. (eds), *Recent Advances of Research in Antinutritional Factors in Legume Seeds*. PUDOC, Wageningen, pp. 342–350.

Schroeder, H.E. (1984) Major albumins of *Pisum* cotyledons. *Journal of the Science of Food and Agriculture* 35, 191–198.

Scott, M.P., Jung, R., Müntz, K. and Nielsen, N.C. (1992) A protease responsible for post-translational cleavage of a conserved Asn–Gly linkage in glycinin, the major seed storage protein of soyabean. *Proceedings of the National Academy of Sciences, USA* 89, 658–662.

Shewry, P.R. and Gutteridge, S. (1992) *Plant Protein Engineering*. Cambridge University Press, Cambridge.

Shirsat, A.H. (1988) A transposon-like structure in the 5' flanking sequence of a legumin gene from *Pisum sativum*. *Molecular and General Genetics* 212, 129–133.

Shirsat, A.H., Wilford, N., Croy, R. and Boulter, D. (1989) Sequences responsible for the tissue specific promoter activity of a pea legumin gene in tobacco. *Molecular and General Genetics* 215, 326–331.

Shirsat, A.H., Meakin, P.J. and Gatehouse, J.A. (1990) Sequences 5' to the conserved 28 bp Leg box element regulate the expression of pea seed storage protein gene *leg*A. *Plant Molecular Biology* 15, 685–693.

Shotwell, M.A. and Larkins, B.A. (1991) Improvement of the protein quality of seeds by genetic engineering. In: Dennis, E.S. and Llewellyn, D.J. (eds), *Molecular Approaches to Crop Improvement*. Springer-Verlag, Vienna, pp. 33–61.

Smith, A.M. (1988) Major differences in isoforms of starch-branching enzyme between developing embryos of round-and wrinkled-seeded peas. *Planta* 175, 170–179.

Smith, A.M. (1990) Evidence that the 'waxy' protein of pea is not the major, starch-granule-bound starch synthase. *Planta* 182, 599–604.

Smith, A.M. and Denyer, K. (1992) Starch synthesis in developing pea embryos. *New Phytologist* 122, 21–33.

Smith, A.M. and Martin, C. (1993) Starch biosynthesis and the potential for its manipulation. In: Grierson, D. (ed.), *Biosynthesis and Manipulation of Plant Products*. Plant Biotechnology, vol. 3. Blackie and Son, Glasgow, pp. 1–54.

Smith, A.M., Bettey, M., and Bedford, I.D. (1989) Evidence that the *rb* locus alters the starch content of developing pea embryos through an effect on ADPglucose pyrophosphorylase. *Plant Physiology* 89, 1279–1284.

Smith, A.M., Quinton-Tulloch, J. and Denyer, K. (1990) Characteristics of plastids responsible for starch synthesis in developing pea embryos. *Planta* 180, 517–523.

Spencer, D. (1984) The physiological role of storage proteins in seeds. *Philosophical Transactions of the Royal Society of London Series B* 304, 275–285.

Spencer, D. and Higgins, T.J.V. (1980) The biosynthesis of legumin in maturing pea seeds. *Biochemistry International* 1, 502–509.

Spencer, D., Chandler, P.M., Higgins, T.J.V., Inglis, A.S. and Rubira, M. (1983) Sequence interrelationships of the subunits of vicilin from pea seeds. *Plant Molecular Biology* 2, 259–267.

Spencer, D., Rerie, W.G., Randall, P.J. and Higgins, T.J.V. (1990) The regulation of pea seed storage protein genes by sulfur stress. *Australian Journal of Plant Physiology* 17, 355–363.

Tanchak, M.A. and Chrispeels, M.J. (1989) Crosslinking of microsomal proteins identifies P-9000, a protein that is co-transported with phaseolin and phytohaemagglutinin in bean cotyledons. *Planta* 179, 495–505.

Thompson, A.J., Evans, I.M., Boulter, D., Croy, R.R.D. and Gatehouse, J.A. (1989) Transcriptional and posttranscriptional regulation of seed storage-protein gene expression in pea (*Pisum sativum* L.). *Planta* 179, 279–287.

Thompson, A.J., Bown, D., Yaish, S. and Gatehouse, J.A. (1991) Differential expression of seed storage protein genes in the pea *legJ* subfamily; sequence of gene *legK*. *Biochemie und Physiologie der Pflanzen* 187, 1–12.

Thomson, J.A. and Schroeder, H.E. (1978) Cotyledonary storage proteins in *Pisum* sativum. II. Hereditary variation in components of the legumin and vicilin fractions. *Australian Journal of Plant Physiology* 5, 281-294.

Thomson, J.A., Schroeder, H.E. and Dudman, W.F. (1978) Cotyledonary storage proteins in *Pisum sativum*. I. Molecular heterogeneity. *Australian Journal of Plant Physiology* 5, 263-279.

Thomson, J.A., Schroeder, H.E. and Tassie, A.M. (1980) Cotyledonary storage proteins in *Pisum sativum*. V. Further studies on molecular heterogeneity in the vicilin series of holoproteins. *Australian Journal of Plant Physiology* 7, 271-282.

Tranbarger, T.J., Franceschi, V.R., Hildebrand, D.F. and Grimes, H.D. (1991) The soybean 94-kilodalton vegetative storage protein is a lipoxygenase that is localized in paraveinal mesophyll cell vacuoles. *Plant Cell* 3, 973-987.

Trowbridge, I.S. (1974) Isolation and chemical characterization of a mitogenic lectin from *Pisum sativum*. *Journal of Biological Chemistry* 249, 6004-6012.

Turner, L., Hellens, R.P., Lee, D. and Ellis, T.H.N. (1993) Genetic aspects of the organization of legumin genes in pea. *Plant Molecular Biology* 22, 101-112.

Turner, S.R., Barratt, D.H.P. and Casey, R. (1990) The effect of different alleles at the *r* locus on the synthesis of seed storage proteins in *Pisum sativum*. *Plant Molecular Biology* 14, 793-803.

Vandekerckhove, J., Van Damme, J., Van Lijsenbettens, M.V., Botterman, J., De Block, M., Vandewiele, M., De Clercq, A., Leemans, J., Van Montagu, M. and Krebbers, E. (1989) Enkephalins produced in transgenic plants using modified 2S seed storage proteins. *Bio/Technology* 7, 929-932.

van Eijsden, R.R., Hoedemaeker, F.J., Diaz, C.L., Lugtenberg, B.J.J., de Pater, B.S. and Kijne, J.W. (1992) Mutational analysis of pea lectin. Substitution of Asn^{125} for Asp in the monosaccharide-binding site eliminates mannose/glucose-binding activity. *Plant Molecular Biology* 20, 1049-1058.

Vick, B.A. and Zimmerman, D.C. (1983) The biosynthesis of jasmonic acid: a physiological role for plant lipoxygenase. *Biochemical and Biophysical Research Communications* 111, 470-477.

Wandelt, C.I., Khan, M.R.I., Craig, S., Schroeder, H.E., Spencer, D. and Higgins, T.J.V. (1992) Vicilin with carboxy-terminal KDEL is retained in the endoplasmic reticulum and accumulates to high levels in the leaves of transgenic plants. *Plant Journal* 2, 181-192.

Wang, T.L. and Hedley, C. (1991) Seed development in peas: knowing your three 'r's (or four, or five). *Seed Science Research* 1, 3-14.

Welland, M.E., Miles, M.J., Lambert, N., Morris, I., Coombs, J.H. and Pethica, J.B. (1989) Structure of the globular protein vicilin revealed by scanning tunnelling microscopy. *International Journal of Biological Macromolecules* 11, 29-32.

Wolswinkel, P. (1992) Transport of nutrients into developing seeds: a review of physiological mechanisms. *Seed Science Research* 2, 59-73.

Wright, D.J. (1987) The seed globulins. In: Hudson, B.J.F. (ed.), *Developments in Food Proteins*, vol. 5. Elsevier Applied Science, London, pp. 81-157.

Wright, D.J. and Bumstead, M.R. (1984) Legume proteins in food technology. *Philosophical Transactions of the Royal Society of London Series B* 304, 381-393.

Yang, L., Domoney, C., Casey, R. and Hall, T.C. (1990) Processing *Pisum sativum* seed storage protein precursors *in vitro*. *Cell Research* 2, 153-162.

Developmental Mutants 7

I.C. MURFET AND J.B. REID
Department of Plant Science, University of Tasmania, GPO Box 252C, Hobart 7001, Australia

Genetic variation in peas (*Pisum sativum* L.) has been known for over 400 years (see Blixt, 1972) and developmental mutants in this species provided the experimental basis for the development of modern genetics (Mendel, 1866). The number of developmental and morphological mutants described in peas exceeds 400, placing this species among an élite group of plant species for which a detailed genetical understanding has been obtained. This group includes, amongst others, maize, tomato, *Arabidopsis*, barley and rice. Much of our understanding of plant growth and development is also based upon this restricted group of species.

In this chapter we provide details of the major developmental mutants that modify the overall structure or architecture of the pea plant. These mutants are grouped according to the developmental process they influence: leaf form (Table 7.1), internode elongation (Table 7.2), flowering (Tables 7.3 and 7.4), branching pattern (Table 7.5), fruit development and senescence. However, while some genes are very specific in the process they influence (e.g. the leaf form genes *Af*, *Tl* and *St*), mutations in other genes result in widespread pleiotropic effects on several developmental processes. Examples of the latter are the length mutant *lv*, which also shows major effects on flowering and branching, and the flowering mutant *sn*, which shows major effects on branching and senescence. Thus some genes are mentioned under more than one heading.

There are many mutants that alter the phenotype of the whole plant but are not considered here. Some modify plant appearance by altering the production of the flower and pod pigments (e.g. *a*, *ar*, *b*, *ce*, *cr*, *gp*, *pu*) or the epidermal waxes (e.g. *wlo*, *wsp*, *wel*, *wa*, *wb*, *was*). Genes modifying chlorophyll levels (e.g. *chi-6*, *alb*, *cov*, *alt*), the necrosis genes (e.g. *dgl*, *len*, *len-2*, *bulf*) and genes reported to alter the resistance to pathogens (e.g. *er*, *en*, *lr*, *sbm*) may also influence the whole plant phenotype but do not affect development *per se*. Genes influencing seed development and nitrogen fixation are discussed in Chapters 5 and 9 and will not be considered here. Finally, the male and female fertility mutants in peas, although frequent, have not received sufficient examination to justify detailed consideration although they have dramatic effects on plant architecture and plant senescence.

For certain of these developmental mutants, the action has been described only at the whole plant level (branching mutants, for example; Arumingtyas *et al.*, 1992) whereas for others, such as flowering and senescence, the whole plant physiology has also received considerable attention (Murfet, 1985b). In the best-understood mutants, the biochemical action of the genes has also been determined, examples of these being the internode elongation genes that control GA biosynthesis (Reid and Ross, 1993). In all cases, detailed molecular studies have not been completed although work has begun on some of the genetic systems controlling these developmental processes in peas. The aim of this chapter is to provide an overview of the classical genetic understanding of these processes and, where possible, their physiological and biochemical control. Some systems such as those affecting flowering, internode length and leaf form are as well, or better, understood than in any other plant species and may provide suitable experimental systems for future molecular studies. While peas may not possess many of the advantages for this type of study that are present in *Arabidopsis*, maize, tomato or *Antirrhinum*, they do possess certain distinct advantages (e.g. the compound leaf structure and large seed size), which will become evident from the case studies outlined.

Leaf Development

A wide range of mutations influence the form and structure of the pea leaf. An excellent review by Marx (1987) deals in detail with the genes responsible and their genetic interactions. The complexity of the compound leaf of peas means that some of these mutations have a dramatic influence on the overall appearance of the plant.

A list of the major developmental mutants influencing leaf form is contained in Table 7.1. The best known are the homoeotic mutants that cause the replacement of one leaf part by another. These mutants may provide the tools necessary to dissect the control of leaf development at the molecular level in the same way that homoeotic mutants have allowed flower development to be explored (Coen and Meyerowitz, 1991).

The first of these homoeotic mutants to be described was the *acacia* mutant, *tl* (Vilmorin and Bateson, 1912; White 1917), which results in the replacement of the tendrils on the wild-type leaf by leaflets (Fig. 7.1). The terminal leaflets on *tl* plants possess normal leaflet anatomy (Gould *et al.*,

Table 7.1. The major mutations influencing leaf development are listed with brief descriptions of their phenotype. Further details regarding most mutations can be found in Marx (1987).

Mutation	Phenotype	Key reference
tl	*acacia*, tendrils replaced by leaflets	Vilmorin and Bateson, 1912
st	stipules reduced to short, very narrow laminae	Pellew and Sverdrup, 1923
st^{bs}	butterfly stipules, stipules intermediate in size between WT and st. $St > st^{bs} > st$	Apisitwanich and Swiecicki, 1992
Td	stipule and leaflet margins dentate, incompletely dominant	Wellensiek, 1925a
uni	*unifoliata*, single leaflet (lamina) replaces multiple leaflets and tendrils	Lamprecht, 1933
uni^{tac}	*tendrilled acacia*, terminal tendril replaced by leaflet. $Uni > uni^{tac} \geq uni$	Sharma, 1972; Marx, 1986b
cri	*crispa*, leaflets and stipules crinkled	Lamm, 1949
coch	*cochleata*, stipules often replaced by stalked leaflet	Wellensiek, 1959
$coch^{het}$	*heterophyllus*, stipules differ in size and often reduced	Rozov *et al.*, 1992
af	*afila*, leaflets replaced by tendrils	Goldenberg, 1965
sil	*sinuate leaf*, leaflet and stipule margins undulate	Marx, 1977
apu	*apulvinic*, leaflets not sessile but on stalks	Harvey, 1979
arg	*argenteum*, leaflets and stipules silver grey due to lack of contact between epidermis and mesophyll cells	Hoch *et al.*, 1980

1986). The development of the leaf primordia in *tl* plants is similar to that of wild-type plants except that marginal and adaxial meristems are active where none exists in wild-type plants (Meicenheimer *et al.*, 1983). The change occurs late in leaf ontogeny.

The *af* mutation results essentially in the reverse of the *tl* mutation at the gross phenotypic level (Kujala, 1953; Goldenberg, 1965). All leaflets are replaced by tendrils of normal anatomy (Fig. 7.1) although the degree of

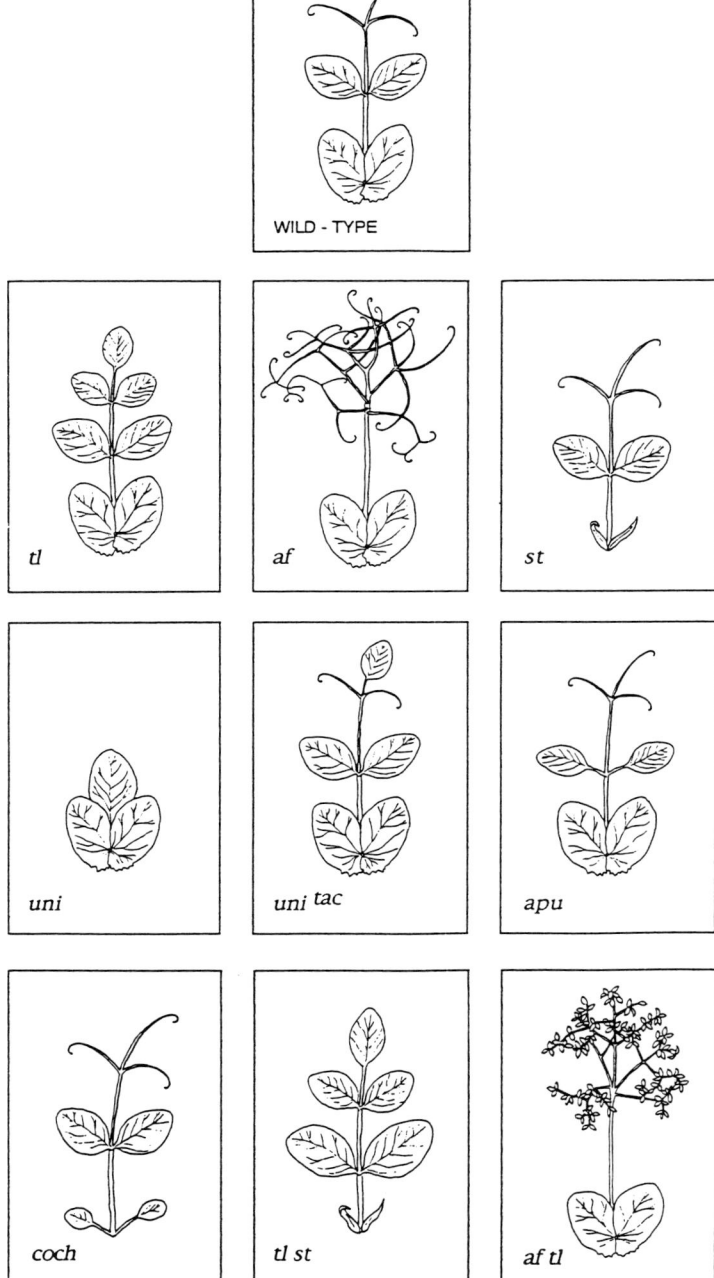

Fig. 7.1. A comparison of the compound leaf of wild-type plants with a range of single and double mutants influencing leaf development.

branching of the tendrils may be substantially greater (Gould et al., 1986, 1991, 1992). This results from the complete lack of leaflet marginal meristems in *af* plants. The 'leaflet' primordia on *af* plants develop secondary branches with radial meristems, characteristic of a radial marginal meristem (Meicenheimer et al., 1983). The *af* allele is of considerable agronomic significance since it has been incorporated into many commercial varieties in order to reduce lodging and provide improved airflow, and consequently reduced humidity, within the stand (Pyke and Hedley, 1985; Snoad, 1985).

The mutation *st* reduces stipule size dramatically, resulting in a narrow, strap-like organ (Fig. 7.1). This results from the early loss of the stipule marginal meristems, and consequently, although these meristems are present in *st* plants, they cease functioning after a shorter time than in wild-type plants (Meicenheimer et al., 1983). An intermediate allele st^{bs} (butterfly stipules) has now been identified (Apisitwanich and Swiecicki, 1992).

The mutations *tl*, *af* and *st* are probably the best studied of the leaf mutations and illustrate an important principle. They are all highly specific in their action and do not show a wide range of pleiotropic effects. For example, the stipules on *af* or *tl* plants are normal whereas *st* has no apparent effects on leaflet or tendril formation. Likewise, none of the mutations alters flower or pod characteristics even though these organs are modified leaves. Consequently the *af*, *st* and *tl* mutations appear to act by controlling specific aspects of leaf development. This may imply that they become activated only at certain points in the developmental sequence, analogous to the *apetala 2*, *apetala 3* and *agamous* mutations controlling flower development in *Arabidopsis* (Bowman et al., 1989; Coen, 1991).

The mutations *uni*, uni^{tac} and *apu* all modify the form of the leaflets or tendrils, like *tl* and *af*. (The mutation formerly known as *tac* was shown by Marx (1986b) to be allelic with *uni* and given the symbol uni^{tac}.) In uni^{tac} plants, the terminal tendril is replaced by a laminate leaflet. Later-formed leaves possess a single pair of subterminal tendrils (Sharma, 1972, Marx, 1986b; Fig. 7.1). The more severe allele, *uni* (Lamprecht, 1933), results in plants possessing a single leaf blade in addition to normal stipules (Fig. 7.1). However, the expression of both alleles is variable both within and between plants (Marx, 1987). Unlike the *tl* and *af* mutations, the *uni* and uni^{tac} alleles have pleiotropic effects on flower development (Sharma, 1972; Marx, 1986b). These are so severe that *uni* plants are sterile (Marx, 1986b). This non-specificity of site of action of the alleles at the *uni* locus may indicate that the product of this locus is expressed in many organs derived from leaves. It may thus be involved at a more basic level of leaf development than either *tl* or *af*. Whether this implies that it influences events at an earlier stage of primordia development warrants examination. The *apu* mutation results in leaflets being borne on stalks or petiolules (Harvey, 1979; Fig. 7.1), perhaps suggesting that some signal required for the change from petiole to lamina growth is delayed in this mutant.

The alleles at the *coch* locus, *coch* and *coch*^{het} (Wellensiek, 1959, 1962; Rozov *et al.*, 1992), like *st*, influence stipule development. Stipules may be missing from the lower nodes of *coch* plants and at higher nodes may be malformed or replaced by structures similar to the leaflets on *apu* plants (Fig. 7.1) or by normal leaves with leaflets and tendrils. The *coch*^{het} plants possess reduced stipules at the basal nodes and stipules of different size at the upper nodes (Swiecicki, 1989). The large differences in the expression of the *coch* alleles suggest a regulatory role in the site of gene expression, as befits a homoeotic mutation.

The *sil*, *cri*, *Arg* and *Td* mutations influence both the stipules and leaflets and do not appear to be homoeotic in their action. While the *sil* mutation causes the leaflet and stipule margins to undulate (Marx, 1977), the *cri* mutation results in a crinkled appearance to leaflets and stipules that are revolute in structure (Lamm, 1949). In addition, in *cri* plants the veins are thick and prominent, the petiole is very short, and flower and fruit abnormalities lead to reduced seed set. The *Arg* mutation causes the leaves to be silvery grey due to air spaces below the epidermis (Hoch *et al.*, 1980). The *sil*, *cri* and *Arg* mutations, therefore, modify normal leaf structure rather than displacing the timing of development or positioning of leaf parts. This may imply an abnormality in cell development or tissue organization rather than an inappropriate developmental programme as seen in the homoeotic mutants. Several mutations influencing other developmental processes (e.g. the internode length mutations *lk* and *lkb*, or the flowering mutation *dm*) also have pronounced effects on normal leaf development by altering leaflet size, shape and overall appearance. These types of mutations are unlikely to be useful for the exploration of the control of organogenesis but may prove useful for understanding cell differentiation and tissue organization.

There is also a class of genes that influence leaf development by affecting the timing of the transition from the relatively simple juvenile leaves to the more complex adult leaves. These changes in leaf complexity have been reported by Barber (1959) and examined in some detail by Young and co-workers (Young, 1983; Gould *et al.*, 1986). Genes influencing this change may be classed as heterochronic genes if the timing is changed relative to reproductive development. While little work has been published on genes modifying this juvenile to adult transition in leaf form, appreciable changes do occur and frequently appear to cause substantial heterochronic shifts in the rate of vegetative relative to reproductive development (e.g. flowering mutations *lf*, *sn*, *dne*, *ppd*, *hr* and *dm*; R.J.E. Wiltshire, I.C. Murfet and J.B. Reid, unpublished). Some of these genes therefore appear to be heterochronic in nature.

Another group of mutations affecting leaf appearance are the mutations altering wax production on the leaf surfaces. While many such mutations exist, such as *wa*, *was*, *wb*, *wel*, *wlo* and *wsp*, their primary relevance here is that they frequently show precise site-specific expression. For example, in *wlo* plants wax is absent only from the adaxial surface of the leaflets and in *wb* plants wax is absent from all leaf parts except the adaxial surface of leaflets. This can be a useful guide as to which set of genes may be activated in a given situation, particularly in combination with the homoeotic mutants. The opposing action of *wlo* and *wb* is of interest because it implies the adaxial surface of the stipules is developmentally different from the adaxial surface of the leaflets and perhaps more akin to the abaxial surface of the leaflets.

Finally, there is a large heterogeneous group of mutants influencing leaf development about which little is known. Some affect leaflet (and/or stipule) area by either increasing (*lat*) or decreasing (e.g. *fo* and *red*) their width. Others influence the number of leaflet pairs (e.g. *up* results in only one pair of leaflets) or the edge of the leaflets (e.g. *ser*: edge serrated, or *ins*: tip incised). Further details regarding these and other similar mutants can be found in Blixt (1972). These mutants may prove useful in the future to test hypotheses on the control of leaf development.

Gene interactions in leaf development

Possibly the most informative work that has been done so far with these leaf mutations has been the exploration of the genetic interactions between them and, in particular, between the homoeotic mutations. These interactions take several forms and are too numerous to mention in detail (see Marx, 1987). Instead, examples of the distinct types of interaction that can occur are described to indicate the power of this type of analysis.

Interactions between homoeotic mutations affecting different organs or domains, e.g. *st*, reduced stipules, and *tl*, acacia leaflets, result in plants where there is an additive effect of the mutant phenotypes (Fig. 7.1). Such an interaction provides little new information but confirms the different domains of activity of the genes involved and their apparent independence of action. Epistatic relationships are more informative. They illustrate that the action of one mutation can override the expression of a mutation at another locus. Such mutations must affect the same domain(s) of the leaf. For example, *af* (*afila*) conceals the expression of the *apu* (*apulvinic*) mutation by turning all leaflets into tendrils (Marx, 1987). This indicates a more basic role for the *af* gene in controlling leaflet initiation than for *apu*, which merely modifies leaflet development once it is initiated. Likewise, the uni^{tac} mutation, by reducing the number of branches on the leaf rachis, may reduce the number of leaflets formed on $tl\ uni^{tac}$ plants or the number of tendrils on $af\ uni^{tac}$ plants (Marx, 1987).

The most informative gene interactions are those in which a novel phenotype is produced in the double recessive combination. The most widely examined of this type of interaction is undoubtedly the interaction of the *af* and *tl* mutations, where the double recessive has a distinctive phenotype in which small laminae occur on the tips of many secondary rachises (Fig. 7.1). Although this distinctive 'parsley leaf' (*pleofila* phenotype) of *af tl* plants is normally considered to be due to the combination of features of both the *af* and *tl* mutants (Marx, 1987), another way to consider this phenotype is to explore the idea of what the ground state of the compound leaf of a pea might be expected to look like. Presumably the tendrils are specialized, possibly relatively recent, adaptations of the leaf to aid the plant in climbing, whereas the leaflets may have resulted from the fusion of divaricating branches of a lateral shoot of finite growth potential. If this is the case, a mutant lacking the function of a gene controlling tendril production (*tl*) and the function of a gene controlling full lamina production (*af*) might be expected to produce a highly branched rachis without tendrils. This is not dissimilar to the 'leaves' (excluding stipules) of *af tl* plants, with the exception that the rachis tips show some small areas of fusion to produce tiny laminate structures. The 'parsley-leaf' phenotype may, therefore, represent some movement towards the 'ground state' for the pea leaf rather than some positive interaction between the characteristics of *af* and *tl* plants. This hypothesis would suggest that the signal coded for by the *Af* gene in particular is of key evolutionary significance, since it may be involved with the primary development from a three-dimensional divaricating branch structure into a fused, two-dimensional leaf lamina. This is consistent with the results of Meicenheimer *et al.* (1983), who showed that *af* plants lack marginal leaf meristems, and the observation of Marx (1987) that the leaves of *af* and *af tl* plants are arranged in three dimensions and not in one plane like wild-type (WT) leaves (see also Cote *et al.*, 1992).

Another informative set of gene interactions involve the *sil* mutation. As a single mutant this allele does not appear to be homoeotic since it simply causes both the leaflets and stipules to undulate (Marx, 1977). The *wlo* mutation reduces wax formation on the upper surface of leaflets (see earlier). In *sil wlo* plants, however, the stipule tips as well as the leaflets are waxless. Likewise *sil af* plants possess deeply incised stipule tips from which tendrils may grow (Marx, 1987). Consequently, *sil* appears to possess a homoeotic action when combined with *wlo* or *af*. This may imply homology between the tip of the stipule and the adaxial leaflet surface (Marx, 1977). Thus *af* in the presence of *sil* may be viewed as able to exert an effect outside its normal domain. In this sense *sil* is modifying the normally distinct sites of action or domains of the leaf development genes *Af* and *Tl*.

Three-way interactions may also be instructive. As indicated earlier, *af tl* plants possess many small laminate structures and *wlo* reduces wax formation on the upper surface of leaflets. It might reasonably be expected that the laminate structures on *af tl wlo* plants might be waxless on the upper surface. However, they are not; they possess the wild-type layer of wax on their upper surface. This may be interpreted as *af* blocking the action of *wlo* (Marx, 1987). Another explanation may be that these laminate structures are not equivalent to leaflets. Rather, as suggested earlier, they may represent the 'ground state' for development of the pea leaf. Therefore genes normally specifically expressed in leaflets are not expressed in these structures due to the lack of a functional *Af* product which is required for leaflet formation.

The final interactions that we wish to mention involve the *cri* mutation. When this gene is combined with mutations such as *af*, *sil*, *tl* or *tac*, disorganized tissues frequently result (Marx, 1987). The *cri* mutation therefore is disrupting the normal developmental patterns that occur in the other mutants. This may suggest that the *cri* mutation operates at a more basic level of tissue organization than the other mutations and that its effects become more pronounced in mutant backgrounds where normal homoeostatic mechanisms are already disturbed.

Internode Length

While genes at well over 20 loci influencing internode length have been described, only 15 have been examined in detail (Table 7.2). Length mutants are the most commonly described class of developmental mutants. This partially reflects their ease of identification, due to their distinct appearance at an early stage of seedling emergence, and the ease of quantifying internode length. However, it also reflects the wide range of processes that impinge on the final internode length of the plant. Some of these processes have a direct role in the control of internode elongation while others are far more indirect, influencing internode elongation only as a result of effects on other processes (e.g. *lm*; Reid and Ross, 1988a). This allows the internode length mutants to be divided into a range of subgroups depending on their possible mode of action, such as mutants with altered levels of the biologically active gibberellin, GA_1, mutants with an altered response to GA_1, and mutants with altered photomorphogenic responses (Table 7.2).

Photomorphogenic mutants

These mutants possess altered responses to light. In peas (Weller and Reid, 1993) and other species (Kendrick and Nagatani, 1991), they have proved useful for exploring the role of the different species of phytochrome in controlling aspects of photomorphogenesis.

Plants homozygous for the mutant allele *lv* possess more elongated internodes than the wild-type plants when grown in white or red light (Nagatani *et al.*, 1990). They also possess a reduced photoperiod response for flowering (Weller and Reid, 1993) and an enhanced response to GA_1 (Reid and Ross, 1988b). They lack the usual shade-avoidance mechanisms and do not respond to end-of-day treatment with far-red (FR) light (Weller and Reid, 1993). This last response has become a bioassay for phytochrome B activity. Since

Table 7.2. The major mutations influencing stem elongation along with their proposed action. They are grouped into GA synthesis types, which block steps in the GA biosynthetic pathway, GA response types, which possess modified responses to applied GA_1, and photomorphogenic types, which alter the response to light. Further details may be found in Reid and Ross (1993).

Mutation	Phenotype and proposed action	Key reference(s)
GA synthesis		
le	dwarf, reduces 3β-hydroxylation of GA_{20} to GA_1	Mendel, 1866; Ingram et al., 1984; Ross et al., 1989
na	nana, blocks ent-7α-hydroxykaurenoic acid to GA_{12}-aldehyde	Wellensiek, 1969; Ingram and Reid, 1987a
ls	dwarf, reduces step GGPP to CPP*	Reid, 1986; Ingram and Reid, 1987a; S. Swain, T. Saito, Y. Kamiya and J.B. Reid, unpublished
lh	dwarf, may block prior to ent-kaurene	Reid, 1986; Ingram and Reid, 1987a; Swain and Reid, 1992
sln	slender, blocks production of GA_{29}-catabolite	Reid et al., 1992; Ross et al., 1993
GA response		
lk	erectoides, very short and may modify ethylene levels	Reid, 1986; Ross and Reid, 1986
lkb	semi-erectoides, short, reduced IAA levels and altered cell-wall properties	Reid and Ross, 1989; Behringer et al., 1990
lka	semi-erectoides, short, altered cell-wall properties	Reid and Ross, 1989; Behringer et al., 1990
lkc	short, normal GA levels	Reid et al., 1991
lkd	short, normal GA levels	Cramp and Reid, 1993
la cry[s]	slender, duplicate genes, possible receptor mutant	de Haan, 1930; Potts et al., 1985
lm	micro, all parts reduced, modified root development	Lindqvist, 1951; Reid and Ross, 1988a
lgr	short, normal GA levels, possible receptor mutant	I.C. Murfet, J. Kusnadi, J.J. Ross and J.B. Reid, unpublished
Photomorphogenic		
lw	short, reduced GA response, enhanced photoperiod response	Jolly et al., 1987; Weller and Reid, 1993
lv	elongated, GA-hypersensitive, reduced photoperiod response	Reid and Ross, 1988b; Nagatani et al., 1990; Weller and Reid, 1993
lip1	short, light-independent photomorphogenesis	Frances et al., 1992

*Geranylgeranyl pyrophosphate to copalyl pyrophosphate.

preliminary immunological and spectrophotometric studies suggested that light-stable phytochrome is not altered in *lv* plants (Nagatani *et al.*, 1990) this mutant may influence the transduction pathway for the phytochrome B signal. However, the level of all the light-stable phytochrome species (phytochromes B, C, D and E) needs to be determined before this can be confirmed.

The other established photomorphogenic mutant is *lw*. In many respects this mutant possesses a phenotype which is the reverse of *lv*. For example, *lw* plants show a reduced internode elongation in white light, a reduced response to GA_1 (Jolly *et al.*, 1987), an enhanced photoperiod response for flowering and an enhanced end-of-day FR response (Weller and Reid, 1993). Subsequent experiments suggest that the mutation confers a hypersensitivity to phytochrome, similar to that observed in the *hp* mutant in tomato (Peters *et al.*, 1989) and a transgenic tobacco line which overexpresses rice phytochrome A (Nagatani *et al.*, 1991).

While further study is necessary to determine the precise action of *lv* and *lw*, present results indicate that phytochrome function influences stem elongation via changes in the ability of the tissue to respond to endogenous GA_1 levels (Jolly *et al.*, 1987; Reid and Ross, 1988b) rather than by modifying GA_1 levels (J.L. Weller, J.J. Ross and J.B. Reid, unpublished). This is the reverse of suggestions made in the case of *Brassica rapa* using the stable-phytochrome-deficient mutant *ein* (Devlin *et al.*, 1992) and the phytochrome-altered mutant ma_3^R in *Sorghum bicolor* (Childs *et al.*, 1991). The conclusion that GA responsiveness is altered in the pea photomorphogenic mutants is also consistent with studies from both light- and dark-grown plants, which suggest that light reduces the responsiveness of pea internodes to applied GA_1 (Reid, 1988) and does not cause a substantial drop in GA_1 levels (Ross and Reid, 1989). Additional photomorphogenic mutants are under examination (J.L. Weller and J.B. Reid, unpublished), which may further clarify the role of the different phytochrome types in stem elongation and flowering.

A recently described mutant, *lip1*, is reported to result in light-independent photomorphogenesis. It has characteristics normally associated with light-grown seedlings even when grown in complete darkness (Frances *et al.*, 1992). For example, dark-grown *lip1* plants have short epicotyls and partially expanded leaves which display clear cellular differentiation and partially developed plastids. The plants possess less than 10% of the wild-type level of spectrally detectable phytochrome due to a reduction in the PHYA polypeptide. *Cab*, *Feb1* and *Rbc5* transcripts are present in the dark-grown *lip1* plants at levels comparable to those in light-grown material (Frances *et al.*, 1992). However, *lip1* also results in a substantial reduction in internode elongation in the light but tests for allelism with the other internode length genes have not been undertaken. Furthermore its morphological characteristics in the dark are somewhat similar to those described for dark-grown *na* plants (Reid, 1983; see section on GA-synthesis mutants; Fig. 7.2). The relationships clearly need clarification and highlight the difficulties encountered when mutants affecting

Fig. 7.2. The phenotype of *na* plants of line NGB1766 when grown in continuous white light (L) or complete darkness (D). The *na* plants show a lack of elongation and apical hook development, and considerable leaf expansion in the dark compared with dark-grown wild-type plants (Reid, 1983). (Photograph by courtesy of J.L. Weller.)

similar processes are approached from entirely different perspectives. This is particularly clear in the case of *lip1*, where the wild-type but not the short mutant was treated with GA (Frances *et al.*, 1992) even though GA_1 deficiency is well established as the cause of many dwarf mutants (Table 7.2; see below).

GA-synthesis mutants

The genes at five loci appear to modify stem elongation by directly altering the level of GA_1. Mutations at four of these loci, *le*, *na*, *ls* and *lh*, result in dwarf to extreme dwarf (nana) phenotypes (Mendel, 1866; Wellensiek, 1969, 1971a; Reid *et al.*, 1983; Reid, 1986) due to a deficiency of the biologically active GA, GA_1 (see Reid and Ross, 1993). In contrast, the *sln* mutation causes an increase in elongation of the basal internodes and *sln* plants possess substantially elevated GA_1 levels in young shoots (Reid *et al.*, 1992).

The *le* mutation reduces the 3β-hydroxylation of GA_{20} to GA_1 in the apical portion of shoots (Ingram *et al.*, 1984). It also reduces the 3β-hydroxylation of GA_9 to GA_4 (J.J. Ross, J.B. Reid and N.L. Lawrence, unpublished), a similar conversion in the apparently minor (for peas) non-13 hydroxylation pathway. A more severe allele, le^d (Ross and Reid, 1987), results in reduced internode lengths and GA_1 levels compared with *le* plants

(Ross et al., 1989). A less severe allele, le^{5839}, has also been identified (Ross and Reid, 1991). The le allele reduces GA_1 levels in the apical bud, leaves, petioles, tendrils (Smith et al., 1992), pods and young developing seeds (Santes et al., 1993). However, while it has frequently been assumed that le results in the production of a defective 3β-hydroxylase, other possible actions have not been excluded (e.g. that Le is a regulatory gene). Indeed, Smith (1992 and unpublished results) has suggested, on the basis of metabolism of GA_{20} by excised sections, that Le controls the production of an inhibitor of 3β-hydroxylation in the apex. While the evidence for such a role is weak, partly due to the poor separation and identification of metabolites, it does highlight the need for identification of the action of the Le gene at the molecular level. The Le gene does not appear to be expressed in relatively mature developing seeds, at or beyond contact point (Gaskin et al., 1985), and hence there is no effect of le in this tissue at this developmental stage.

The primary effect of the GA_1 deficiency in le plants is a reduction in stem elongation, and compelling evidence has been obtained using this mutant that GA_1 acts as an endogenous regulator of internode elongation in wild-type pea plants (Ross et al., 1992). However, the reduction in GA_1 levels in pods, leaves and developing seeds is not correlated with major phenotypic changes in these organs (Smith et al., 1992; Santes et al., 1993). Further work is therefore required to clarify the role of GA_1 in the growth of these organs.

Gene Ls appears to be required for the conversion of geranylgeranyl pyrophosphate (GGPP) to copalyl pyrophosphate (CPP) since *in vitro* feeding studies showed reduced conversion of [^{3}H]GGPP to [^{3}H]*ent*-kaurene by ls plants while the conversion of [^{3}H]CPP to [^{3}H]*ent*-kaurene was not inhibited compared with Ls plants (S. Swain, T. Saito, Y. Kamiya and J.B. Reid, unpublished). This is consistent with earlier application experiments that had shown that ls plants responded as well to applied *ent*-kaurene as wild-type plants (Ingram and Reid, 1987a). A second mutant allele at this locus, ls^{M26}, has been described (Reid, 1986) but its effects have not been examined in detail.

Gene Lh is thought to be required to allow the conversion of GGPP to *ent*-kaurene, based on experiments involving the application of intermediates in the GA biosynthetic pathway such as *ent*-kaurene (Ingram and Reid, 1987a). However, results of experiments with paclobutrazol show an interesting interaction with alleles at the lh locus (Swain and Reid, 1992) and indicate that metabolic studies are required to confirm the site of action. Two mutant alleles, lh (Reid, 1986) and lh^i (Swain and Reid, 1992), both of which result in dwarf phenotypes, have been identified at this locus. They appear to possess differential tissue specificity, since lh causes a greater reduction in GA levels in the young shoot (Swain and Reid, 1992) whereas lh^i is more severe in developing seeds (Swain et al., 1993). The lh^i allele has allowed us to demonstrate that GAs may be important in controlling seed development as well as stem elongation in peas since homozygous lh^i plants display substantial seed abortion when self-pollinated but substantially reduced abortion if crossed with *Lh* (wild-type) pollen. GA levels in the heterozygous *Lhlh^i* seeds were also restored to at least wild-type levels (Swain *et al.*, 1993).

The *na* allele appears to block the conversion of *ent*-7α-hydroxykaurenoic acid to GA_{12}-aldehyde since *na* plants do not elongate in response to the application of precursors prior to GA_{12}-aldehyde whereas *Na* plants respond significantly (Ingram and Reid, 1987a). Further, *na* plants do not metabolize [^{3}H]*ent*-kaurenoic acid to substances co-eluting with C19 GAs, in contrast to *Na* plants. The block to GA biosynthesis caused by *na* appears to be tissue-specific, influencing shoot and pod tissue, but not developing seeds (Potts and Reid, 1983; Potts, 1986; S. Swain, unpublished). Mutations at the *na* locus appear to have a more severe effect in shoots than mutations at the *le*, *ls* or *lh* loci, and they result in a reduction of GA_1 levels by up to 50 times compared with wild-type plants (Ingram *et al.*, 1984; Proebsting *et al.*, 1992; S. Swain, unpublished). This makes the *na* mutation ideal for the study of the role of GA_1 in the photomorphogenic control of stem elongation and leaf expansion. Plants possessing *na* are substantially dwarfed in both the light and the dark (Reid, 1983). They also show appreciable leaf development and reduced apical hook development in the dark (Reid, 1983; Fig. 7.2), not unlike the *det* and *cop* mutations in *Arabidopsis* (Chory *et al.*, 1989; Deng and Quail, 1992). While chlorophyll formation and full leaf development are not present in dark-grown *na* plants, the results suggest that GA_1 is required to allow normal etiolation processes to proceed and raise the possibility that some of the constitutive mutations that lack photomorphogenic control may influence GA levels or the GA response pathway.

The *sln* mutation blocks the production of GA_{29}-catabolite in developing seeds and testas (Ross *et al.*, 1993). This is associated with a dramatic build-up of GA_{20} and, to a lesser extent, GA_{29} in dry seeds. On imbibition, the GA_{20} is converted to GA_1, causing a two- to threefold increase in the length of the basal internodes (Reid *et al.*, 1992; Ross *et al.*, 1993). Feeding with [^{13}C ^{3}H]GA_{29} shows that the step GA_{29} to GA_{29}-catabolite is blocked in *sln* plants but the results do not indicate why GA_{20} levels are also elevated in *sln* plants (Ross *et al.*, 1993). Treatment with prohexadione-Ca on germination reduces elongation, indicating that it is indeed the 3β-hydroxylation of the GA_{20} carried over in the seed which is responsible for the rapid early growth of *sln* plants. Paclobutrazol treatment at this time has no effect but paclobutrazol does reduce seedling elongation if the seeds are treated at an early stage of development on the maternal plant (Reid *et al.*, 1992; Ross *et al.*, 1993). On the basis of phenotypic characteristics, the *Sln* gene may also be tissue-specific, acting in developing seeds but not the mature shoot; metabolic and genetic studies are required to verify this observation. This proposed site of action for *Sln* is the reverse of the apparent situation for *Na* and may suggest that both genes are regulatory, rather than structural, genes. Although alternative explanations exist, the apparent tissue specificity of these genes suggests that there is a much tighter control in specific tissues over GA metabolism, and, subsequently, GA levels, than appeared to be the case a few years ago.

The sln mutation also displays a unique pattern of inheritance for a gene controlling a growth process in peas. In crosses between homozygous Sln and sln plants, the F_1 and F_2 are all wild-type. In the F_3, one quarter of progenies breed true for the slender (sln) phenotype (Reid et al., 1992). This appears to occur because GA_{20} levels do not build up unless both the embryo and testa are homozygous sln. In F_2 seed, the testa is from the maternal F_1 plant and hence has a heterozygous Sln sln genotype. The testa has previously been shown to be a major site of catabolism of GA_{20} to GA_{20}-catabolite (Sponsel, 1983; Zhu et al., 1991).

GA-response mutants

These mutants possess an altered response to the application of GA_1 and display a range of internode-length phenotypes from extremely short (e.g. lk) to elongated, slender types (e.g. the gene combination la cry^s). This group of mutants is also extremely diverse in other phenotypic characters (Reid and Davies, 1992), indicating that they are probably a heterogeneous group of mutants that do not necessarily have underlying similarities at the biochemical and molecular levels. Indeed, when underlying mechanisms are identified for such mutants, they have frequently been shown to fall into another category; lv and lw, for example, both possess altered GA_1 responses due to their effects on photomorphogenesis.

Some of the mutants may act at, or close to, perception of the GA_1 signal. Such mutants should be true phenocopies of the GA-synthesis mutants. This means that not only should all GA-mediated processes be altered but also these mutants should lack the additional pleiotropic effects present in many GA-response mutants. Two such mutants exist. One is the slender mutant conferred by the gene combination la cry^s at the duplicate loci La and Cry (de Haan, 1927, 1930; Reid et al., 1983; Potts et al., 1985). On seedling emergence, la cry^s plants are true phenocopies of the GA_1-overproducing mutant sln (Reid et al., 1992, Ross et al., 1993). The la cry^s slender mutant behaves as if saturated with GA_1 regardless of the endogenous GA_1 levels since na la cry^s plants are still slender in phenotype (Potts et al., 1985; Ingram and Reid, 1987b). A second mutant with the desired phenotype is the short internode mutant lgr, which is morphologically similar to the ls, lh and le GA_1-deficient mutants (I.C. Murfet, J. Kusnadi, J.J. Ross and J.B. Reid, unpublished). This mutant responds only weakly to applied GA_1 but can be promoted to an equivalent length to wild-type plants if treated continuously with high doses of GA_1. The lgr plants also possess normal levels of GA_1. The sequence of action of these two mutants has not been determined but the phenotype of the lgr la cry^s gene combination may indicate which of the two acts closer to the point of GA reception.

Another group of GA_1-response mutants possesses a phenotype which suggests that they influence the ability of the plant to respond to GA_1 by partially blocking some step well down the transduction pathway leading from GA_1 perception to elongation. These mutants show a reduction in internode length, leaf area, petiole length, peduncle length and tendency to branch (Reid, 1986; Reid and Ross, 1989). The stems are also frequently ridged and there appears to be a reduction in wax production on the leaves and stem (Ross and Reid, 1986). This group includes relatively severe mutants at the *lk*, *lka* and *lkb* loci (the so-called erectoides and semi-erectoides phenotypes). In these mutants, treatment with GA_1 never results in elongation comparable to similarly treated wild-type plants. The mutations isolated at the *lkc* and *lkd* loci have an effect similar to, but less dramatic than, *lk*, *lka* and *lkb*, and they have received less attention (Reid *et al.*, 1991; Cramp and Reid, 1993).

Work on *lka* and *lkb* suggests that these mutations reduce elongation in both the light and dark by altering the ability of the cell wall to elongate. The *lka* and *lkb* mutants possess elevated osmotic and turgor pressures, a change not apparent in *ls*, a GA_1-deficient mutant of similar stature. In addition, the wall yield threshold is substantially elevated compared with wild-type plants and is greater than in *ls* plants (Behringer *et al.*, 1990). These results suggest that *lka* and *lkb* alter wall relaxation by altering its chemorheological properties. The biochemical basis for these changes is unknown, but mutant *lkb* also has reduced levels of IAA and shows pronounced elongation when treated with auxin transport inhibitors (M.J. McKay, J.J. Ross, N.L. Lawrence and J.B. Reid, unpublished). The most extreme of these mutants, *lk*, also appears to possess elevated ethylene levels under certain growing conditions (Ross and Reid, 1986). Whether the changed levels of auxin and ethylene in this group of mutants are the primary cause of the mutant phenotype (or at least major aspects of the phenotype) or are merely secondary consequences deserves examination, particularly since there is a paucity of mutants influencing the levels of these hormones in higher plant species and the role of auxin in internode elongation of intact plants is still uncertain.

Finally, there are mutants that alter elongation in a purely indirect way by limiting normal growth. One example is *lm*, which results in a reduction in the size of all parts of the vegetative shoot. The first visible effect of the mutant is to cause abnormal root development (Reid and Ross, 1988a).

Flowering

The control of flowering in pea has been reviewed on several occasions (Haupt, 1969; Murfet, 1977, 1985b, 1989a, 1990b; Murfet and Reid, 1985). Continuous variation for the trait denied Mendel (1866) the opportunity to identify any major flowering genes and some papers have analysed the trait in terms of quantitative genetic control (e.g. Clay, 1935; Rowlands, 1964). Quantitative systems certainly play a role (Murfet, 1985b) but a considerable number of major flowering genes have been identified (Tables 7.3 and 7.4). In addition,

Table 7.3. Details of the origin, authors, type lines and chromosomal location for mutant alleles at 13 major flowering loci. The genotype of the wild-type (WT) is arbitrarily designated *Lt E Sn Dne Ppd hr Gi Fsd Veg-1 Veg-2 Det Dm Fds*. Most domestic cultivars used as initial lines in mutation programmes have this genotype.

Mutant allele	Author of symbol	Type line	Chromosome*	Author of linkage	Source of mutant, initial line/author
lf	White, 1917	L58	1	Hoshino, 1915	Nature
lf^a	Murfet, 1975	L7	1	As above	Nature
Lt^d	Murfet, 1975	WL1771	1	As above	Nature
sn	Barber, 1959	L58	7 (2)	Weeden et al., 1988	Nature
e	Murfet, 1971a	L58	6	Murfet, 1971b	Nature
Hr	Murfet, 1973	L64	3	Murfet, 1973	Nature
veg-1	Gottschalk, 1979	L172	–	–	Dippes Gelbe Viktoria/W. Gottschalk
fds	Gottschalk, 1983	R20E	4	Murfet, 1990b	Dippes Gelbe Viktoria/W. Gottschalk
dne	King and Murfet, 1985	K218	3	King and Murfet, 1985	Torsdag/K. Sidorova
det	Marx, 1986a	L89m	5 (7)	Marx, 1986a	Several
dm	Murfet, 1989a	L89m	–	–	Natural causes/C. Duchêne
gi	Murfet, 1989a	L158	5	Murfet, 1990a	Virtus/M. Vassileva
ppd	Arumingtyas and Murfet, 1993	M2/137	1	Arumingtyas and Murfet, 1993	Borek/N. Naidenova
veg-2	I.C. Murfet, M.J. McKay and W.K. Swiecicki, unpublished	Wt16005	1	I.C. Murfet and M.J. KcKay, unpublished	Kaliski/W.K. Swiecicki
veg-2^{inc}	I.C. Murfet, M.J. McKay and W.K. Swiecicki, unpublished	Wt16123	1	As above	Kaliski/W.K. Swiecicki
fsd	I.C. Murfet and S.A. Taylor, unpublished	Wt16015	–	–	Porta/W.K. Swiecicki

*Chromosomal location according to the revised pea map of Weeden et al. (1991). The numbers in parentheses refer to the chromosomes of Blixt's (1972) map.

Table 7.4. Summary of the proposed action and category of 13 flowering genes in pea and the phenotypic effect of the mutant allele in a wild-type (WT) background. The WT genotype is arbitrarily designated *Lf E Sn Dne Ppd hr Gi Fsd Veg-1 Veg-2 Det Dm Fds*. Most domestic cultivars used in mutation programmes have this genotype. In nature the WT almost certainly carries the dominant allele *Hr*.

Mutant allele	Dom (D) Rec (R)	Response[a] to Photoperiod	Response[a] to Vernalization	Action, category and phenotypic effect of mutant allele. Key references on action
WT		+	+	Late, quantitative long-day plant with vernalization response
lf	R	–[b]	–	Sensitivity to flowering signal, heterochronic. Early flowering; day-neutral for initiation but photoperiodic for other traits. Murfet (1971c)
e	R	+[c]	+	Synthesis, inhibitor, heterochronic. Increases *Sn Dne Ppd* activity in cotyledons. Murfet (1971c)
sn	R	–	–	Synthesis, inhibitor, heterochronic. Early-flowering day-neutral plant. Promotes transition from 2 to > 2 leaflets per leaf. Barber (1959), Murfet (1971c)
dne	R	–	–	As for *sn*. Leaky mutant. King and Murfet (1985)
ppd	~R	–	–	As for *sn*. Incompletely recessive. Arumingtyas and Murfet (1993)
Hr	D	++	++	Synthesis, inhibitor, heterochronic. *Hr* prolongs *Sn Dne Ppd* activity in shoot. Reid and Murfet (1977a)
gi	R	++	++	Synthesis, floral stimulus. Flowering delayed. C.A. Beveridge and I.C. Murfet (unpublished)

Gene			Description
fsd	R	++[d]	+[d]	Synthesis, floral stimulus. Flowering delayed in short days, does not flower in long days. S.A. Taylor and I.C. Murfet (unpublished)
veg-1	R	−	−	Sensitivity, homoeotic. Mutant never flowers; all inflorescence sites occupied by vegetative branches. Reid and Murfet (1984)
veg-2	R	−	−	Appears similar to, but not allelic with, veg-1. Responses not fully tested. I.C. Murfet (unpublished)
veg-2^{inc}	R	++	±	Sensitivity, homoeotic. Flowering delayed, inflorescence converted to leafy lateral branch, flowers often malformed. I.C. Murfet (unpublished).
det	R	+	NT	Heterochronic, shoot meristem activity terminated soon after the onset of flowering. Singer et al. (1991)
dm	R	NT	NT	Heterochronic, large to indefinite delay in flowering, shorter internodes, reduced leaf size, transition to >2 leaflets per leaf extensively delayed, female-sterile. Murfet (1989a)
fds	R			Development of flower buds suppressed in short days in certain genetic backgrounds.[e] Gottschalk (1983); Murfet (1990b)

[a] Response as measured by change in node of flower initiation; ++ large response, + moderate response, − no response, NT not tested.
[b] Flower initiation is early and unaffected by environment but short-day conditions increase the time to first open flower and length of the reproductive phase.
[c] Allele e has no phenotypic effect in a WT background but in an lf background lf E is early flowering and lf e is late (WT phenotype).
[d] fsd plants do not flower in long days (18–24 h) even when vernalized. Effect of vernalization in short days not yet tested.
[e] Effect of fds in a WT background not yet explored.

several internode length genes have substantial effects on flowering behaviour.

The flowering genes may be viewed and classified in several ways (Table 7.4). Some determine response to environmental factors like photoperiod whereas others do not. Some control events such as the induction process in the leaves, organs distant from the shoot meristem, whereas others act in the apical bud to determine reception of the flowering signal and transduction events leading to development of floral organs and expression of the floral programme. Some genes control the synthesis of mobile substances that act as growth regulators whereas others appear to control response to, or reception of, such substances. Lastly, some mutant alleles alter the timing of developmental events and thus may be viewed as heterochronic mutants, whereas others replace one organ with another and therefore qualify as homoeotic mutants.

For reference purposes, we have arbitrarily defined genotype *Lf E Sn Dne Ppd hr Gi Fsd Veg-1 Veg-2 Det Dm Fds* as wild-type (WT). This genotype is commonly encountered among domestic cultivars and most cultivars used in mutation programmes have this genotype (e.g. Dippes Gelbe Viktoria, Kaliski, Paloma, Parvus, Porta and Torsdag). In some cases the effect of a mutant allele is only known in this background. We believe the true WT in nature probably carries the dominant allele *Hr*. For example, a wild *P. fulvum* accession from Asia Minor (WL1256) has responses indicative of genotype *Hr*, but flowering behaviour needs to be examined in a range of wild accessions. The WT as defined has a late-flowering, quantitative long-day habit and a phenotype known as K-type (Marx, 1968) or L-type (Murfet, 1971a). WT plants usually flower around node 15 under long days (18–24 h) and about 10 nodes later under short days (8–10 h). However, a WT phenotype does not necessarily imply a full WT genotype since genotypes *e* (e.g. Hobart line 24) and *lf e* (e.g. Hobart line 53) also have an L-type phenotype (Murfet, 1971b).

The Lf *locus*

Lf (late flowering) was the first major flowering gene identified in pea (Hoshino, 1915; White, 1917). *Lf* is of great practical and evolutionary importance since alleles at this locus govern the length of the juvenile phase (Murfet, 1985b, 1990b). The alleles Lf^d, *Lf*, *lf* and lf^a specify minimum flowering nodes of 15, 11, 8 and 5, respectively (Murfet, 1978, 1985b) and they represent an excellent example of heterochrony. In lf^a plants, flower initiation can occur at nodes already laid down in the embryo and which bear a leaf with only one pair of leaflets, whereas WT peas normally possess leaves with two or more pairs of leaflets before flowering commences. Thus lf^a provides an example of paedomorphy since reproduction commences while leaf morphology is still in the juvenile form. Moreover, in lf^a plants flower initiation is possible before the plant is properly geared to reproduction. As a consequence, flower buds at several nodes begin to develop at about the same time, leading to simultaneous flower opening at two, three or even four nodes on the one day (Marx, 1972; Duchêne, 1984; Murfet, 1985b). In contrast, in WT (*Lf*) plants flowers open at consecutive nodes at about 2-day intervals (Murfet, 1985b). In Lf^d plants, flowering is delayed for so long that the plant is developmentally old in other respects, e.g. the leaves have two or more pairs of leaflets, internode length has decreased and apical dominance has decreased to such an extent that the bulk of the yield is borne on aerial laterals (Murfet, 1985b; Murfet and Reid, 1985). On the basis of grafting studies, it is proposed that the *Lf* alleles determine the threshold level of flowering signal necessary to trigger flower initiation at the shoot apex, with Lf^d specifying the highest threshold and lf^a the lowest (Murfet, 1971c, 1975). There is evidence that callus cultures of genotype *lf* grow faster than those of genotype *Lf* at low IAA (indole-3-acetic acid) levels while the reverse applies at high IAA levels (Ezhova and Kovalenko, 1992). The significance, if any, of this result in relation to flowering has not been established.

The *Lf* gene is by far the most mutable of the flowering genes in pea and the great majority of induced flowering mutants, including those symbolized *efr*, *no* and *pra*, have been traced to this locus (e.g. Murfet, 1978, 1982b; Uzhintseva and Sidorova, 1979, 1988; Murfet and Ezhova, 1991; Arumingtyas and Murfet, 1992). Most of the induced mutant alleles tend to match fairly closely the strength of *lf* or lf^a but we suspect that molecular evidence will eventually reveal a multitude of different alleles at this locus with subtly different strengths not detectable by classic methods. However, the fact that mutational changes appear to occur in close to quantum jumps is in itself an interesting question. Is the observation an artefact resulting from the threshold nature of flower initiation or are certain sites in the gene more prone to mutation?

Photoperiod response: genes Sn, Dne *and* Ppd, *and modifiers* E *and* Hr

The existence of both photoperiodic (long-day habit) and day-neutral types among pea cultivars was first demonstrated by Doroshenko and Rasumov (1929). The ability to respond to photoperiod in pea requires the joint presence of the dominant alleles for three complementary genes: *Sn* (sterile nodes; Barber, 1959; Murfet, 1971b), *Dne* (day-neutral; Murfet, 1982b; King and Murfet, 1985) and *Ppd* (photoperiod response; Arumingtyas and Murfet, 1993). In a WT background, mutation to the recessive allele at any one of these three loci results in much earlier flowering and a day-neutral habit. It is proposed that these three genes govern steps in the synthesis of a substance that functions as a graft-transmissible flower inhibitor since flower initiation in day neutral (e.g. *sn*) shoots is significantly delayed by grafting to photoperiodic stocks (Barber and Paton, 1952; Murfet, 1971c; Murfet and Reid, 1973; King and Murfet, 1985). The primary role of the *Sn Dne Ppd* product may be to direct assimilate flow (Reid and Murfet, 1984; Murfet, 1985b) since a loss of system activity, either through mutation (e.g. *Sn* to *sn*) or because of environmental conditions (e.g. long days), not only promotes flower initiation, but also leads to a series of other consequences. The tendency to form basal branches (secondary stems) is diminished, the growth of flower buds relative to leaf growth is promoted (flowers open closer to the apical bud in *sn* than in *Sn* plants), and the onset of apical arrest and senescence is accelerated so that the duration of the reproductive phase and seed yield are decreased and maturity date brought forward. Other consequences are a decrease in the number of flowers per inflorescence, peduncle length and flower lifespan (Murfet, 1971b, 1977, 1982a, 1985b; Duchêne, 1984; Murfet and Reid, 1985). The suggestion that the *Sn Dne Ppd* product directs assimilate flow is supported by recent results for sweetpeas (Beveridge *et al.*, 1992) in which photoperiodic plants were shown to direct a greater proportion of labelled assimilates in a basipetal direction than day neutral counterparts. Perhaps more importantly, within the apical bud itself photoperiodic sweetpea plants distributed significantly more assimilate into leaflets and less into petioles and internodes than day-neutral plants. These results support the nutrient diversion theory of flowering control (e.g. Sachs and Hackett, 1969; Bernier, 1988).

In a WT background, mutation of *Sn*, *Dne* or *Ppd* to the recessive allele lowers both the node of flower initiation and, to a smaller extent, the node of transition from two to more than two leaflets per leaf (Barber, 1959; R.J. Wiltshire, I.C. Murfet, and J.B. Reid, unpublished). These mutants therefore provide an example of heterochrony since there is an acceleration in the rate of change to both the reproductive state and the adult leaf form. Such day-neutral mutants also largely lose the ability to respond to vernalization which is present in the WT (Barber, 1959; Murfet and Reid, 1974). Vernalization appears to promote flowering in pea by at least two mechanisms – one by a reduction in the activity of the *Sn Dne Ppd* system, i.e. by reducing the synthesis of a graft-transmissible inhibitor, and the other by rendering the apex more responsive to the flowering signal (Reid and Murfet, 1975). In WT plants inhibitor synthesis occurs in the cotyledons and leaves under short-day conditions. Long-day conditions promote flowering because activity of the *Sn Dne Ppd* system is shut down by light through two phytochrome-mediated mechanisms (Murfet and Reid, 1974; Reid and Murfet, 1977b).

The genes *E* (early; Murfet, 1971a) and *Hr* (high response; Murfet, 1973) influence expression of the *Sn Dne Ppd* system during different stages of ontogeny. The dominant allele *E* seems to reduce activity of the system in the cotyledons and it therefore acts during the early stages of seedling growth (Murfet, 1971c), whereas *Hr* acts later in ontogeny to maintain *Sn Dne Ppd* activity and inhibitor production by the shoot (Reid and Murfet, 1977a). The dominant allele *Hr* enhances the capacity to respond to photoperiod and in the presence of an otherwise WT background results in the phenotype known as G-type (Marx, 1968) or LHR (Murfet, 1971a). An allelic difference at the *E* locus is not expressed phenotypically in a WT background and this locus is not easy to work with. However, if the background is changed to *lf*, *E* plants have an early photoperiodic habit (class early initiating = EI; Murfet, 1971a) and *e* plants a WT phenotype. Photoperiod does not affect the node of flower initiation in EI plants but the time to first open flower and duration of the reproductive phase are increased under short-day conditions. The extended duration of the reproductive phase (delayed apical arrest) is particularly marked in the genotype *lf E Hr* (*Sn Dne Ppd*, etc.) which has the phenotype described by Marx (1968) as G2.

The mutant allele *sn* appears largely inactive (an amorph) but *dne* is undoubtedly leaky. This leakiness is manifest by the fact that, when the background is shifted from WT to Lf^d *Hr*, *dne* plants display a very large response to photoperiod, i.e. an LHR phenotype (Murfet, 1989a). The mutant allele *ppd* is incompletely recessive (Arumingtyas and Murfet, 1993).

Mutants gi *and* fsd

The *gi* (gigas; Murfet, 1989a) mutant flowers later than its WT initial line (Virtus) under a range of conditions and the plants are often taller as a consequence of having more internodes. Internode length is not increased. Our studies (C.A. Beveridge, and I.C. Murfet, unpublished) show that the normal long-day photoperiod response is still apparent, indeed enhanced, in the *gi* mutant but with distinct peculiarities. In one experiment the mutant plants flowered at node 54 in a 12-h photoperiod and 28 in a 24-h photoperiod (8 h daylight + 16 h fluorescent light at 50 μmol m^{-2} s^{-1}) while the WT initial line flowered at nodes 20 and 15, respectively. However, *gi* plants generally, but not invariably, remained vegetative indefinitely under a 24-h photoperiod consisting of 8 h daylight and 16 h of weak (3 μmol m^{-2} s^{-1}) incandescent light. *gi* plants sometimes also failed to flower in an 18-h photoperiod comprising a natural day extended at each end with mixed incandescent/fluorescent light (25 μmol m^{-2} s^{-1}). Reciprocal grafts between the mutant and its initial line have clearly established that *gi* affects synthesis rather than sensitivity. The graft results indicate that *gi* plants are deficient in some substance essential for flowering, that the missing substance (or precursor) can be supplied from the WT across a graft union, and that the substance is certainly produced in the shoot but is also supplied by the cotyledons and/or root. Donor WT shoots with as little as three foliage leaves were effective at inducing flowering in mutant *gi* graft partners. Our graft studies gave no evidence that *gi* plants produced more flower inhibitor than WT plants.

Vernalization strongly promoted flowering in *gi* plants and the difference between the mutant and its WT initial line was reduced greatly by this treatment. Vernalized *gi* plants flowered in both short-day and long-day conditions whereas unvernalized *gi* plants did not flower in some long-day conditions. Reciprocal grafts (epicotyl/epicotyl) between vernalized and unvernalized *gi* plants showed that the promotive effect of vernalization in this genotype occurred in the shoot apical bud, not the cotyledons and rootstock. This result is consistent with vernalization effecting an increase in sensitivity to the floral signal.

The *gi* allele has now been transferred from a WT to a day-neutral (*sn*) background. The *sn gi* plants flowered as early as node 11 in short days (8 h) but they showed erratic and unstable flowering behaviour and some plants remained vegetative indefinitely. These non-flowering *sn gi* plants in short days are similar in phenotype to non-flowering *Sn gi* plants in long-day conditions. The sporadic flowering pattern was evidenced by the occurrence of one or two inflorescences on the main shoot followed by vegetative reversion and a similar pattern on lateral shoots. This pattern suggests that the hormonal situation in *sn gi* plants hovers close to the threshold for flowering for much of the time and the net result is sporadic flowering and a failure to form sufficient flowers and fruits to effect monocarpic senescence; a lingering death ensues. The non-flowering *sn gi* segregants were clearly distinguishable from vegetative *Sn gi* segregants in short days on the basis of internode length, leaf size, diminished vigour and other non-reproductive features in much the same way as *Sn veg* and *sn veg* segregants were distinguished previously (Reid and Murfet, 1984).

The recessive *fsd* (flowering short days) mutant (Tables 7.3 and 7.4) flowers in short days (8–14 h) but much later than the WT initial line. However, mutant plants do not flower in long days (18–24 h). Technically the *fsd* mutant behaves as a short-day plant, which contrasts with the quantitative long-day habit of WT peas. Graft results (S.A. Taylor and I.C. Murfet, unpublished) indicate that *fsd* plants are deficient in some substance essential for flowering, since mutant plants will flower belatedly in long days if grafted to a WT donor comprising a rootstock and a shoot with healthy leaves. Thus the *fsd* mutant has features in common with the *gi* mutant. Both mutations may block synthesis of the floral stimulus. However, unlike *gi* plants, *fsd* plants do not flower in a 24-h photoperiod when vernalized, or in 18-h conditions when grafted to a WT rootstock without leaves. Again, in contrast to *gi* plants, the internode length of *fsd* plants declined well before flowering commenced in short days. A test for allelism has not wholly resolved the genetic relationship between *fsd* and *gi*. All F_1 hybrids ($n = 12$) from the cross *fsd* × *gi* flowered in long days (18 h) ahead of the *gi* controls. Some *gi* controls and all *fsd* controls remained vegetative. Nevertheless, the phenotype of the hybrid plants was not fully WT. In many cases the F_1 plants showed only transient flowering followed by reversion to the vegetative state and their phenotype was not unlike that of the *gi* controls, which also showed transient flowering. The question of whether *fsd* and *gi* are allelic or non-allelic is therefore not resolved and further genetic tests are under way.

The mutations *gi* and *fsd* both substantially delay the onset of reproduction but do not appear to have any effect on the node of transition from two to more than two leaflets per leaf (R.J.E. Wiltshire, I.C. Murfet and J.B. Reid, unpublished). They may therefore be considered a further example of heterochronic action.

Loci Veg-1 *and* Veg-2

Reid and Murfet (1984) found that plants homozygous for the recessive allele *veg-1* (vegetative; Gottschalk, 1979) would not flower under any of the circumstances tested. They tried, unsuccessfully, a range of environmental conditions, treatments such as grafting or application of growth regulators, and manipulation of the background for the other flowering loci known at that time (*Lf*, *E*, *Sn* and *Hr*). However, Singer and Maki (1993) have recently obtained some evidence that flowers are produced by the double mutant *det veg-1*, but at a higher node than in *det* or WT plants. The *veg-1* mutant is homoeotic, in that axillary sites occupied by inflorescences with flowers in WT are occupied by vegetative lateral branches (inflorescences without flowers?) in *veg-1* plants. These branches in turn bear second-and third-order vegetative laterals in a reiterative pattern terminated only by the unnatural lingering death of the plant because normal monocarpic senescence does not occur in pea in the absence of fruiting. Whether *veg-1* blocks both inflorescence and flower initiation, or just flower initiation, is open to interpretation (Coen, 1991; Singer and Maki, 1993).

Two recessive mutants induced in the WT cv. Kaliski by Prof. W.K. Swiecicki have been traced to a new flowering locus symbolized *Veg-2* (vegetative; I.C. Murfet, M.J. McKay and W.K. Swiecicki, unpublished). In the Kaliski background, *veg-2* plants have remained wholly vegetative in long days or short days and whether vernalized or unvernalized. They therefore have a phenotype similar to the *veg-1* mutant (Gottschalk, 1979; Reid and Murfet, 1984). However, they have not yet been subjected to the full range of tests and conditions which were applied to *veg-1* (Reid and Murfet, 1984; Singer and Maki, 1993). Twenty five F_1 plants derived from crosses between *Veg-1/veg-1* and *Veg-2/veg-2* heterozygotes flowered in a similar manner to WT. Thus the probability that *veg-1* and *veg-2* are alleles is less than 0.001. Other tests showed *veg-2* is not allelic with *gi* or *fsd*. Plants homozygous for the second mutant allele, *veg-2^{inc}*, showed a 1.5- to 2.1-fold delay in node of flower initiation, compared with the WT initial line, in photoperiods of 24, 18, 14 and 8 h. However, the cardinal feature of the *veg-2^{inc}* mutant is that the inflorescence converts to a leafy lateral branch instead of terminating in a stub after the production of one or two flowers (Fig. 7.3). The flowers of *veg-2^{inc}* plants can be perfect but are often malformed; petals may be deleted, with the standard, or standard and one or two wings, missing; some petal expression may occur in the sepal whorl; petals may be replaced by leafy

Fig. 7.3. Axillary inflorescences of WT *Veg-2* (left) and mutant *veg^{inc}* (right). The WT inflorescence terminated in a stub after producing two flowers. The mutant inflorescence was converted into a leafy, flower-bearing, lateral shoot of indeterminate growth. The first flower was malformed and ineffective. Not all *veg-2^{inc}* inflorescences proliferate to the extent shown here; some produce only a single aberrant flower with bract and several unexpanded nodes before growth is arrested.

bracts; and some 'flowers' are principally recognizable as such by the presence of a small nest of anthers in green bract surrounds. A small number of *veg-2^{inc}* plants have failed to produce any recognizable flowers in an 18 h photoperiod.

An interaction of considerable developmental interest occurs in plants homozygous for veg-2^{inc} and the recessive *fasciata* allele *fa* (White, 1917). The *fa* mutation interferes with normal expression of the vegetative programme and the stem eventually expands into a ribbon-like structure, leaf insertion becomes irregular and the stem may fork into two or more shoots. However,

Fig. 7.4. Developmental confusion in the double mutant fa veg-2^{inc}: two carpels appear in the place of a leaflet.

the flowers of *fa* plants are usually unaffected. Conversely, *veg-2^{inc}* disrupts expression of the floral programme but vegetative growth seems unaffected. In the double mutant *fa veg-2^{inc}* developmental confusion ensues with the onset of the reproductive phase. For example, carpels are formed in place of leaflets (Fig. 7.4) or at the tip of leaflets. In one case a whorl of anthers occurred in one leaf axil and a cluster of carpels occurred in the next leaf axil as if consecutive stem nodes were interpreted as succeeding whorls of a flower. These developmental mutants warrant detailed study. Crosses between the *veg-2^{inc}* mutant and *Veg-2/veg-2* heterozygotes produced about 50% of plants with a WT phenotype and 50% with a *veg-2^{inc}* phenotype. Thus *veg-2^{inc}* and *veg-2* are alleles and the dominance order is *Veg-2* > *veg-2^{inc}* > *veg-2*. As befits homoeotic mutants neither *veg-1* nor *veg-2* appears to alter the node of transition from two to more than two leaflets per leaf (R.J. Wiltshire, I.C. Murfet and J.B. Reid, unpublished).

The recently described recessive mutant *pim* (proliferating inflorescence meristem; Maki *et al.*, 1993; Singer and Maki, 1993) has some features in common with *veg-2^{inc}*. The *pim* mutant shows extensive indeterminate growth of axillary inflorescences and aberrant flowers but, unlike *veg-2^{inc}*, it does not alter the node of flower initiation. *pim* and *veg-2^{inc}* should be tested for allelism.

Mutants dm, det and fds

Plants homozygous for the recessive mutant allele *dm* (diminutive) show a twofold to indefinite flowering delay depending on the genotype for the other flowering genes, internode length is reduced 50–65% wholly by a reduction in cell number, leaf size is reduced 40–50%, leaf shape is more ovate and the flowers are female sterile (Murfet, 1989a). *Dm* is thus not specifically a flowering gene and the recessive allele *dm* may cause a major defect in some process required for normal growth and development. The *dm* mutation also provides a striking example of heterochrony. In an *lf* but otherwise WT background, *dm* delays flower initiation by around 20 nodes and the transition from two to more than two leaflets per leaf by about 35 nodes (R.J.E. Wiltshire, I.C. Murfet and J.B. Reid, unpublished). Thus *dm* plants are paedomorphic. They represent an example of neoteny since the transition in leaf form is slowed relative to the onset of the reproductive state.

The recessive mutant allele *det* (determinate) causes the shoot to cease growth after formation of a small number of reproductive nodes (Marx, 1986a). The *det* mutation does not appear to alter the node of flower initiation but it can reduce the time to first open flower by several days in what is probably a compensatory response (Murfet, 1989b). The *det* plants are not determinate in the botanical sense since the 'terminal' flower arises from an axillary meristem rather than the apical meristem, which simply ceases to grow (Singer *et al.*, 1991). Thus the main shoot is changed from an organ of indeterminate growth to one of limited growth terminating in a stub, much like a normal inflorescence in pea (see Fig. 7.3). This is the converse of the effect of the *veg-2^{mc}* and *pim* mutations, which cause the inflorescence to grow in an indeterminate pattern. Examination of the interaction between *det* and other genes affecting inflorescence and floral development should help elucidate the role of these genes and the sequence of gene action. The expression of *det* is influenced by the genotype at the *Lf* locus; in *lf^d* plants, more reproductive nodes are formed before it is expressed (Murfet, 1989b, 1992). Whether this is a consequence of very early flower initiation in *lf^a* plants or a true interaction with *lf^d* remains to be determined. However, it suggests that *Lf* may act before *Det*. The *veg-1* mutation may block the transition from an inflorescence to a floral meristem, since the double mutant *det veg-1* has shorter inflorescences and is able to produce flowers (Singer and Maki, 1993). The double mutant *det pim* produces only a few flowers at each floral node compared with dozens in *pim* plants (Singer and Maki, 1993). Thus *Pim* appears to control inflorescence to floral transition in pea and it is hypostatic to *det*. The double mutants *det veg-2* and *det veg-2^{mc}* have yet to be examined. The late Dr G.A. Marx found that *det* could be combined with multiple flowers per inflorescence. Thus the *det* locus is of continuing interest both from a developmental viewpoint and for its practical potential to alter plant architecture.

Gottschalk (1983) reported that the recessive mutant allele *fds* (flower development suppressor) was responsible for the failure of flower buds to develop in several of his lines (e.g. R20E). We have not been able to obtain unequivocal evidence for *fds*, but flower bud development is strongly suppressed in R20E under short day conditions; the breeding results are consistent with the hypothesis that the trait is monogenic recessive and only expressed in a clear manner under short-day conditions and in plants which are both fasciated (*fa*) and photoperiodic (*Sn Dne Ppd*) (Murfet, 1990b).

The effect of internode-length genes on flowering

Mutant alleles that shorten internode length by blocking gibberellin biosynthesis (*le*, *lh*, *ls*, *na*; Table 7.2) all reduce the rate at which new leaves are produced, for example, *le* by 15% and *na* by 25% (Murfet and Reid, 1987). As a consequence the time to first open flower can be increased significantly in these gibberellin-deficient genotypes (Rasmusson, 1935; Marx, 1975; Murfet and Reid, 1987). Nevertheless, even double mutants (e.g. *lh ls*), which cannot emerge above ground without assistance and which attain a mature height of 2 cm, still produce flowers (Reid, 1986; S. Swain, unpublished). If gibberellin is required for flower initiation in pea, the level must be extremely low (Murfet and Reid, 1987).

The *lka* mutation results in a short internode, semi-erectoides phenotype (Reid and Ross, 1989). In a WT flowering background the mutant alleles *lka* and *lka*^{11242} cause an increase of some 6–14% in the node of flower initiation (Kusnadi *et al.*, 1992).

Weller and Reid (1993) proposed that the length mutant *lv* has a reduced response to light-stable phytochrome whereas the *lw* mutant may be hypersensitive to phytochrome action. In plants with a WT flowering genotype, both mutations had a substantial effect on flowering behaviour; the *lv* mutation almost eliminated the normal increase in node of flower initiation under short days whereas *lw* substantially increased the magnitude of the flowering delay in short days. Photomorphogenic mutations of this kind enhance the prospect of understanding how phytochrome mediates the flowering response to photoperiod in pea.

Branching

Genes that influence branching in pea may be subdivided into two categories: those that have been identified primarily because of their effect on branching (Table 7.5) and those identified for their effect on some other trait but which nevertheless have a marked influence on branching (Table 7.6). Genes in the first category may determine the synthesis, transport or reception of substances directly involved in the control of apical dominance and lateral bud outgrowth,

Table 7.5. List of pea genes identified primarily because of an effect on branching habit.

Gene	Author of symbol	Type line	Chromosome	Author of linkage	Phenotype
Asc	Lamprecht, 1951	WL936	–	–	*Ascendens*: stem branches semi-prostrate
fr	Lamprecht, 1950		3	Blixt, 1968	*fructicosa*: number of secondary stems
fru	Lamprecht, 1950		4	Blixt, 1968	*fructicosa*: number of secondary stems
ho	Lamprecht, 1958	WL1180	–	–	*horizontalis*: lateral branches growing horizontally
pro	Lamprecht, 1963	WL851	4	Lamprecht, 1963	*procumbens*: stem branches first horizontal then ~45°
ram	Monti and Scarascia Mugnozza, 1967	Monti P745 d-r	2 (now 7)	Monti, 1970 (Weeden et al., 1991)	*ramosus*: profuse branching, poorly fertile
rms-1	Blixt, 1976	WL5237	3	Blixt, 1976	*ramosus*: increased branching
rms-2	Arumingtyas et al., 1992	K524	–	–	*ramosus*: increased branching
rms-3	Arumingtyas et al., 1992	K487	–	–	*ramosus*: increased branching
rms-4	Arumingtyas et al., 1992	K164	–	–	*ramosus*: increased branching
rms-5	Apisitwanich et al., 1992 Arumingtyas et al., 1992	Wt15244	5	Apisitwanich et al., 1992	*ramosus*: increased branching

Table 7.6. List of flowering and internode-length genes which influence branching.

Trait	Genes	Effect on branching
Flowering	sn, dne, ppd	Reduced basal branching
	Hr	Increased basal branching
	Lf^d, gi, e	Increased aerial branching
	det	Increased branching
	veg-1, veg-2	Inflorescence converted into leafy lateral branch
	$veg\text{-}2^{inc}$	Inflorescence converted into leafy lateral branch with flowers
Internode length	le, lh, ls, na, lgr	Short internodes, increased basal branching
	la + cry^s, lv	Long internodes, reduced branching
	lk, lka, lkb, lkc, lkd	Short internodes, reduced branching

and mutations in such genes may therefore provide a valuable insight into the control of branching.

We have arbitrarily chosen common domestic cultivars with tall stature and a late flowering, quantitative long day habit to represent the normal or wild-type (WT) branching habit. These cultivars have WT stature and flowering behaviour as defined in previous sections. Examples include Torsdag, Parvus, Weitor and Kaliski, which are the initial lines for a number of induced branching mutants. These WT lines differ somewhat in branching habit due to differences in their genetic background. However, environmental conditions exert a major influence over the degree of branching (Lamprecht, 1950; Blixt, 1968) and the branching pattern (Arumingtyas *et al.*, 1992) actually displayed by such lines; for example, good field growing conditions and wide spacing between plants increase the number of secondary stems produced (Lamprecht, 1950). Likewise, photoperiod has a major effect (Nakamura, 1965; Arumingtyas *et al.*, 1992). The latter authors found Torsdag did not branch in a 24-h photoperiod but branched from both the upper and lower nodes in an 8-h photoperiod. Thus it is possible to designate a WT genotype, but the branching phenotype displayed is highly subject to environmental influence.

Arumingtyas *et al.* (1992) recognized five patterns of branching among pea cultivars and branching mutants: total absence of branches (nil); branches from the upper nodes only (aerial); branches from the basal nodes only (basal); branches from the upper and lower nodes separated by a region devoid of branches (gap); and branches from all, or almost all, vegetative nodes (complete). Second- and high-order branches may occur in a reiterative pattern. In plants with the basal pattern, one or more basal laterals may develop into secondary stems, which rival or even surpass the primary shoot in certain genotypes and environmental conditions. In plants with the gap pattern, the gap frequently occurred in the region of nodes 4–8 (counting the first scale leaf as node 1) but the gap sometimes occurred further up the stem. Some lines displayed up to three patterns depending on the conditions; for example, the pattern for Torsdag varied from nil to aerial to gap as the photoperiod varied from 24 to 16 to 8 h. Hence it was not possible to assign most lines to a particular pattern category.

Branching genes

Loci identified specifically because of their effect on branching are listed in Table 7.5. Two genes, *Fr* and *Fru* (*fructicosa*), were proposed to control the number of secondary stems in pea (Lamprecht, 1950; Blixt, 1968). Continuous unimodal F_2 distributions were partitioned arbitrarily into a 15 : 1 ratio (*Fr Fru, Fr fru, fr Fru* low : *fr fru* high) by Lamprecht and a 9 : 7 ratio (*Fr Fru* low : *Fr fru, fr Fru, fr fru* high) by Blixt. Lamprecht considered the genes unsuitable for linkage analysis because of the high variability of expression, but Blixt succeeded in placing *Fr* on chromosome 3 and *Fru* on chromosome 4. (See Chapter 2 regarding the identification and numbering of linkage groups and chromosomes.)

Six *ramosus* loci have been identified among 16 induced, single-gene recessive, branching mutants: *ram* (Monti and Scarascia Mugnozza, 1967), *rms-1* (Blixt, 1976), *rms-2*, *rms-3* and *rms-4* (Arumingtyas *et al.*, 1992) and *rms-5* (Apisitwanich *et al.*, 1992; Arumingtyas *et al.*, 1992). The *rms-1* locus is on chromosome 3 (Blixt, 1976), *rms-5* is on chromosome 5 (Apisitwanich *et al.*, 1992) and *ram* is on a chromosomal segment formerly assigned to the upper section of chromosome 2 (Monti, 1970) but now considered part of chromosome 7 (Weeden *et al.*, 1991). All 16 mutants exhibit increased branching compared with their initial lines and clear segregation has been demonstrated at each of the six loci. In some cases and conditions a qualitative difference occurred between mutant and non-mutant segregants, since the former branched and the latter did not (e.g. *rms-1*, *rms-3*, *rms-4*), and in other cases a clear segregation could be demonstrated using a one-way plot of total lateral length divided by total length of the main shoot (e.g. *rms-5*), or a two-way plot of these two variables (e.g. *rms-2*) (Arumingtyas *et al.*, 1992).

The *ram* mutant is characterized by profuse branching and a large number of poorly fertile flowers (Monti and Scarascia Mugnozza, 1967). The phenotype of *ram* plants is distinctly different from that of any of the *rms* mutants since the shoots show a progressive decline in meristem function and normal stem and leaf organogenesis, culminating in failure of continued growth, while the shoots of the *rms* mutants appear to retain normal meristem activity (Arumingtyas *et al.*, 1992). With the possible exception of *rms-2*, Arumingtyas *et al.* (1992) found no indication that mutation at any of the five *rms* loci resulted in a distinct branching pattern. The most profuse branching occurred with mutants *rms-1* and *rms-4* which displayed a complete pattern in some circumstances. As with normal lines, branching pattern and intensity varied with photoperiod, but the mutant *rms-2* appeared the least susceptible to this influence and tended to produce a gap pattern in all photoperiods tested (24, 18, 16 and 8 h).

We are currently studying the five *rms* mutants by grafting experiments, application of growth regulators, and analysis of phytohormone levels. C.A. Beveridge, R.S. Floyd, M.J. Gregory, I.C. Murfet and J.J. Ross (unpublished) have shown that the branching phenotype of *rms-1*, *rms-2* and *rms-5* shoots becomes more normal when grafted on to a normal stock, while *rms-4* shoots retain a mutant phenotype when grafted on a normal stock. The *rms-3* shoots showed some evidence of a shift toward normality but largely retained the mutant phenotype. These results indicate that the mutant alleles *rms-1*, *rms-2* and *rms-5* may block synthesis or transport of a mobile substance that suppresses lateral outgrowth, while *rms-4* appears to act in the shoot, possibly by interfering with response to, or reception of, a regulatory substance. The action of *rms-3* remains unresolved. Reciprocal grafts between the *rms-1*, *rms-2* and *rms-5* mutants may enable us to determine whether the three genes act on the same biosynthetic pathway and, if so, in what order. The shoots of *rms-2* plants have elevated levels of the auxin indole acetic acid (IAA), and decapitated *rms-2* and *rms-4* shoots both require higher levels of applied IAA to suppress lateral bud outgrowth than is the case for their normal counterparts (C.A. Beveridge, J.J. Ross and I.C. Murfet, unpublished). Clearly mutations at the five *rms* loci offer considerable potential for gaining an insight into the control of branching. On the other hand, profuse branching in *ram* plants appears to result from a decline in meristem function and this mutant may be more useful as a tool to study processes in the meristem itself.

Effect of the flowering genes on branching

Most flowering genes in pea have a significant effect on branching habit (Table 7.6). Day-neutral lines have a much lower tendency to produce basal laterals (secondary stems) than photoperiodically WT lines (Doroshenko and Rasumov, 1929; Murfet and Reid, 1985). The long-day habit of WT peas is conferred by the gene combination *Sn Dne Ppd* and, as detailed in the flowering section, there is evidence that the *Sn Dne Ppd* product increases assimilate flow in a basipetal direction. The mutations *sn*, *dne* and *ppd* cause loss of system function and result in a day-neutral habit and reduced basal branching. Light also shuts down the activity of the *Sn Dne Ppd* system (Murfet and Reid, 1974; Reid and Murfet, 1977b) and this can explain why the tendency of photoperiodic lines to produce basal laterals diminishes as the photoperiod increases (Nakamura, 1965; Floyd and Murfet, 1986; Arumingtyas *et al.*, 1992). The flowering gene *Hr*, which prolongs the activity of the *Sn Dne Ppd* system (Murfet, 1973; Reid and Murfet, 1977a), further enhances the capacity of photoperiodic lines to produce basal laterals (Ross, 1983). It is not clear whether the outgrowth of basal laterals in photoperiodic plants is stimulated directly by the greater availability of nutrients in the lower stem region or whether a stimulatory signal is relayed from the roots.

Many primitive accessions of *P. sativum* subsp. *humile*, *P. sativum* subsp. *elatius* and *P. fulvum* display an enhanced branching response to photoperiod when compared with the domestic cultivars arbitrarily chosen here to represent WT. These primitive accessions produce a dense mat of first-, second- and third-order basal laterals under short-day conditions such as a photoperiod of 8–12 h. Accession JI2055 (*P. sativum* subsp. *elatius*) was identified as among the most profusely branching of all the John Innes Institute accessions (M. Ambrose, personal communication). In tests at Hobart, (I.C. Murfet, unpublished), accession JI2055 branched only very weakly under continuous light. However, individual plants of JI2055 grown in 14-cm pots under greenhouse conditions from 24 April to 25 September (photoperiod 10–12 h) produced over 65 secondary stems with a total lateral length in excess of 31 m and a main shoot height of around 1 m. Such profuse basal branching is presumed to be of evolutionary advantage to those wild ecotypes that are adapted to growing vegetatively through the cool, moist conditions of winter and that flower with the advent of long days in the spring. There may be similar advantages to those adapted for winterhardiness (Murray and Swensen, 1991). Flowering in JI2055 is strongly delayed under short days and the accession probably carries the dominant alleles of all the major flowering genes.

In pea, a leaf normally subtends either an inflorescence or a vegetative lateral bud but not both. Thus any gene that increases the number of vegetative nodes also increases the number of potential sites at which outgrowth of a lateral branch can occur. Thus allele Lf^d and mutations *gi*, *fsd* and *e* (in some circumstances) all increase the occurrence of aerial laterals (Murfet and Reid, 1985; Floyd and Murfet, 1986; Murfet, 1989a, 1990b). The effect of the *Lf* alleles on branching is readily apparent in an *sn* background, where lf^a and *lf* plants are usually devoid of laterals but *Lf* and Lf^d plants generally possess aerial laterals (see cover of *Pisum Newsletter*, vol. 22, 1990). In the genotype Lf^d *sn*, the bulk of the yield may be borne on short aerial laterals (Murfet and Reid, 1985). The homoeotic mutants *veg-1* and *veg-2* do not produce flowers, and all inflorescence sites are occupied by vegetative lateral branches. These branches may in turn produce second- and third-order laterals, leading to a dense canopy of leaves. A similar phenotype occurs in *fsd* plants in long days and in the rarely encountered non-flowering Lf^d *sn* segregants. In *veg-2^{mc}* the inflorescences are converted into leafy lateral branches with flowers (Fig. 7.3).

Shoots homozygous for the *det* allele produce only a small number of reproductive nodes, usually one or two, before growth is terminated. The resulting seed crop is frequently insufficient to effect monocarpic senescence, and an outgrowth of lateral shoots occurs. The exception is the genotype lf^a *sn det*, where the primary shoot may bear three to six reproductive nodes and the gross phenotype resembles that of lf^a *sn Det* plants (Murfet, 1989b).

Again, genetically male sterile pea plants cannot undergo normal monocarpic senescence, and display a proliferation of lateral branches and a lingering death.

Effect of the internode length genes on branching

The majority of internode-length genes have a pronounced effect on branching in pea (Table 7.6). The GA_1-deficient short internode mutants *le*, *lh*, *ls* and *na* all display increased basal branching (Floyd and Murfet, 1986; J.B. Reid, unpublished). Conversely, the long-internode two-gene combination *la cry*^{s} (slender), which behaves as if saturated with GA (Potts *et al.*, 1985), generally does not branch at all, but may occasionally produce a weak lateral at one of the higher nodes (de Haan, 1927; Floyd and Murfet, 1986). Again, the *lv* mutant, which has long internodes, an etiolated appearance and an enhanced response to GA (Reid and Ross, 1988b), likewise shows reduced branching (Fig. 7.5). It appears that these effects on branching, associated with real or perceived differences in GA level, reflect compensatory changes: the more resources used on internode elongation, the less available for lateral outgrowth. However, it has been proposed that *lv* reduces the response to light-stable phytochrome (Weller and Reid, 1993); a direct effect of phytochrome B on branching may therefore be possible, and be part of the normal shade-avoidance mechanism. In contrast to the GA-deficient dwarf-internode types, the short-internode erectoides (*lk*) and semi-erectoides (*lka*, *lkb*) mutants show increased apical dominance (Reid, 1986; Reid and Ross, 1989; Kusnadi *et al.*, 1992). These mutants have approximately normal GA levels but show a reduced response to applied GA_1. However, there is reason to believe that *lk*, *lka* and *lkb* act some distance down the transduction pathway beyond the point

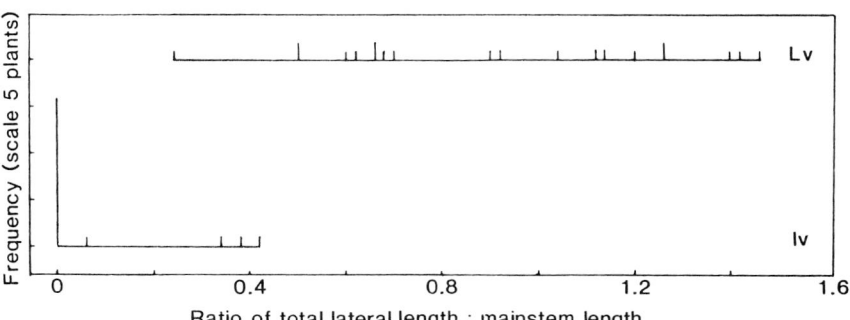

Fig. 7.5. Reduced branching in the *lv* mutant as shown by the distribution of the ratio of total lateral length to total length of the main shoot for plants of line 80 (*Lv*) and its isoline 80 m (*lv*^{80m}) (see Weller *et al.*, 1992) grown in an 8-h photoperiod. The total length of *lv* plants was approximately double that of *Lv* plants. (Data provided by I.C. Murfet and J.L. Weller.)

of GA reception (Lawrence *et al.*, 1992; Reid and Davies, 1992). The short-internode mutant *lgr* may act close to the point of GA reception and is of particular interest since it has the phenotype of a GA-deficient mutant, including increased branching, but, like the erectoides mutants, it has 'normal' GA levels and a reduced response to GA. The increased branching of the *lgr* mutant shows that the lack of branching in the erectoides mutants is not a consequence of reduced response to GA_1. The mutations that lead to an erectoides phenotype may interfere directly with the branching control system to block the compensatory increase in branching that we would expect to find in short internode types. The epistasis of the erectoides type of mutation (e.g. *lkc*) is well illustrated by the double mutant line K202 (*ls lkc*) (Reid, 1986; Reid *et al.*, 1991). Although deficient in GA_1 as a result of the *ls* mutation (Reid and Potts, 1986), this short-internode mutant branches significantly less than its normal initial line, Torsdag (Floyd and Murfet, 1986).

The phenotype of the double mutant *lkb rms-2* is not yet known, but it would be of interest to examine the interaction between the mutant alleles at these two loci. The non-branching, semi-erectoides mutant, *lkb*, has reduced IAA levels (Reid and Davies, 1992) whereas the branching mutant *rms-2* has increased levels of free IAA (C.A. Beveridge, J.J. Ross and I.C. Murfet unpublished). This is the reverse of that expected from traditional theories (Phillips, 1975) on the role of IAA in apical dominance. These two mutants may therefore help to elucidate the role of IAA in the control of both branching and internode elongation.

Angle of growth of lateral branches

Lamprecht has identified three loci that influence the angle of growth of stem branches (Table 7.5). The dominant allele *Asc* (*Ascendens*) causes branches to grow semi-prostrate (Lamprecht, 1951), the recessive *ho* (*horizontalis*) allele causes lateral branches to grow horizontally (Lamprecht, 1958) and the *pro* (*procumbens*) allele is said to cause stem branches to grow at first horizontal but subsequently at approximately 45° (Lamprecht, 1963).

The angle at which lateral branches emerge is also influenced by photoperiod and is generally further from the horizontal in long-day than short-day conditions (Table 7.7).

Plant Senescence

Peas have been used for many years as a model system for studying monocarpic senescence (e.g. Marx, 1968; Malik and Berrie, 1975; Proebsting *et al.*, 1976; Reid, 1980). They are well suited for this due to their limited lifespan, their large fruit size, the strong environmental control of senescence displayed and

Table 7.7. Effect of photoperiod on the angle of growth (from the horizontal) of the strongest lateral emerging from node two of 16-day-old seedlings. All plants received 8 h of daylight followed by 16 h of darkness (8-h photoperiod) or 16 h of weak (3 μmol m^{-2} s^{-1}) incandescent light (24-h photoperiod).

	Angle (from horizontal) of lateral at node 2						
	24-h photoperiod			8-h photoperiod			Significance of
Line	Mean	SE	n	Mean	SE	n	difference
31	53.7	1.0	6	27.5	1.9	6	$P < 0.001$
80	58.2	3.5	5	45.0	3.5	6	$P < 0.05$
173	53.7	4.4	3	24.0	2.3	3	$P < 0.01$

SE: standard error. n: number of plants.

the considerable genetic variation available. Much of this work has focused on senescence in the G2 type of peas (Marx, 1968), where detailed studies on the role of fruits on senescence have been carried out (e.g. Gianfagna and Davies, 1981). In this genotype the node of flower initiation is largely unaffected by photoperiod, but apical senescence (or apical arrest: Wang and Woolhouse, 1982) is dramatically delayed by short photoperiods, making it an excellent model system (Marx, 1968; Kelly and Davies, 1988). However, the genes involved with this response are those that influence the synthesis of the flower inhibitor, *Sn*, *Dne*, *Hr* and *E* (Murfet, 1973, 1978, 1985b; Reid, 1979), and it would appear that the product of these genes may influence both senescence and flowering by altering the assimilate distribution within the plant (Reid and Murfet, 1984; Kelly and Davies, 1988; Beveridge *et al.*, 1992). This can result from both a direct action on assimilate distribution, as seen from the effect of these genes in defruited plants, and an indirect effect due to a nutrient diversion to fruit growth (Gianfagna and Davies, 1981; Reid, 1980). These genes can therefore be considered to be 'true' senescence genes.

Many other genes described in this review, e.g. *veg-1*, *ar*, *n*, *lh'*, also modify plant senescence. However, their action appears to be indirect, due to their effects on the size of the reproductive sinks (Murfet, 1985a; Swain and Reid, 1992). This illustrates one difficulty in using developmental mutants to explore the control of certain basic developmental processes. The mutants provide maximum benefit if they can be shown to act directly on the process under consideration but, for processes such as senescence and yield, where many correlative factors interact to control the process, it may be necessary to screen the mutants very strictly to identify genes specifically regulating the desired process. This may partly explain why molecular advances have yet to be made for such processes.

Inflorescence and Pod Development

One to several flowers are borne on axillary racemes in the pea. Most domestic cultivars with a WT flowering habit usually produce two flowers per raceme, but multipodded types have been selected which produce three to six or more flowers. The trait is influenced by the environment (Lamprecht, 1947; Hole and Hardwick, 1976) and no clear agreement exists concerning the genetic control (Murfet, 1985b). Some reports have proposed a polygenic control (e.g. Clay, 1935; Ibarbia and Bienz, 1970; Snoad and Arthur, 1973), whereas others propose control by two genes, *Fn* and *Fna*, with *Fn Fna* having one flower per raceme, *Fn fna* and *fn Fna* two flowers, and *fn fna* three or more flowers (White, 1917; Lamprecht, 1947). The trait is certainly influenced by some major flowering genes. Murfet (1985b) described a cross segregating for *Sn-sn* and *Hr-hr* in which *sn* segregants had one or two flowers, *Sn hr* plants two or three flowers, and *Sn Hr* generally three or four flowers per raceme at the most productive nodes. The presence of *sn* can account for the observations of Meunissier (1922) and Lamprecht (1947) that early cultivars often have only one flower per raceme. The mutations $veg\text{-}2^{inc}$ (Tables 7.3 and 7.4) and *pim* (Maki et al., 1993; Singer and Maki, 1993) radically alter inflorescence morphology and can markedly increase the number of flowers per raceme. Blixt (1972) lists a number of other mutations that alter inflorescence traits. Some flowering genes have a very large influence on peduncle length, which can range from 1 mm in genotype Lf^d *sn* to over 200 mm in tall, early photoperiodic lines. In a similar background, peduncle length decreases with the allelic sequence lf^a, *lf*, *Lf*, Lf^d (Murfet, 1990b). In some circumstances, peduncle length can be used to clearly distinguish *Sn* and *sn* segregants where flowering node or time fail (Murfet, 1985b). The internode-length mutations, *lka* and *lkb*, substantially reduce peduncle length (Reid and Ross, 1989; Kusnadi et al., 1992). As detailed in the section on flowering, all inflorescence sites in the mutants *veg-1* and *veg-2* are occupied by vegetative lateral branches, or inflorescences without flowers, depending on interpretation.

Several mutations have a clear and specific effect on pod development. The pods of primitive peas dehisce at maturity, scattering the seeds, whereas the pods of domestic peas are non-dehiscent. This difference is determined by the alleles and *Dpo* (dehiscent)-*dpo* (non-dehiscent) (Marx, 1971). The complementary genes *P* and *V* interact to control formation of a tough, sclerenchymatous membrane lining the inner pod wall: *P V* has a complete membrane, *P v* has patches of sclerenchyma, *p V* has a band of sclerenchyma along the region near the ovule-bearing suture and *p v* has no membrane (White, 1917). Pods lacking part or all of this membrane can be eaten prior to pod inflation and are variously known as non-parchmented, sugar or snow peas. The *n* mutation results in a thick, fleshy pod wall (Wellensiek, 1925b), the pods are shorter and narrower, and seed size is reduced (Wehner and Gritton, 1981; Murfet, 1985a; Reid, 1989). Pea pods have a string of lignified sclerenchymatous fibres along both sutures. Two recessive mutations resulting in stringless pods have been reported: *sin* (Lamprecht, 1938; a mutant that has been lost) and *sin-2* (Wellensiek, 1971b; McGee and Baggett, 1992). Expression of *sin-2* is dependent on high temperature (Wellensiek, 1971b). McGee and Baggett (1992) have shown that *sin-2* pollen grows more slowly than pollen with the *Sin-2* allele and they suggest this accounts for the deficiency of stringless plants found in segregating F_2 populations. The combination *p v n sin-2* results in a snap pea with a pod that can be eaten, even when fully inflated.

The shape of the pod apex is controlled by the alleles *Bt* (blunt) and *bt* (pointed) (White, 1917). The *te* mutation reduces pod breadth by about 25% (Lamprecht, 1953). Several mutations influence pod curvature, and complex interactions occur, for example, with *n*. The *con* mutation (Lamprecht, 1936) causes convex pods, and the *cp* mutation (Wellensiek, 1925b) results in concavely curved pods. The *twp* mutation (Marx, 1973) causes twisting of the immature pod, due to arrested growth in patches of tissue in the central region of the valve; the symptoms diminish as the pod matures, and expression is variable. Blixt (1972) lists several other mutations that influence pod shape and breadth.

The mutations *ar* and *def* influence development of the funiculus. The *ar* mutation results in funiculi of reduced diameter, the parenchyma cells are smaller and less densely packed, and the vascular system appears disorganized compared with that of *Ar* plants; seed size is also substantially reduced (Reid, 1989). The *def* (development funiculus) mutation results in a funiculus that remains attached to the seeds when they are harvested (Khangildin and Khangildin, 1969).

The dominant allele *Np* causes a pustule-like growth on the external surface of the pod; the expression of this allele is influenced by light quality (Nuttal and Lyall, 1964; Dodds and Matthews, 1966). Snoad (1969) found the neoplasms to be highly polyploid, and to result from induced mitotic activity in the subsidiary cells that surround the stomata. Berdnikov *et al.* (1992) suggested that *Np* may confer a measure of resistance to pea weevil, since there is a high frequency of the *Np* allele among accessions from regions where pressure from the pest is high. They found that pustules formed on *Np* plants in response to oviposition by the beetle, and also following application of NaCl solution containing the homogenized remains of a beetle.

Duchêne (1984) compared the growth of pea pods in an early photoperiodic *Sn* line and in its day-neutral *sn* isoline. Under short day conditions, the pods on *Sn* plants were shorter and the maximum elongation rate was less than in *sn* plants. The time taken to achieve maximum growth rate was about 2 days longer in *Sn* pods, due to a corresponding delay in flower senescence and in the onset of pod elongation. Pod inflation was also delayed, and maximum pod breadth was narrower and attained later in *Sn* than in *sn* plants. However, the *Sn–sn* difference did not significantly affect the maximum rate of inflation of the pod. Differences between the isolines were largely eliminated in long days. The allele present at the *Lf* locus also influences pod growth (Duchêne, 1984; Murfet and Reid, 1985). For example, in a segregating progeny and background *sn* genotype, Duchêne found that the pods of *lf/lf* segregants reached their maximum elongation rate 1 day later than Lf^d/Lf^d pods and their maximum growth rate was only 60% that of the Lf^d/Lf^d pods.

Other Processes

Winterhardiness

Considerable variation in cold tolerance has been identified in peas (see Murray and Swensen, 1991) but single genes controlling this process have not yet been recognized. It is reported to be a quantitatively inherited trait, although genes for winterhardiness have been linked to *pl* and *en* (Auld *et al.*, 1985; Liesenfeld *et al.*, 1986). Given environments that maximize the phenotypic differences present, cold tolerance, like flowering, may prove amenable to genetic analysis; for many years, the latter was also considered to be quantitatively inherited.

Leaf wilting

The *wil* mutation results in plants that wilt when exposed to even minor water stress (Marx, 1976). These plants possess a lower percentage of water, water potential and diffusive resistance than wild-type plants (Donkin *et al.*, 1983). Leaves from drought-stressed *wil* plants also contain less abscisic acid than wild-type plants (Wang *et al.*, 1984). These results suggest that the *wil* mutation influences water relations through the control of ABA levels, in a similar manner to the well-examined mutants *flc*, *sit* and *not* in tomatoes and the *droopy* mutant in potatoes (Taylor and Tarr, 1984; Parry and Horgan, 1992). De Bruijn and Vreugdenhil (1992) recently used *wil* to suggest that abscisic acid does not influence the growth rate of pea seeds.

Root development

Genes controlling aspects of root development have not been examined extensively, due to the need for specific screening systems to identify mutants of interest. However, one mutant, *lm*, described principally as an internode-length mutant (Lindqvist, 1951; Reid and Ross, 1988a) would perhaps better be described as a root-growth mutant since the first visible effects of the 'micro' phenotype caused by *lm* are observed in the roots. The roots spiral and the cortex degenerates at an early age in *lm* plants (Reid and Ross, 1988a), possibly resulting in a reduction in size of all vegetative characters of the shoot.

A second mutant altering root development is *ageotropum*. The *age* mutation results in the main root growing upwards and protruding from the soil surface (Blixt, 1970), due to a breakdown in the normal gravitropic mechanisms. Subsequent work by Olsen and Iversen (1980a, b) has confirmed the inability of this mutant to respond in a normal manner to gravity. Jaffe *et al.* (1985) reported that *age* roots are neither gravitropic nor phototropic, but do respond tropically to a moisture gradient. This effect seemed to need an intact root cap, and it was suggested that the mutant would be a suitable vehicle for studying hydrotropism, unencumbered by phototropic and gravitropic responses.

Stem form

A range of mutants modifying the structure of the stem have been described (Blixt, 1972). They include mutations for fasciation (*fa*, Mendel, 1866; White, 1917), which result in the stem becoming flattened, with irregular insertion of leaves and the proliferation of many small shoot apices. These mutations may modify the structure and organization of the apical meristem and will be of use in exploring the control of meristem organization. The mutant alleles show variable expressivity. The mutation *art-1* (Marx, 1981) modifies the structure of the plant in the nodal regions, resulting in swollen nodes, undulate leaves and shortening of the upper internodes, due to a reduction in cell number (Rothwell and Murfet, 1987). Its effects may also result from abnormal meristem activity. The *creep* mutation (Sidorova, 1975) causes the vegetative shoots to be prostrate, but the stems grow vertically from the onset of flowering. In contrast, etiolated *creep* seedlings show normal vertical growth from the outset (Scott, 1988). This mutation may be of use in modifying plant architecture, since the growth of mutant plants is quite normal in other respects. The product of the *Creep* gene could be of widespread evolutionary relevance, because the gene converts growth from the prostrate to the upright habit (Scott, 1988).

Concluding Comments

This chapter has illustrated some of the diverse array of genetic mutants available in pea, and which may be valuable for investigating the control of developmental processes. Some of these, for example the internode-length mutants, have already provided much information at both the physiological and biochemical level. An extensive physiological model has been built up around the flowering mutants, but our efforts to identify the action of these genes at the biochemical level remain frustrated. Many of the other examples of this wealth of genetic variation are ripe for exploitation; the mutants influencing plant architecture, e.g. leaf form and branching pattern, and some clearly expressed mutations influencing pod development are particularly noteworthy in this respect. In most of these cases there is a need for a molecular analysis of the phenotypes.

The pea has many advantages as a model species for the study of plant development. The large seeds are ideal for studies of seed storage materials and seed development (see Chapters 5 and 6). The size of the plant is an advantage for many purposes, for example, for biochemical analysis of the levels of particular substances in various organs, but can also be a disadvantage, for example, tall, late-flowering genotypes can be a problem in growth chambers. The leaves show clear genetic differences, which make them suitable for developmental analysis, and the caulescent growth pattern enables clear measurements of internode length; again, the number of nodes can easily be counted as a measure of developmental age. The pea is well suited to, and has been used widely for, physiological studies. For example, it can be easily grafted and this technique has been used to study several developmental processes, including stem elongation, flowering, branching, and root nodule development. Indeed, the last process cannot be studied in many of the other model species, such as cereals, crucifers and members of the Solanaceae. Genetically pure lines abound in pea, the plant is easily crossed and the generation time can be as short as 9–10 weeks.

It is our view that the most rapid and efficient means of understanding the control of plant development will be to focus simultaneously on several model species, and then capitalize on the advantages of each species by establishing gene homologies at the molecular level. Thus information from one species may complement that from another, leading to a synergistic increase in knowledge.

Recent years have seen some very rapid and exciting progress in our understanding of the molecular events underlying the control of floral organogenesis in *Antirrhinum majus* and *Arabidopsis thaliana* (Coen and Meyerowitz, 1991) and of the physiological genetics of flowering, stem elongation and photomorphogenesis in the latter species (e.g. Martinez-Zapater and Somerville, 1990; Koornneef *et al.*, 1991; Zagotta *et al.*, 1992; Araki and Komeda, 1993). Homology between some flowering genes in *Arabidopsis* and *Antirrhinum*, e.g. that between *LFY* and *FLO* (Weigel *et al.*, 1992), has already been established. Certain genes in pea and *A. thaliana* are particularly worthy of comparison at the molecular level: the flowering genes *Veg-1* and *Veg-2* in pea with *LFY* and *AP1* in *A. thaliana*; the length genes *Ls* and *Le* with *GA1* and *GA4*; *Wil* with *ABA*; and the photomorphogenic gene *Lv* with *HY3*. Lastly, whereas developmental steps are ultimately controlled by specific events at the molecular level, changes in a single gene can have far-reaching effects on several developmental processes. Again, these processes may be influenced in a correlative manner, and by several interacting factors. A knowledge of events at the molecular level is crucial to gaining an understanding of developmental processes; however, an integrated approach, combining physiological and biochemical studies of the whole plant, genetic analysis and molecular studies, is necessary if a full understanding of plant development is to be achieved.

References

Apisitwanich, S. and Swiecicki, W.K. (1992) A new allele, st^{bs}, at the *St* locus. *Pisum Genetics* 24, 13.
Apisitwanich, S., Swiecicki, W.K. and Wolko, B. (1992) A new *ramosus* gene on chromosome 5. *Pisum Genetics* 24, 14-15.
Araki, T. and Komeda, Y. (1993) Analysis of the role of the late flowering locus, *GI*, in the flowering of *Arabidopsis thaliana*. *Plant Journal* 3, 231-239.
Arumingtyas, E.L. and Murfet, I.C. (1992) The genetic basis of early flowering in mutant line I/178. *Pisum Genetics* 24, 32-36.
Arumingtyas, E.L. and Murfet, I.C. (1993) Flowering in *Pisum*. A further gene controlling response to photoperiod. *Journal of Heredity* (in press).
Arumingtyas, E.L., Floyd, R.S., Gregory, M.J. and Murfet, I.C. (1992) Branching in *Pisum*: inheritance and allelism tests with 17 *ramosus* mutants. *Pisum Genetics* 24, 17-31.
Auld, D.L., Field, L.A. and Crock, J.C. (1985) *University of Idaho, Experimental Station Progress Report* No. 228.
Barber, H.N. (1959) Physiological genetics of *Pisum* II. The genetics of photoperiodism and vernalisation. *Heredity* 13, 33-60.
Barber, H.N. and Paton, D.M. (1952) A gene controlled flowering inhibitor in *Pisum*. *Nature* 169, 592.
Behringer, F.J., Cosgrove, D.J., Reid, J.B. and Davies, P.J. (1990) The physical basis for altered stem elongation rates in internode length mutants of *Pisum*. *Plant Physiology* 94, 166-173.
Berdnikov, V.A., Trusov, Y.A., Bogdanova, V.S., Kosterin, O.E., Rosov, S.M., Nedel'kina, S.V. and Nikulina, Y.N. (1992) The neoplastic pod gene (*Np*) may be a factor for resistance to the pest *Bruchus pisorum* L. *Pisum Genetics* 24, 37-39 and cover.
Bernier, G. (1988) The control of floral evocation and morphogenesis. *Annual Review of Plant Physiology and Plant Molecular Biology* 39, 175-219.
Beveridge, C.A., Ross, J.J. and Murfet, I.C. (1992) Mutant *dn* influences dry matter distribution, assimilate partitioning and flowering in *Lathyrus odoratus* L. *Journal of Experimental Botany* 43, 55-62.
Blixt, S. (1968) Linkage studies in *Pisum*. XII. Linkage relations of the genes *Fr* and *Fru* determining ramification. *Agri Hortique Genetica* 26, 136-141.
Blixt, S. (1970) The *ageotropum* mutant. *Pisum Newsletter* 2, 11-12.
Blixt, S. (1972) Mutation genetics in *Pisum*. *Agri Hortique Genetica* 30, 1-293.
Blixt, S. (1976) Linkage studies in *Pisum*. XV. Establishing the *Rms* gene and linkage of *Rms* and *Fas* in chromosome 3. *Agri Hortique Genetica* 34, 83-87.
Bowman, J.L., Smyth, D.R. and Meyerowitz, E.M. (1989) Genes directing flower development in *Arabidopsis*. *Plant Cell* 1, 37-52.

Childs, K.L., Pratt, L.H. and Morgan, P.W. (1991) Genetic regulation of development in *Sorghum bicolor*. VI. The ma_3^R allele results in abnormal phytochrome physiology. *Plant Physiology* 97, 714–719.

Chory, J., Peto, C., Feinbaum, R., Pratt, L. and Ausubel, F. (1989) *Arabidopsis thaliana* mutant that develops as a light-grown plant in the absence of light. *Cell* 58, 991–999.

Clay, S. (1935) Notes on the inheritance of quantitative characters in a cross between two varieties of garden pea (*Pisum sativum* L.). *Journal of Pomology and Horticultural Science* 13, 149–189.

Coen, E.S. (1991) The role of homoeotic genes in flower development and evolution. *Annual Review of Plant Physiology and Plant Molecular Biology* 42, 241–279.

Coen, E.S. and Meyerowitz, E.M. (1991) The wars of the whorls: genetic interactions controlling flower development. *Nature* 353, 31–37.

Cote, R., Gerrath, J.M., Pasluszny, U. and Grodzinski, B. (1992) Comparative leaf development of conventional and semileafless peas (*Pisum sativum*). *Canadian Journal of Botany* 70, 571–580.

Cramp, R.E. and Reid, J.B. (1993) Internode length in *Pisum*. Gene *lkd*. *Plant Growth Regulation* 12, 141–147.

de Bruijn, S.M. and Vreugdenhil, D. (1992) Abscisic acid and assimilate partitioning to developing seeds. I. Does abscisic acid influence growth rate of pea seeds? *Journal of Plant Physiology* 140, 201–206.

de Haan, H. (1927) Length factors in *Pisum*. *Genetica* 9, 481–497.

de Haan, H. (1930) Contributions to the genetics of *Pisum*. *Genetica* 12, 321–439.

Deng, X.-W. and Quail, P.H. (1992) Genetic and phenotypic characterisation of *cop1* mutants of *Arabidopsis thaliana*. *Plant Journal* 2, 83–95.

Devlin, P.F., Rood, S.B., Somers, D.E., Quail, P.H. and Whitelam, G.C. (1992) Photophysiology of the elongated internode (*ein*) mutant of *Brassica rapa*. *ein* mutant lacks a detectable phytochrome B-like polypeptide. *Plant Physiology* 100, 1442–1447.

Dodds, K.S. and Matthews, P. (1966) Neoplastic pod in the pea. *Journal of Heredity* 57, 83–85.

Donkin, M.E., Wang, T.L. and Martin, E.S. (1983) An investigation into the stomatal behaviour of a wilty mutant of *Pisum sativum*. *Journal of Experimental Botany* 34 (144), 825–834.

Doroshenko, A.V. and Rasumov, V.I. (1929) Photoperiodism of some cultivated forms in connection with their geographical origin. *Trudy po Prikladnoi Botanike Genetike i Selektsii* 22, 219–276.

Duchêne, Ch. (1984) Reproductive development in *Pisum*: the role of genes *Sn* and *Lf*. MSc thesis, University of Tasmania, Hobart.

Ezhova, T.A. and Kovalenko, O.V. (1992) Manifestation of the *Lf* locus in tissue culture of pea. *Pisum Genetics* 24, 40–43.

Floyd, R.S. and Murfet, I.C. (1986) Branching in *Pisum*: effect of the flowering and length genes. *Pisum Newsletter* 18, 12–15.

Frances, S., White, M.J., Edgerton, M.D., Jones, A.M., Elliott, R.C. and Thompson, W.F. (1992) Light characterisation of a pea mutant with light-independent photomorphogenesis. *Plant Cell* 4, 1519–1530.

Gaskin, P., Gilmour, S.J., MacMillan, J. and Sponsel, V.M. (1985) Gibberellins in immature seeds and dark-grown shoots of *Pisum sativum*. Gibberellins identified

in the tall cultivar Alaska in comparison with those in the dwarf Progress No. 9. *Planta* 163, 283-289.
Gianfagna, T.J. and Davies, P.J. (1981) The relationship between fruit growth and apical senescence in the G2 line of peas. *Planta* 152, 356-364.
Goldenberg, J.B. (1965) *Afila*, a new mutation in pea (*Pisum sativum* L.). *Boletin Genetico* 1:27-28.
Gottschalk, W. (1979) A *Pisum* gene preventing transition from the vegetative to reproductive stage. *Pisum Newsletter* 11, 10.
Gottschalk, W. (1983) A *Pisum* gene preventing flower development under short day conditions. *Egyptian Journal of Genetics and Cytology* 12, 385-396.
Gould, K.S., Cutter, E.G. and Young, J.P.W. (1986) Morphogenesis of the compound leaf in three genotypes of the pea, *Pisum sativum*. *Canadian Journal of Botany* 64, 1268-1276.
Gould, K.S., Cutter, E.G. and Young, J.P.W. (1991) Modification of pea leaf morphology by 2,3,5-triiodobenzoic acid. *Botanical Gazette* 152, 133-138.
Gould, K.S., Young, J.P.W. and Cutter, E.G. (1992) L-system analysis of compound leaf development in *Pisum sativum* L. *Annals of Botany* 70, 189-196.
Harvey, D.M. (1979) Evolution of an apulvinic foliar mutation in *Pisum sativum* L. *John Innes Institute Seventieth Annual Report* 34.
Haupt, W. (1969) *Pisum sativum* L. In: Evans, L. (ed.) *The Induction of Flowering: Some Case Histories*, Macmillan, Melbourne, pp. 393-408.
Hoch, H.C., Pratt, C. and Marx, G.A. (1980) Subepidermal air spaces: basis for the phenotype expression of the *Argenteum* mutant of *Pisum*. *American Journal of Botany* 67, 905-911.
Hole, C.C. and Hardwick, R.C. (1976) Development and control of the number of flowers per node in *Pisum sativum* L. *Annals of Botany* 40, 707-722.
Hoshino, Y. (1915) On the inheritance of the flowering time in peas and rice. *Journal of the College of Agriculture Hokkaido Imperial University* 6, 229-288.
Ibarbia, A.E. and Bienz, D.R. (1970) The inheritance of numbers of pods per node in peas, *Pisum sativum* L. *Journal of the American Society of Horticultural Science* 95, 137-139.
Ingram, T.J. and Reid, J.B. (1987a) Internode length in *Pisum*. Gene *na* may block gibberellin synthesis between *ent*-7α-hydroxykaurenoic acid and gibberellin A_{12}-aldehyde. *Plant Physiology* 83, 1048-1054.
Ingram, T.J. and Reid, J.B. (1987b) Internode length in *Pisum*. Biochemical expression of the *le* and *na* mutations in the slender phenotype. *Journal of Plant Growth Regulation* 5, 235-243.
Ingram, T.J., Reid, J.B., Murfet, I.C., Gaskin, P., Willis, C.L. and MacMillan, J. (1984) Internode length in *Pisum*. The *Le* gene controls the 3β-hydroxylation of gibberellin A_{20} to gibberellin A_1. *Planta* 160, 454-463.
Jaffe M.J., Takahashi, H. and Biro, R.L. (1985) A pea mutant for the study of hydrotropism in roots. *Science* 230, 445-447.
Jolly, C.J., Reid, J.B. and Ross, J.J. (1987) Internode length in *Pisum*. Action of gene *lw*. *Physiologia Plantarum* 69, 489-498.
Kelly, M.O. and Davies, P.J. (1988) Photoperiodic and genetic control of carbon partitioning in peas and its relationship to apical senescence. *Plant Physiology* 86, 978-982.
Kendrick, R.E. and Nagatani, A. (1991) Phytochrome mutants. *Plant Journal* 1, 133-139.

Khangildin, V.K. and Khangildin, V.V. (1969) Nekotorye razul'taty genetcheskikh issledovanii s gorokhom. *Trudy Bashkirskogo NIISKh* 3, 40–60.
King, W.M. and Murfet, I.C. (1985) Flowering in *Pisum*: a sixth locus, *Dne*. *Annals of Botany* 56, 835–846.
Koornneef, M., Hanhart, C.J. and van der Veen, J.H. (1991) A genetic and physiological analysis of late flowering mutants in *Arabidopsis thaliana*. *Molecular and General Genetics* 229, 57–66.
Kujala, V. (1953) Felderbse, bei welcher die ganz Blattspreite in Ranken umgewandelt ist. *Archivum Societatis Zoologiche Botanicae Fennicae 'Vanamo'* 8, 44–45.
Kusnadi, J., Gregory, M.J., Murfet, I.C., Ross, J.J. and Bourne, F.C. (1992) Internode length in *Pisum*: phenotypic characterisation and genetic identity of short internode mutant Wt11242. *Pisum Genetics* 24, 64–74.
Lamm, R. (1949) Contributions to the *Gp*-chromosome of *Pisum*. *Hereditas* 35, 203–214.
Lamprecht, H. (1933) Ein *unifoliata* – Typus von *Pisum* mit gleichzeitiger Pistilloidie. *Hereditas* 18, 56–64.
Lamprecht, H. (1936) Genstudien an *Pisum sativum*. I. Über den Effekt der Genpaare *Con-con* und *S-s*. *Hereditas* 22, 336–360.
Lamprecht, H. (1938) Über Hülseneingenschaften bei *Pisum*, ihre Vererbung und ihre Züchterischer Wert. *Der Züchter* 10, 150–157.
Lamprecht, H. (1947) The inheritance of the number of flowers per inflorescence and the origin of *Pisum* illustrated by polymeric genes. *Agri Hortique Genetica* 5, 16–25.
Lamprecht, H. (1950) The degree of ramification in *Pisum* caused by polymeric genes. *Agri Hortique Genetica* 8, 1–6.
Lamprecht, H. (1951) Genanalytische Studien zur Artberechtigung von *Pisum humile* Boiss et Noë. *Agri Hortique Genetica* 9, 107–134.
Lamprecht, H. (1953) Ein Gene für schmale Hülsen bei *Pisum* und seine Koppelung. *Agri Hortique Genetica* 11, 15–27.
Lamprecht, H. (1958) Über grundlegende Gene für die Gestaltung höherer Pflanzen sowie über neue und bekannte Röntgenmutanten. *Agri Hortique Genetica* 16, 145–192.
Lamprecht, H. (1963) Die Vererbung des ascendens-Merkmals von *Pisum* Genmanifestation und Chromosomenzugehörigkeit. *Agri Hortique Genetica* 21, 87–110.
Lawrence, N.L., Ross, J.J., Mander, L.N. and Reid, J.B. (1992) Internode length in *Pisum*. Mutants *lk*, *lka* and *lkb* do not accumulate gibberellins. *Journal of Plant Growth Regulation* 11, 35–37.
Liesenfeld, D.R., Auld, D.L., Murray, G.A. and Swensen, J.B. (1986) Transmittance of winterhardiness in segregated populations of peas. *Crop Science* 26, 49–54.
Lindqvist, K. (1951) The mutant 'micro' in *Pisum*. *Hereditas* 37, 389–420.
Maki, S.L., Horan, A.D. and Singer, S.R. (1993) Characterization of a floral mutant in pea affecting the transition from inflorescence to flower development. *Plant Physiology* 102, *Supplement* 124.
Malik, N.S.A. and Berrie, A.M.M. (1975) Correlation effects of fruits and leaves in senescence of pea plants. *Planta* 124, 169–175.
Martinez-Zapater, J.M. and Somerville, C.R. (1990) Effect of light quality and vernalization on late-flowering mutants of *Arabidopsis thaliana*. *Plant Physiology* 92, 770–776.

Marx, G.A. (1968) Influence of genotype and environment on senescence in peas, *Pisum sativum* L. *BioScience* 18, 505–506.

Marx, G.A. (1971) New linkage relations for chromosome 3 of *Pisum*. *Pisum Newsletter* 3, 18–19.

Marx, G.A. (1972) Simultaneous flowering in peas: a character of possible value in breeding. *Pisum Newsletter* 4, 28–29.

Marx, G.A. (1973) A new gene in *Pisum* affecting the contour of the pod. *Pisum Newsletter* 5, 25–26.

Marx, G.A. (1975) The *Le* locus: its influence on flowering time. *Pisum Newsletter* 7, 30–31.

Marx, G.A. (1976) 'Wilty': a new gene of *Pisum*. *Pisum Newsletter* 8, 40–41.

Marx, G.A. (1977) A genetic syndrome affecting leaf development in *Pisum*. *American Journal of Botany* 64, 273–277.

Marx, G.A. (1981) Arthritic (*art*): a new gene on chromosome 6. *Pisum Newsletter* 13, 38–39.

Marx, G.A. (1986a) Linkage relations of *Curl*, *Orc* and '*Det*' with markers on chromosome 7. *Pisum Newsletter* 18, 45–48.

Marx, G.A. (1986b) *Tendrilled acacia* (*tac*): an allele at the *uni* locus. *Pisum Newsletter* 18, 49–52.

Marx, G.A. (1987) A suite of mutants that modify pattern formation in pea leaves. *Plant Molecular Biology Reporter* 5, 311–335.

McGee, R.J and Baggett, J.R. (1992) Unequal growth rate of pollen tubes from normal and stringless pea genotypes. *HortScience* 27, 833–834.

Meicenheimer, R,D., Muehlbauer, F.J., Hindman, J.L. and Gritton, E.T. (1983) Meristem characteristics of genetically modified pea (*Pisum sativum*) leaf primordia. *Canadian Journal of Botany* 61, 3430–3437.

Mendel, G. (1866) Versuche über Pflanzenhybriden. *Verhandlungen des Naturforschenden Vereines in Brünn* 4, 3–47.

Meunissier, A. (1922) Observations sur l'hérédité du caractère 'Pois à trois cosses' et caractère 'Pois Chenille'. *Genetica* 4, 279–320.

Monti, L.M. (1970) Linkage studies on four induced mutants of peas. *Pisum Newsletter* 2, 21–22.

Monti, L.M. and Scarascia Mugnozza, G.T. (1967) Mutazioni per precocitá e ramositá indotte in pisello. *Genetica Agraria* 21, 303–312.

Murfet, I.C. (1971a) Flowering in *Pisum*: three distinct phenotypic classes determined by the interaction of a dominant early and a dominant late gene. *Heredity* 26, 243–257.

Murfet, I.C. (1971b) Flowering in *Pisum*: a three-gene system. *Heredity* 27, 93–110.

Murfet, I.C. (1971c) Flowering in *Pisum*: reciprocal grafts between known genotypes. *Australian Journal of Biological Sciences* 24, 1089–1101.

Murfet, I.C. (1973) Flowering in *Pisum*: Hr, a gene for high response to photoperiod. *Heredity* 31, 157–164.

Murfet, I.C. (1975) Flowering in *Pisum*: multiple alleles at the *lf* locus. *Heredity* 35, 85–98.

Murfet, I.C. (1977) The physiological genetics of flowering. In: Sutcliffe, J.F. and Pate, J.S. (eds), *The Physiology of the Garden Pea*. Academic Press, London. pp. 385–430.

Murfet, I.C. (1978) The flowering genes, *Lf, E, Sn* and *Hr* in *Pisum*: their relationship with other genes, and their descriptions and type lines. *Pisum Newsletter* 10, 48–52.

Murfet, I.C. (1982a) Flowering in the garden pea: expression of gene *Sn* in the field and use of multiple characters to detect segregation. *Crop Science* 22, 923–926.

Murfet, I.C. (1982b) The present state of physiological and biochemical gene mapping. In: Blixt, S. and Williams, J.T. (eds), *Documentation of Genetic Resources: A Model*. IBPGR, Rome, pp. 45–51.

Murfet, I.C. (1985a) The influence of genes *ar* and *n* on senescence in *Pisum sativum* L. *Annals of Botany* 55, 675–685.

Murfet, I.C. (1985b) *Pisum sativum*. In: Halevy, A.H. (ed.), *CRC Handbook of Flowering*, vol. IV. CRC Press, Boca Raton, Florida, pp. 97–126.

Murfet, I.C. (1989a) Flowering genes in *Pisum*. In: Lord, E.M. and Bernier, G. (eds), *Plant Reproduction: From Floral Induction to Pollination*. American Society of Plant Physiologists, Rockville, Maryland. pp. 10–18.

Murfet I.C. (1989b) Interaction of the *det* (determinate) mutant with other flowering genes in *Pisum*. *Pisum Newsletter* 21, 44–47.

Murfet, I.C. (1990a) The *gi* locus shows linkage with *gp*, *r* and *tl*. *Pisum Newsletter* 22, 38–40.

Murfet, I.C. (1990b) Flowering genes in pea and their use in breeding. *Pisum Newsletter* 22, 78–86.

Murfet, I.C. (1992) Expression of *det* (determinate) in genotypes Lf^d, *Lf*, *lf* and lf^a. *Pisum Genetics* 24, 75–78.

Murfet, I.C. and Ezhova, T.A. (1991) Differences in flowering behaviour between cultivar Ranny Zeleny and regenerant lines developed from callus cultures are attributable to variation at the *Lf* and *Sn* loci. *Pisum Genetics* 23, 19–25.

Murfet, I.C. and Reid, J.B. (1973) Flowering in *Pisum*: evidence that gene *Sn* controls a graft-transmissible inhibitor. *Australian Journal of Biological Sciences* 26, 675–677.

Murfet, I.C. and Reid, J.B. (1974) Flowering in *Pisum*: the influence of photoperiod and vernalising temperatures on the expression of genes *Lf* and *Sn*. *Zeitschrift für Pflanzenphysiologie* 71, 323–331.

Murfet, I.C. and Reid, J.B. (1985) The control of flowering and internode length in *Pisum*. In: Hebblethwaite, P.D., Heath, M.C. and Dawkins, T.C.K. (eds), *The Pea Crop*. Butterworths, London, pp. 67–80.

Murfet, I.C. and Reid, J.B. (1987) Flowering in *Pisum*: gibberellins and the flowering genes. *Journal of Plant Physiology* 127, 23–29.

Murray, G.A. and Swensen, J.B. (1991) Winterhardiness of pea. *Pisum Genetics* 23, 4–13.

Nagatani, A., Reid, J.B., Ross, J.J., Dunnewijk, A.E. and Furuya, M. (1990) Internode length in *Pisum*. The response to light quality, and phytochrome type I and II levels in *lv* plants. *Journal of Plant Physiology* 135, 667–674.

Nagatani, A., Kay, S.A., Deak, M., Chua, N. and Furuya, M. (1991) Rice type I phytochrome regulates hypocotyl elongation in transgenic tobacco seedlings. *Proceedings of the National Academy of Sciences USA* 88, 5207–5211.

Nakamura, E. (1965) Studies on branching in *Pisum sativum* L. *Special Report of Laboratory of Horticulture*, Shiga Agricultural College, Kusatsu.

Nuttal, V.W. and Lyall, L.H. (1964) Inheritance of neoplastic pod in the pea. *Journal of Heredity* 55, 184–186.
Olsen, G.M. and Iversen, T. (1980a) Growth and curvature in seedlings of *Pisum sativum* and an ageotropic mutant. *Physiologia Plantarum* 50, 269–274.
Olsen, G.M. and Iversen, T. (1980b) Ultrastructure and movements of cell structures in normal pea and an ageotropic mutant. *Physiologia Plantarum* 50, 275–284.
Parry, A.D. and Horgan, R. (1992) Abscisic acid biosynthesis in higher plants. In: Karssen, C.M., Van Loon, L.C. and Vreugdenhil, D. (eds), *Plant Growth Regulation*. Kluwer Academic Publishers, Dordrecht. pp. 160–178.
Pellew, C. and Sverdrup, A. (1923) New observations on the genetics of peas. *Journal of Genetics* 13, 125–131.
Peters, J.L., van Tuinen, A., Adamse, P., Kendrick, R.E. and Koornneef, M. (1989) High pigment mutants of tomato exhibit high sensitivity for phytochrome action. *Journal of Plant Physiology* 134, 661–666.
Phillips, I.D.J. (1975) Apical dominance. *Annual Review of Plant Physiology* 26, 342–367.
Potts, W.C. (1986) Gibberellins in light-grown shoots of *Pisum sativum* L. and the influence of reproductive development. *Plant and Cell Physiology* 27, 997–1003.
Potts, W.C. and Reid, J.B. (1983) Internode length in *Pisum*. III. The effect and interaction of *Na/na* and *Le/le* gene differences on endogenous gibberellin metabolism. *Physiologia Plantarum* 57, 448–454.
Potts, W.C., Reid, J.B. and Murfet, I.C. (1985) Internode length in *Pisum*. Gibberellins and the slender phenotype. *Physiologia Plantarum* 63, 357–364.
Proebsting, W.M., Davies, P.J. and Marx, G.A. (1976) Photoperiodic control of apical senescence in a genetic line of peas. *Plant Physiology* 58, 800–802.
Proebsting, W.M., Hedden, P., Lewis, M.J., Croker, S.J. and Proebsting, L.N. (1992) Gibberellin concentration and transport in genetic lines of pea. *Plant Physiology* 100, 1354–1360.
Pyke, K.A. and Hedley, C.L. (1985) Growth and photosynthesis of different pea phenotypes. In: Hebblethwaite, P.D., Heath, M.C. and Dawkins, T.C.K. (eds), *The Pea Crop*. Butterworths, London, pp. 297–306.
Rasmusson, J. (1935) Studies on the inheritance of quantitative characters in *Pisum*. I. Preliminary note on the genetics of time of flowering. *Hereditas* 20, 161–180.
Reid, J.B. (1979) Red–far-red reversibility of flower development and apical senescence in *Pisum*. *Zeitschrift für Pflanzenphysiologie* 93, 297–301.
Reid, J.B. (1980) Apical senescence in *Pisum*: a direct or indirect role for the flowering genes? *Annals of Botany* 45, 195–201.
Reid, J.B. (1983) Internode length in *Pisum*. Do the internode length genes effect growth in etiolated plants? *Plant Physiology* 72, 759–763.
Reid, J.B. (1986) Three further loci, *lh*, *ls* and *lk*. *Annals of Botany* 57, 577–592.
Reid, J.B. (1988) Internode length in *Pisum*. Comparison of genotypes in the light and dark. *Physiologia Plantarum* 74, 83–89.
Reid, J.B. (1989) Seed development in *Pisum*. The influence of genes *ar* and *n*. *Journal of Plant Physiology* 134, 751–755.
Reid, J.B. and Davies, P.J. (1992) The genetics and physiology of gibberellin sensitivity mutants in peas. In: Karssen, C., Van Loon, L.C. and Vreugdenhil, D. (eds), *Progress in Plant Growth Regulation*. Kluwer, Dordrecht, pp. 214–225.

Reid, J.B. and Murfet, I.C. (1975) Flowering in *Pisum* and possible mechanisms of the vernalisation response. *Journal of Experimental Botany* 26, 860-867.

Reid, J.B. and Murfet, I.C. (1977a) Flowering in *Pisum*: the effect of genotype, plant age, photoperiod and number of inductive cycles. *Journal of Experimental Botany* 28, 811-819.

Reid, J.B. and Murfet, I.C. (1977b) Flowering in *Pisum*: the effect of light quality on genotype *lf e Sn Hr*. *Journal of Experimental Botany* 28, 1357-1364.

Reid, J.B. and Murfet, I.C. (1984) Flowering in *Pisum*: a fifth locus, *veg*. *Annals of Botany* 53, 369-382.

Reid, J.B. and Potts, W.C. (1986) Internode length in *Pisum*. Two further mutants, *lh* and *ls*, with reduced gibberellin biosynthesis, and a gibberellin insensitive mutant, *lk*. *Physiologia Plantarum* 66, 417-426.

Reid, J.B. and Ross, J.J. (1988a) Internode length in *Pisum*. Further studies on the 'micro' gene, *lm*. Physiologia Plantarum 72, 547-554.

Reid, J.B. and Ross, J.J. (1988b) Internode length in *Pisum*. A new gene, *lv*, conferring an enhanced response to gibberellin A_1. *Physiologia Plantarum* 72, 595-604.

Reid, J.B. and Ross, J.J. (1989) Internode length in *Pisum*. Two further gibberellin-insensitivity genes, *lka* and *lkb*. *Physiologia Plantarum* 75, 81-88.

Reid, J.B. and Ross, J.J. (1993) A mutant based approach, using *Pisum sativum*, to understanding plant growth. *International Journal of Plant Science* 154, 22-34.

Reid, J.B., Murfet, I.C. and Potts, W.C. (1983) Internode length in *Pisum*. II. Additional information on the relationship and action of loci *Le*, *La*, *Cry*, *Na* and *Lm*. *Journal of Experimental Botany* 34, 349-364.

Reid, J.B., Ross, J.J. and Hasan, O. (1991) Internode length in *Pisum*: gene *lkc*. *Journal of Plant Growth Regulation* 10, 11-16.

Reid, J.B., Ross, J.J. and Swain, S.M. (1992) Internode length in *Pisum*. A new slender mutant with elevated levels of C_{19} gibberellins. *Planta* 188, 462-467.

Ross, J.J. (1983) Flowering and branching in the sweet pea. PhD thesis. University of Tasmania, Hobart.

Ross, J.J. and Reid, J.B. (1986) Internode length in *Pisum*. The involvement of ethylene with the gibberellin insensitive erectoides phenotype. *Physiologia Plantarum* 67, 673-679.

Ross, J.J. and Reid, J.B. (1987) Internode length in *Pisum*. A new allele at the *le* locus. *Annals of Botany* 59, 107-109.

Ross, J.J. and Reid, J.B. (1989) Internode length in *Pisum*. Biochemical expression of the *le* mutation in darkness. *Physiologia Plantarum* 76, 164-172.

Ross J.J. and Reid, J.B. (1991) Internode length in *Pisum*: le^{5839} is a less severe allele than Mendel's *le*. *Pisum Genetics* 23, 29-34.

Ross, J.J., Reid, J.B., Gaskin, P. and MacMillan, J. (1989) Internode length in *Pisum*. Estimation of GA_1 levels in genotypes *Le*, *le* and *le^d*. *Physiologia Plantarum* 76, 173-176.

Ross, J.J., Reid, J.B. and Dungey, H.S. (1992) Ontogenetic variation in the levels of gibberellin A_1 in *Pisum*. Implications for the control of stem elongation. *Planta* 186, 166-171.

Ross, J.J, Reid, J.B. and Swain, S.M. (1993) Control of stem elongation by gibberellin A_1. Evidence from genetic studies including the slender mutant *sln*. *Australian Journal of Plant Physiology* (in press).

Rothwell, R.N. and Murfet, I.C. (1987) Internode length in *Pisum*: effect of the gene *art* (arthritic). *Pisum Newsletter* 19, 59–60.

Rowlands, D.G. (1964) Genetic control of flowering in *Pisum sativum* L. *Genetica* 35, 75–94.

Rozov, S.M., Gorel, F.L. and Berdnikov, V.A. (1992) *coch* and *het* are allelic. *Pisum Genetics* 24, 82.

Sachs, R.M. and Hackett, W.P. (1969) Control of vegetative and reproductive development in seed plants. *HortScience* 4, 103–107.

Santes, C.M., Hedden, P., Sponsel, V.M., Reid, J.B. and Garcia-Martinez, J.L. (1993) Expression of the *le* mutation in young ovaries of *Pisum sativum* and its effect on fruit development. *Plant Physiology* 101, 759–764.

Scott, I.M. (1988) Stem growth habit and starch statolith content of the *creep* pea mutant. *Journal of Plant Physiology* 133, 103–106.

Sharma, B. (1972) *Tendrilled acacia*, a new mutation controlling tendril formation in *Pisum sativum*. *Pisum Newsletter* 4, 50.

Sidorova, K.K. (1975) Induced pea mutant with creeping stem. *Pisum Newsletter* 7, 57–58.

Singer, S.R. and Maki, S.L. (1993) Interactions of meristem identity genes affecting floral architecture in pea. *Plant Physiology* 102, *Supplement* 121.

Singer, S.R., Hsiung, L.P. and Huber, S.C. (1991) Determinate (*det*) mutant of *Pisum sativum* L. (Leguminosae: Papilionoideae) exhibits an indeterminate growth pattern. *American Journal of Botany* 77, 1330–1335.

Smith, V.A. (1992) Gibberellins A_1 biosynthesis in *Pisum sativum* L. II. Biological and biochemical consequences of the *le* mutation. *Plant Physiology* 99, 372–377.

Smith, V.A., Knatt, G.J., Gaskin, P. and Reid, J.B. (1992) The distribution of gibberellins in vegetative tissues of *Pisum sativum* L. I. Biological and biochemical consequences of the *le* mutation. *Plant Physiology* 99, 368–371.

Snoad, B. (1969) Neoplastic pod of pea. *John Innes Institute 60th Annual Report* 17.

Snoad, B. (1985) The need for improved pea-crop plant ideotypes. In: Hebblethwaite, P.D., Heath, M.C. and Dawkins, T.C.K. (eds), *The Pea Crop*. Butterworths, London. pp. 31–41.

Snoad, B. and Arthur, A.E. (1973) Genetical studies of quantitative characters in pea. I. A seven parent diallel cross of cultivars. *Euphytica* 22, 237–337.

Sponsel, V.M. (1983) The localization, metabolism and biological activity of gibberellins in maturing and germinating seeds of *Pisum sativum* cv. Progress No. 9. *Planta* 159, 454–468.

Swain, S.M. and Reid, J.B. (1992) Internode length in *Pisum*. A new allele at the *Lh* locus. *Physiologia Plantarum* 86, 124–130.

Swain, S.M., Reid, J.B. and Ross, J.J. (1993) Seed development in *Pisum*. The lh^i allele reduces gibberellin levels, and increases seed abortion. *Planta* (in press).

Swiecicki, W.K. (1989) A new gene heterophylus (*het*) on chromosome 7. *Pisum Newsletter* 21, 75–76.

Taylor, I.B. and Tarr, A.R. (1984) Phenotypic interactions between abscisic acid deficient tomato mutants. *Theoretical and Applied Genetics* 68, 115–119.

Uzhintseva, L.P. and Sidorova, K.K. (1979) Genetic nature of early flowering pea mutants. *Geneticka* 15, 1076–1082.

Uzhintseva, L.P. and Sidorova, K.K. (1988) Genetics of early flowering pea mutants. *Pisum Newsletter* 20, 39–40.

Vilmorin, P. de, and Bateson, W. (1912) A case of gametic coupling in *Pisum*. *Proceedings of the Royal Society B*, 84, 9–11.
Wang, T. and Woolhouse, H.W. (1982) Hormonal aspects of senescence in plant development. In: Jackson, M.B., Grout, B. and Mackenzie, I.A. (eds), *Growth Regulators in Plant Senescence*. British Plant Growth Regulation Group Monograph 8, Wantage, pp. 5–25.
Wang, T.V., Donkin, M.E. and Martin, E.S. (1984) The physiology of a wilty pea: abscisic acid production under water stress. *Journal of Experimental Botany* 35, 1222–1232.
Weeden, N.F., Kneen, B.E. and Murfet, I.C. (1988) Mapping of the *Sn* locus to chromosome 2. *Pisum Newsletter* 20, 49–51.
Weeden, N.F., Ambrose, M. and Swiecicki, W.K. (1991) Cover. *Pisum Genetics* 23.
Wehner, T.C. and Gritton, E.T. (1981) Effect of the *n* gene on pea pod characteristics. *Journal of the American Society of Horticultural Science* 106, 181–183.
Weigel, D., Alvarez, J., Smyth, D.R., Yanofsky, M.F. and Meyerowitz, E.M. (1992) *LEAFY* controls floral meristem identity in *Arabidopsis*. *Cell* 69, 843–859.
Wellensiek, S.J. (1925a) Genetic monograph on *Pisum*. *Bibliographia Genetica* 2, 343–476.
Wellensiek, S.J. (1925b) *Pisum* crosses I. *Genetica* 7, 1–64.
Wellensiek, S.J. (1959) Neutronic mutations in peas. *Euphytica* 8, 109–215.
Wellensiek, S.J. (1962) The linkage relations of the *cochleata* mutant in *Pisum*. *Genetica* 33, 145–153.
Wellensiek, S.J. (1969) The physiological effects of flower forming genes in peas. *Zeitschrift für Pflanzenphysiologie* 60, 388–402.
Wellensiek, S.J. (1971a) The localization of some new mutants. *Pisum Newsletter* 3, 46.
Wellensiek, S.J. (1971b) Lamprecht's gene *sin* for stringless. *Pisum Newsletter* 3, 48.
Weller, J.L. and Reid, J.B. (1993) Photoperiodism and photocontrol of stem elongation in two photomorphogenic mutants of *Pisum sativum* L. *Planta* 189, 15–23.
Weller, J.L., Murfet, I.C. and Reid, J.B. (1992) Internode length in *Pisum*. Two further *lv* mutants. *Pisum Genetics* 24, 86–89.
White, O.E. (1917) Studies of inheritance in *Pisum*. II. The present state of knowledge of heredity and variation in peas. *Proceedings of the American Philosophical Society* 56, 487–588.
Young, J.P.W. (1983) Pea leaf morphogenesis: a simple model. *Annals of Botany* 52, 311–316.
Zagotta, M.T., Shannon, S., Jacobs, C. and Meeks-Wagner, D.R. (1992) Early flowering mutants of *Arabidopsis thaliana*. *Australian Journal of Plant Physiology* 19, 411–418.
Zhu, Y., Davies, P.J. and Halinska, A. (1991) Metabolism of gibberellin A_{12} and A_{12}-aldehyde in developing seeds of *Pisum sativum* L. *Plant Physiology* 97, 26–33.

Molecular Approaches to the Study of Diseases of Pea

8

A. VIVIAN

Faculty of Applied Sciences, University of the West of England, Coldharbour Lane, Bristol, BS16 1QY, UK

Introduction

Peas are attacked by all the major groups of plant pathogens, including fungi, viruses and bacteria. The interactions between the pea host and particular pathogens vary from those showing very clearly defined race–cultivar reactions (e.g. the *Fusarium* wilts and bacterial blight) to those where a given cultivar may tolerate a wide range of strains (e.g. *Fusarium* root rot) (Hagedorn, 1984). These interactions have allowed an analysis of the contribution of the pathogen, an approach which has had some success, particularly in the case of bacterial diseases. However, the basis of host resistance and susceptibility remains a mystery, largely because of the difficulty of working with a genome the size of that of pea, a size which is typical of many important crop plants. The biochemistry and physiology of infected hosts have been extensively documented and the similarities in the responses of the plant to those induced by stresses such as mechanical wounding noted. In spite of this slow progress, it is possible to visualize experimental routes to the elucidation of the molecular basis of plant responses to pathogen attack. Here, two main approaches to studying the basis of resistance are outlined.

Physiological studies have established that upon infection the host reacts by activating genes, that are assumed to play a role in resisting attack or in limiting damage to the host. Such responses include phytoalexin synthesis and the production of phenolic compounds. Consequently, one approach has been to prepare cDNA libraries from host infected tissue and by differential screening to select clones from mRNA transcripts that are specifically expressed in infected tissue. Such genes, it is argued, are most likely to play a role in the resistance of the plant to pathogen attack and disease; the term pathogenesis-related (PR) proteins (van Loon, 1985) was originally coined to describe a class of such proteins.

A second approach has been used in an attempt to circumvent the problems associated with the generally large size of plant genomes. The common garden weed, *Arabidopsis thaliana*, is a flowering plant with a genome comprising only about one-tenth of the DNA present in pea. The recent discovery that an avirulence gene cloned from race 2 of the pea bacterial pathogen, *Pseudomonas syringae* pv. *pisi*, causes an ecotype-specific response in *A. thaliana* when introduced into *P. syringae* pv. *maculicola* (a pathogen of *A. thaliana*) has opened the possibility of a novel route to the cloning of the *R2* resistance gene of pea (Dangl *et al.*, 1992a). The Flor hypothesis of matching genes in host and pathogen (Flor, 1956) (see later) has been established by co-segregational studies both for the resistance gene *R2* in pea (Vivian *et al.*, 1989) and for the resistance gene *RPM1* in *A. thaliana* (Dangl *et al.*, 1992a). One can envisage that, if a strict gene-for-gene situation pertains, the products of the resistance genes in the two plants may well be homologous or share homologous domains. It has already been possible to identify a 270-kb region of chromosome 3 of *A. thaliana* which carries the *RPM1* gene, and its eventual isolation and sequence determination are close to realization (Dangl *et al.*, 1992b). Homology of the resistance genes, at either the DNA or the peptide level, could be exploited to identify the corresponding gene in pea.

Phytoalexins and Pathogenesis-related (PR) Proteins

Phytoalexins are compounds that have antimicrobial activity and are synthesized by plants in response to pathogenic infection or stress. They are usually uniform within a group of plants but show considerable diversity within the plant kingdom. The Leguminosae produce isoflavonoids, a group of compounds that only occur sporadically outside this family (Deverall, 1982). The major phytoalexin found in pea is pisatin, a pterocarpan (Ingham, 1982). Hadwiger and Webster (1984) investigated the levels of pisatin in five pea cultivars used as host differentials upon inoculation with races 1, 2 and 3 of the bacterial pathogen, *P. syringae* pv. *pisi*. They concluded that the accumulation of pisatin was absent or lower in compatible interactions, i.e. where there was a successful infection of the plant, than in incompatible combinations; this result is consistent with a role for the induction of pisatin synthesis in race-specific resistance. Subsequently, using pisatin accumulation and the detection of five specific mRNA transcripts as a measure of host response, Daniels *et al.* (1988) showed that complete bacterial cell envelopes which had been fragmented and pasteurized at 70°C induced these responses in almost all combinations of races and cultivars. Isolated lipopolysaccharide (LPS), lipoprotein-peptidoglycan or outer-membrane fractions, however, failed to induce accumulation of pisatin. These results indicate a lack of specificity in the induction of host response by the cell envelope of the bacterium and the authors concluded that some additional specific inducer or a suppressor molecule, generated by the bacterium once in contact with the pea, might be responsible for the observed race-specificity.

A crucial step in the biosynthesis of pisatin is the introduction at a late stage of a chiral centre, resulting in (+) optical stereochemistry. Sun et al. (1991) recently characterized the enzyme responsible, 7,2′-dihydroxy-4′,5′-methylenedioxyisoflavone oxidoreductase, and showed peas to be unique among agriculturally important legumes in producing a (+) stereoisomer. Similar 5-deoxyisoflavone oxidoreductases in soyabean (*Glycine max*) (Fischer *et al.*, 1990) and chickpea (*Cicer arietinum*) (Tiemann *et al.*, 1987) produce pterocarpans of (−) stereoisomeric form. The identification and characterization of these enzymes raise the possibility of investigating whether stereoisomerism plays a significant role in the ability of the host to resist pathogen attack (Sun *et al.*, 1991).

Fungal pathogens have been shown to detoxify the phytoalexins of their host (Schaefer *et al.*, 1989) and some vary in their ability to detoxify the optical isomers of pterocarpans (Van Etten *et al.*, 1989). Detoxification of pisatin is mediated by a cytochrome P-450 monooxygenase, which may be important for the pathogenicity of the fungus *Nectria haematococca* (the perfect stage of the fungus *Fusarium solani* f. sp. *pisi*) on pea. Weltring *et al.* (1988) isolated a gene for pisatin demethylation by screening library transformants of *Aspergillus nidulans* for their ability to tolerate pisatin. It is likely that this gene, which appears to act as a virulence factor, may belong to a superfamily of P-450 type genes (Whitlock, 1986).

Fusarium solani is a root-rotting fungus, showing some degree of host specificity. Pea plants resist infection by *F. solani* f. sp. *phaseoli*, a pathogen of *Phaseolus* bean (i.e. it is an incompatible reaction), but are susceptible to *F. solani* f. sp. *pisi*, a pathogen of pea (i.e. it is a compatible reaction). Incompatibility in this system is an active response in which growth of the pathogen is suppressed. Riggleman *et al.* (1985) constructed a cDNA library from poly-adenylated RNA isolated from pea tissue challenged with the incompatible pathogen. They then used differential colony hybridization with mRNA isolated from untreated and from the fungus-induced tissue to identify clones harbouring inducible genes, which they termed disease resistance response genes (DRRG) (Fristensky *et al.*, 1988). Such genes were assumed to be involved in non-host resistance and might include enzymes associated with phytoalexin synthesis and fungal-wall-hydrolysing enzymes such as chitinase. The nine unique cDNA clones obtained from this work were further used as probes to determine the rates of synthesis and accumulation of the corresponding mRNAs when challenged with either the compatible or incompatible strain of *F. solani*. This confirmed that seven of the nine clones behaved in a manner expected of DRRGs (Fristensky *et al.*, 1985). Daniels *et al.* (1987) showed that three races of *P. syringae* pv. *pisi* induced five of these DRRG

RNAs in a race-specific fashion. cDNA clones corresponding to two of the five RNAs were sequenced; these two, DNA clones pI49 and pI176, were 95% homologous (Fristensky et al., 1988). In recognition of this, the genes were redesignated DRRG49-a (pI49) and DRRG49-b (pI176) by Chiang and Hadwiger (1990). These authors also isolated from a pea genomic library a gene, designated DRRG49-c, which contained a single intron of 88 bp separating two exons. The amino acid sequence of DRRG49-c was 85% homologous to the DRRG49-a and DRRG49-b cDNAs (Chiang and Hadwiger, 1990).

At the time the sequences of the DRRGs were published, the authors could find no evidence of similarity with known DNA or peptide sequences in the data bases, but since then a number of apparently related genes have been characterized. BetvI is an allergenic protein, with an M_r of 17,000, which is a constituent of the pollen of white birch (*Betula verrucosa*) and causes allergic rhinoconjunctivitis and allergic bronchial asthma. Sequence analysis of a cDNA clone coding for BetvI has revealed that it shares 55% peptide sequence identity with DRRG49-a (Breiteneder et al., 1989). Two PR proteins, whose sequences have been deduced from elicitor-induced transcripts from bean (*Phaseolus vulgaris*) suspension cultures, also showed 64% identity to DRRG49-a (Walter et al., 1990). Recently, Crowell et al. (1992) have characterized two members of a family of stress-induced, developmentally regulated genes from soyabean whose nucleotide sequences are 86% identical; both share 70% identity at the amino acid level to the pea DRRG proteins described above. These results clearly establish a class of conserved defence-related proteins, which appear to be widely distributed in a number of plant families (Walter et al., 1990). Several indicate complexity in terms of gene families and it will be important in the future to attempt to discover the function of this class of genes.

Chalcone synthase (CHS) is the key regulatory enzyme of flavonoid and isoflavonoid biosynthesis in plants (Dixon et al., 1983). Studies with cultured cells of bean (*Phaseolus vulgaris*) have shown that the level of chalcone synthase mRNA transiently increases after treatment with a fungal elicitor (Ryder et al., 1984). Since the isoflavonoids include phytoalexins such as pisatin, it is likely that this enzyme could play an important role in the ability of legumes to resist infection. A multigene family encoding CHS has been identified in pea, together with two regulatory loci that control differential expression of these genes in a spatially specific manner. The deduced N-terminal amino acid sequences for two of these, CHS1 and CHS2, showed good homology with the consensus CHS sequence (Harker et al., 1990). The nucleotide sequence of three further highly homologous CHS cDNA clones has been elucidated (Ichinose et al., 1992).

This is an appropriate point for a word of caution with regard to conclusions about other substances that might be involved in pathogenicity. Cutin is the main structural component of plant cuticle (Kolattukudy, 1980). The production of cutinase by the pea footrot agent *Nectria haematococca* was considered to be important for penetration of the root during infection (Woloshuk and Kolattukudy, 1986). The insertion of a cloned cutinase gene from this wound pathogen of pea into *Mycosphaerella* spp., wound pathogens of papaya, conferred upon the transformants the ability to infect unwounded papaya fruit (Dickmann et al., 1989). However, this type of experiment provided only indirect evidence for the role of cutinase. Recently Stahl and Schafer (1992) directly evaluated the role of cutinase during infection of pea by *N. haematococca* using gene replacement to construct a cutinase-deficient mutant. Their results demonstrated clearly that cutinase is not required for pathogenicity in this fungus on its natural host. This work also confirmed the value of gene disruption techniques for establishing the involvement of specific genes in disease.

Fungal Pathogens and Pathotype Classification

A preliminary step toward any molecular genetic analysis is to identify suitable race/pathotype variation which can serve as a basis for studies of host specificity. Some recent examples are given here. Vascular wilt disease caused by *Fusarium oxysporum* f. sp. *pisi* infection of pea roots is very destructive. Four races, designated 1, 2, 5 and 6, have been identified by pathogenicity testing on a series of pea differential cultivars (Haglund and Kraft, 1979). Recently a novel variant of race 6, comprising three strains isolated in Denmark, has been identified; evidence from a number of observations indicates that it should be considered a subgroup of race 6 (Bodker et al., 1993).

Restriction fragment length polymorphism (RFLP) analysis has been used to characterize the races of *F. oxysporum* f. sp. *pisi*, using three DNA probes derived from race 1. Two of the probes contained moderately repetitive DNA while the third contained a ribosomal DNA repeat. Isolates of races 1, 2 and 5 were successfully distinguished, although the single North American race 6 isolate was indistinguishable from race 1. Race 2 could also be separated into two subgroups, 2a and 2b. The whole process from isolating the fungus from a plant to obtaining a fingerprint pattern was performed in 10 days compared with 6 weeks for a standard pathogenicity test (Whitehead et al., 1993).

Downy mildew caused by *Peronospora viciae* (Berk.) Casp. is one of the most widespread and damaging diseases of pea crops. *P. viciae* cannot be grown axenically and inoculum has to be produced on living tissue. In a recent study, 11 pathotypes were distinguished on four standard pea differentials, which might serve as a useful basis for molecular genetic investigation of genes responsible for this pattern of interactions (Taylor et al., 1989b).

Bacteria

Isolation of avirulence genes in P. syringae *pv.* pisi

Race structure

Bacterial blight is the most important bacterial disease of pea and was first described by Sackett (1916). The causal organism is *P. syringae* pv. *pisi* (hereafter referred to as *P. s. pisi*), one of a large number of closely related pathovars that are distinguished chiefly by their host range. There are a number of reasons why the interaction between *P. s. pisi* and pea constitutes an ideal model for the investigation of the molecular genetic basis of pathogenicity and host specificity. First, the host has been in cultivation for centuries, during which time a number of well-defined cultivars have been produced, many of which are now known to differ in their reaction to the disease; it is an annual that will produce up to three generations per year under glasshouse and growth chamber conditions; it is naturally self-fertile, permitting the ready selection of genetically uniform stocks; and it has well-established genetics, including maps of the seven chromosomes (see Chapter 2). Second, the pathogen grows well on a simple defined medium; it exhibits a well-defined race structure; it is amenable to genetic manipulation and, because it is a member of the 'syringae' group of pathovars, information gained with this system is potentially applicable to other pathovars and a range of diseases and host plants.

Isolates of *P. s. pisi* can be grouped into different races by the pattern of their interaction with a range of differential cultivars. Taylor (1972) described the occurrence of two distinct races in New Zealand and since that time there has been a steady accumulation of new races; currently some eight races have been identified by their interactions with eight differential cultivars (Taylor *et al.*, 1989a; J.D. Taylor, personal communication). The work of H.H. Flor in the 1930s and 1940s with *Melampsora lini* and flax (*Linum usitatissimum*) culminated in what has come to be known as the gene-for-gene theory (Flor, 1956; Ellingboe, 1984). This theory assumes that specificity resides in matching genes in the host and pathogen and in its most common form this appears to involve the interaction of a single dominant resistance (R) gene in the host with a matching dominant avirulence (A) gene in the pathogen, the outcome of the interaction being a resistant response in the plant. In the case of the *P. s. pisi*/pea interaction, the resistant response is typically a rapid, localized, necrotic collapse of the tissue at the site of inoculation, termed a hypersensitive reaction (HR) (Klement, 1982). Table 8.1 shows the current race structure based on the gene-for-gene hypothesis and the assumption of five matching genes as the minimum number that can account for the observed pattern of interactions. It is worth noting that, in this scheme, uncertainty remains over the presence of some genes in some pathovars or cultivars.

Groups at the University of the West of England, Bristol and Horticulture

Table 8.1. Gene-for-gene relationship between pea cultivars and races of *Pseudomonas syringae* pv. *pisi*, confirmed by genetic analysis. (J.R. Bevan, J.D. Taylor, I.R. Crute, P.J. Hunter and A. Vivian, unpublished.)

		Races (with their avirulence genes)							
		1	2	3	4	5	6	7	8*
		1.34..	2.....	3.....	4..2.	45.....	.234.	1..4.	
Differential cultivars	Resistance genes								
Kelvedon Wonder	.....	+	+	+	+	+	+	+	+
Early Onward	.2...	+	-	+	+	-	+	-	+
Belinda	..3..	-	+	-	+	+	+	-	+
Hurst Greenshaft	1..4.	-	+	+	-	-	+	-	-
Partridge	1.345	-	+	-	-	-	+	-	-
Sleaford Triumph	?2.4?	-	-	+	-	-	+	-	-
Vinco	123..	-	-	-	+	-	+	-	-
Fortune	12345	-	-	-	-	-	+	-	-

* = genetically modified strain derived from race 1
+ = compatible (susceptible); - = incompatible (resistant);
? = unconfirmed; . = absence of resistance or avirulence gene

Research International in Wellesbourne have sought to confirm the genetic basis of the race structure and to identify, isolate and characterize initially the avirulence genes and ultimately the corresponding resistance genes. On the host side, crosses between the differential cultivars have been used to confirm the operation of single dominant resistance genes towards particular races and will be used to map these R genes.

GENE LIBRARY CONSTRUCTION

The approach pioneered by Staskawicz *et al.* (1984) has been used to clone avirulence genes. This involves the construction of gene libraries in the broad-host-range cosmid cloning vector, pLAFR3 (Staskawicz *et al.*, 1987) and their introduction into other races, which can then be screened on a suitable pea cultivar for the acquisition of a particular A gene. The choice of library progenitor was mainly influenced by the availability of suitable recipient races: the races of *P. s. pisi* differ dramatically in their ability to inherit plasmids of the same type as the cosmid vector (incompatibility group P1; Moulton, 1991; Fig. 8.1). The race 1 strain, 299A, was chosen as a recipient and to permit counterselection of the donor it was necessary to obtain a spontaneous rifampicin-resistant mutant. Such mutants vary in the extent to which they retain the pathogenic characteristics of the progenitor and consequently a number of mutants were tested to ensure that no observable change of phenotype had occurred *in planta*.

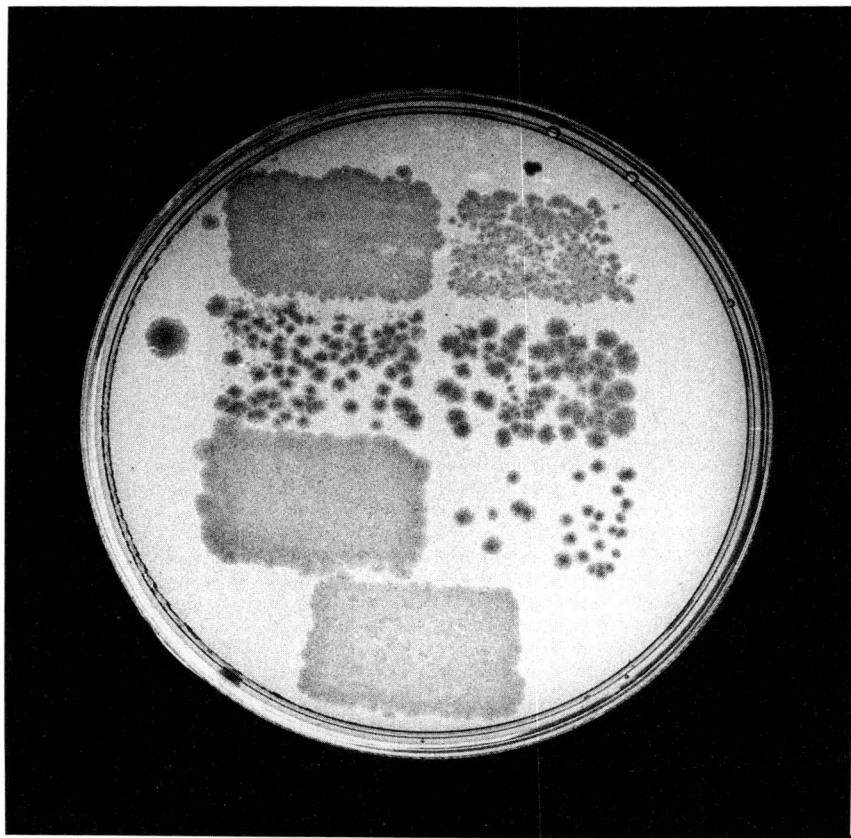

Fig. 8.1. Variation in the ability of races of *P. syringae* pv. *pisi* to inherit plasmids from an *Escherichia coli* donor. Patches of isolates representative of the sevén races of *P. syringae* pv. *pisi* were cultured and replica-plated to a nutrient agar plate (not shown), previously spread with a dense suspension of the *E. coli* donor, harbouring a gene library cosmid (vector pLAFR3), and a second *E. coli* strain, harbouring a mobilizing plasmid (which effects transfer of the cosmid). Following incubation of the resulting mating plate [not shown, but on which there is a cross between each race (patch) and the donor strain], it is replica-plated to a plate selecting for the growth of cosmid transconjugants of each race (from Moulton, 1991).

An estimate of total genome size is an important parameter in the construction of a gene library because this determines the number of clones required for a representative library. Precise estimates of total genome size are not available for *P. syringae* pathovars, but recent estimates, based on pulsed-field gel electrophoresis (PFGE), are 5933 kb for *P. aeruginosa* (Romling *et al.*, 1989) and 6720 kb for *P. tolaasii* (Rainey *et al.*, 1993). Holloway (1992) communicated unpublished estimates of 6920 kb for *P. putida* (R. Saffery and A.

Morgan) and 4925 kb for *P. solanacearum* (D. Escuadra and B. Holloway) without indicating the method of determination. Staskawicz *et al.* (1984) estimated gene frequency in a gene library constructed from race 6 of *P. syringae* pv. *glycinea* by assaying for the expression of a specific gene amongst the *E. coli* clones. The gene chosen was that for ice-nucleation detected by freeze-replica assay (Lindow *et al.*, 1978). Three ice-nucleation-positive clones were detected among more than 1800 tested. This indicated that with an average insert size of about 25 kb, approximately 600 clones would constitute an assumed 'complete' library, and from this one can calculate the size of the *P. syringae* pv. *glycinea* genome to be about 3270 kb. Clearly there is considerable discrepancy between this value and the values for related bacteria determined by PFGE. Pragmatically, we have always doubled the calculated number of clones as a safeguard to try to ensure the likelihood of including all genes (Vivian *et al.*, 1989).

RACE 2 AVIRULENCE GENE (A2)
The hypothetical avirulence gene A2 (Table 8.1) was predicted to be present in race 2 and absent from race 1. Consequently some 1100 clones were conjugated into a race 1 recipient strain and inoculated on the pea cv. Early Onward, postulated to harbour a single resistance gene, R2. A single clone designated pAV270 conferred incompatibility on the recipient strain toward cv. Early Onward, but not toward cv. Kelvedon Wonder, confirming the specificity of the cloned DNA. To confirm the gene-for-gene nature of the specificity, the two cultivars were crossed and a sample of the F_2 progeny was tested for co-segregation of resistance to race 2 and to race 1 harbouring pAV270. These experiments provided the first definitive evidence for the cloning of a specific avirulence allele involved in a gene-for-gene relationship in a *P. syringae* pathovar (Vivian *et al.*, 1989).

The gene responsible on pAV270 was localized through a combination of subcloning and transposon mutagenesis (Vivian *et al.*, 1989; Dangl *et al.*, 1992a). Subsequent DNA sequencing identified a single large open reading frame (ORF) comprising 660 bp and coding for a hypothetical polypeptide of 220 residues. The predicted M_r of the product of the avirulence gene, designated *avrPpiA1*, was 28,000; its predicted hydrophilicity suggested a cytosolic location for the gene product. Searches of the EMBL, GenBank and Swiss-Prot data bases failed to identify any sequence similarity to previously sequenced genes, including the many known avirulence genes from other bacteria. However, functional homologues (designated *avrPmaA1* from strain m2 and *avrPmaA2* from strain 791) were cloned from *P. syringae* pv. *maculicola*, each of which conferred an identical phenotype in *P. syringae* pv. *pisi* toward pea cultivars. Sequence analysis showed the two homologues to be identical to one another, and to be 97% identical to *avrPpiA1* in nucleotide sequence (Dangl *et al.*, 1992a). All three genes when introduced into *P. syringae* pv. *phaseolicola* conferred avirulence toward the *Phaseolus* bean cultivar Canadian Wonder and a delayed hypersensitive response (HR) toward the cultivar 'Seafarer' (Dangl et al., 1992a; Fillingham et al., 1992). Introduction of the three genes into the virulent strain m4 of P. syringae pv. maculicola showed that they conferred identical ecotype-specificity toward A. thaliana (Dangl et al., 1992a).

The similarities observed in the avirulence genes offers the intriguing possibility that the corresponding resistance genes in pea and Phaseolus bean may show some structural similarity to the RPM1 gene of A. thaliana. The gene-for-gene theory envisages matching single genes for avirulence and resistance and it could therefore be possible that the avrPpiA1 gene matches homologous genes in all three host plants. The comparatively small (70,000 kb) genome of A. thaliana should facilitate the isolation of RPM1 and this could provide either a DNA probe to search for a homologue in pea and bean or an immunological probe that might permit the detection of the corresponding R gene product in these host plants. However, it is by no means certain that the avirulence gene matches only a single type of resistance gene; as we shall see below, it has already been shown that avrPpiA1 interacts with at least two resistance genes in Phaseolus bean (Fillingham et al., 1992).

RACE 3 AVIRULENCE GENE (A3)

Early work with P. syringae pv. pisi attempted to isolate non-pathogenic mutants using the transposon, Tn5. This led to the isolation of a mutant of race 1, designated PF24, which had lost one of two indigenous plasmids present in the parent strain, 299A (Malik et al., 1987). Subsequently, Bavage et al. (1991) showed that the plasmid loss in PF24 was associated with a loss of avirulence toward pea cultivars such as Belinda, which harbour the resistance gene R3. This fortuitously enabled the cloning of a second avirulence gene from P. s. pisi. This avirulence gene now designated avrPpiB1 (Vivian and Mansfield, 1993), which corresponds to the A3 gene in Table 8.1, was isolated from a race 3 gene library by testing 1100 transconjugants of PF24 on pea cv. Puget (from the same phenotypic group as Belinda, Table 8.1). Five cosmid clones consistently restored the A3 phenotype in PF24 and all were shown to possess overlapping inserts derived from plasmid DNA in the race 3 strain (Bavage et al., 1991). Further analysis, including subcloning and mutagenesis with the transposon Tn3HoKmGus (Bonas et al., 1989), has localized the gene to a region overlapping two EcoRI fragments, which are currently being sequenced (J.D. Orvis, A. Astuto and A. Vivian, unpublished). This observation, that at least one of the avirulence genes in P. syringae pv. pisi is plasmid-borne, is interesting in the context of some experiments reported by Mazarei and Kerr (1991), from which it appeared that the functions of hypersensitivity and pathogenicity to pea were also plasmid-borne.

SPECIFICITY IN NON-HOST RESISTANCE

Following the demonstration that the cultivar-specific avirulence gene, *avrPpiA1*, from *P. syringae* pv *pisi* could also apparently interact with genes for resistance in *Phaseolus* bean if transferred to *P. syringae* pv. *phaseolicola* (Fillingham et al., 1992), it was of interest to see if novel avirulence genes (perhaps without any detectable avirulence function toward their homologous host) could be identified by transfer from one pathovar to another. This would involve screening on a compatible cultivar of the corresponding host, e.g. a gene library from pv. *phaseolicola* could be transferred to pv. *pisi* and screened on the universally susceptible pea cv. Kelvedon Wonder. This approach has been tried and a total of seven clones from a library of race 4 *P. s.* pv *phaseolicola* were found to confer avirulence to pea (J.R. Wood, P.J. Hunter, J.D. Taylor, J.W. Mansfield and A. Vivian, unpublished). Five of these clones show some degree of DNA homology and demonstrate differential reactions to pea cultivars, enabling the genetic basis of the resistance shown by the host to be analysed through an appropriate crossing programme. Evidence from work with other related bacterial pathogens has indicated that genes involved in cultivar specificity may also be involved in non-host resistance and at least in some cases this has permitted the demonstration of gene-for-gene relationships (Whalen et al., 1988).

Viruses

The molecular biology of virus–host plant interaction is currently being investigated in a range of systems, very few of them being pathogens of pea. Clearly these interactions involve recognition and responses between the partners. Fraser (1990) stressed that these events could perhaps occur at three levels: successful transmission by an animal or fungal vector; direct interaction controlled by genetic factors in plant and virus; and the co-evolution of host and pathogen in populations.

Five major virus diseases of pea in the United States have been recognized together with a number of minor importance (Hampton, 1984a), one of which, pea early browning virus (PEBV), is common elsewhere. Of these, only two, pea seed-borne mosaic virus (PSbMV) and PEBV, have been studied in any detail at a molecular level.

Pea seed-borne mosaic virus

It has been suggested that PSbMV perhaps spread initially in breeding stock from Northern India to Europe (Hampton, 1984b). It was first reported from Czechoslovakia and has subsequently been found worldwide (Khetarpal and Maury, 1987). It is aphid-transmitted and a member of the potyvirus group.

It comprises a flexuous particle, varying slightly from 750 to 780 nm with a positive-sense single-stranded RNA genome. There is evidence for the existence of strains in PSbMV from the work of Hampton *et al.* (1981) and that of Alconero *et al.* (1986), which showed that responses of pea differentials varied from immunity to whole-plant necrosis with different pathotypes.

Timmerman *et al.* (1990) determined the nucleotide sequence of the viral coat protein gene and showed it to comprise an open reading frame (ORF) of 1189 nucleotides (nt). Subsequently the complete nucleotide sequence of the entire viral genome was determined, showing it to be 9924 nt in length (Johansen *et al.*, 1991). Potyvirus genomes typically contain one large ORF with the potential to encode a polyprotein, which is subsequently proteolytically cleaved into at least eight proteins (Dougherty and Carrington, 1988). In some cases the functions of these products are known; for example, the helper-component proteinase (HC-Pro) is necessary for aphid-transmission (Thornbury *et al.*, 1985) and has proteolytic activity (Carrington *et al.*, 1989; Carrington and Herndon, 1992). The small nuclear inclusion protein (49K-Pro), or an N-terminal part of it, is the virus-encoded protein (VPg) which is linked to the 5' terminus of the RNA genome (Murphy *et al.*, 1990).

Analysis of the deduced amino acid sequence of the PSbMV genome indicated that it had a similar organization to those of other previously sequenced potyviruses. It also indicated the presence of an additional proteolytic cleavage site in the 49K-Pro sequence. Similar internal cleavage sites are also found in other potyviruses (Johansen *et al.*, 1991). In the related potyvirus, tobacco etch virus, analysis of naturally occurring cleavage products of the 49,000 M_r nuclear inclusion protein identified an internal cleavage site producing 21,000 M_r and 27,000 M_r products, probably associated with the VPg and proteinase domains of the polyprotein (Dougherty and Parks, 1991).

In a study of 25 UK pea cultivars, all were found to be susceptible to PSbMV, although most dry-pea cultivars had a lower virus content than garden and vining cultivars. Seed transmission was studied in 20 cultivars and found to range from 0% in Maro, Princess and Progreta to 74% in cv. Vedette. However, wide variation in percentage seed transmission was observed between individuals of a single cultivar. The authors concluded that there was no obvious relationship between virus content and the efficiency of seed transmission in different cultivars (Wang *et al.*, 1993). Seed transmission is the end result of a complex interaction between host and virus. Pollen transmission of the virus was shown not to occur in pea; PSbMV infected the floral tissues of both high- and zero-transmission cultivars, but was not detected in the ovules prior to fertilization (Wang and Maule, 1992). Studies on the incidence and concentration of virus in seeds at different stages of development showed that seed transmission of PSbMV resulted from direct invasion of immature pea embryos by the virus and that the block to seed transmission in the non-permissive cultivar probably occurred at this step (Wang and Maule, 1992).

Pea early browning virus

Pea early browning virus (PEBV) is a member of the tobravirus group. It has a genome comprising two positive-sense RNA molecules, each separately encapsulated in rod-shaped particles (Harrison and Robinson, 1986). The virus is transmitted by soil-inhabiting nematodes. A characteristic of tobraviruses is their ability to produce so-called NM-infections in which the larger RNA1 replicates and spreads systemically throughout the affected plant in the absence of RNA2. Since RNA2 encodes the coat protein, RNA1 is unencapsidated in NM-infections. Infections where the full encapsidated genome is present are termed M-infections. The complete sequences for the two RNAs have been determined in the British strain SP5. RNA1 comprises 7300 nt and codes for two putative replicase proteins of 201,000 M_r and 141,000 M_r (the larger being a readthrough product from the smaller). Additionally, RNA1 encodes a protein of M_r 30,000 with homology to the tobacco mosaic virus (TMV) cell-to-cell spread protein (MacFarlane et al., 1989). RNA2 has 3374 nt and contains three ORFs, which are thought probably to function as subgenomic RNAs (Goulden et al., 1990). Full-length cDNA clones for RNA1 and RNA2 have been obtained from which transcripts have been synthesized in vitro. These produce a systemic infection when inoculated into plants, producing symptoms similar to those of the wild-type virus (MacFarlane et al., 1991).

The 3'-proximal portion of the gene encoding the M_r 201,000 protein of PEBV can be expressed separately as a product of M_r 54,000. Tobacco plants transformed with the corresponding ORF (54K) are resistant to infection by PEBV. This resistance could be abolished by premature termination of the translation of the 54K ORF, implying that an intact 54,000 product is necessary for full protection. During the course of these experiments two variants of PEBV were isolated, which were able to overcome the 54K ORF-mediated resistance. Although the authors were not able to rule out the possibility that the variants were contaminants of the virus preparation used, the future characterization of such variants should shed light on the resistance mechanism itself as well as the process of normal virus replication (MacFarlane and Davies, 1992).

Mapping of resistance genes

Mapping of genes for resistance to various types of pathogen has been an important feature in pea, because of the availability of well-defined markers. Such information, apart from its obvious value to breeders, may in due course have value for the isolation and cloning of the genes responsible. Although a comprehensive review of this topic is not appropriate here, a recent example is included to draw attention to the potential value of such information in a molecular context. A single recessive gene in linkage group 2 confers resistance to pea mosaic virus and is closely linked to other recessive genes for resistance to bean yellow mosaic, clover yellow vein (CYVV), the lentil strain of PSbMV and bean common mosaic viruses (BCMV-NL8) (Provvidenti, 1990, 1991). A further cluster of recessive genes for resistance to PSbMV and CYVV was located on chromosome 6 (Provvidenti, 1991). As pointed out by Fraser and Gerwitz (1987), recessive genes for resistance imply the absence of some factor in the resistant plant which is required for pathogenesis (the negative model); however, many single resistance genes for resistance to viruses are dominant (positive model).

Detection

One important aspect of viral diseases is the detection and identification of the agents in plant tissue (Salazar and Querci, 1992; Torrance 1992), and particularly in seed for those that are transmitted by this route. The development of biotechnological approaches to detection owes much to the derivation of methods from fundamental studies involving genetic manipulation. Lange *et al.* (1989) developed a dot immunobinding technique, which is very rapid and can be carried out without the use of specialized equipment. It has been used to detect PSbMV, PEBV and bean common mosaic virus (BCMV) directly in seed. Recently Ball and Reeves (1991) reviewed the techniques, both current and in prospect, for the detection of seed-borne pathogens. They particularly emphasized the potential of various techniques involving nucleic acids, such as DNA amplification by polymerase chain reaction and RFLP analysis, which offer improved speed, specificity and economy in the field of pathogen detection (Vivian, 1992).

Monoclonal antibodies (MAbs) are potentially useful, not only as tools for the specific detection of viruses (Torrance, 1992) and other pathogens (Dewey, 1992; Stead, 1992), but also for investigating the processes of virus–host interaction. Jordan and Hammond (1991) describe MAbs raised against a mixture of potyviruses, including PSbMV.

Conclusions and Prospects

In spite of the considerable recent increase in our knowledge of both plant pathogens and their hosts, thanks to the powerful analytical tools provided by molecular biology, we remain largely ignorant of many of the biological processes that are involved in the interaction of host and pathogen. Many bacterial avirulence genes have been cloned and sequenced, but so far only one acts to produce a detectable elicitor of the specific resistance response, and its precise function remains to be elucidated (Keen, 1991). We know even less about plant resistance genes, although here the prospects are good for progress in isolating such genes from *A. thaliana* in the near future. However, will the familiar 'clone and sequence' merely tell us that this is an entirely novel type of gene? An outcome that may be as frustrating as the avirulence gene story to date!

These thoughts, however, demonstrate the continuing need for progress to be made in dissecting the biological events and that this must proceed alongside the molecular genetic approaches if we are to fully understand the basis of plant diseases. The importance of this endeavour cannot be too highly stressed, against a background of rising world populations and increasing dependence on crops for survival in many parts of that world (Sequeira, 1992).

References

Alconero, R., Provvidenti, R. and Gonsalves, D. (1986) Three pea seedborne mosaic virus pathotypes from pea and lentil germ plasm. *Plant Disease* 70, 783–786.

Ball, S.F.L. and Reeves, J.C. (1991) The application of new techniques in the rapid testing for seed-borne pathogens. *Plant Varieties and Seeds* 4, 169–176.

Bavage, A.D., Vivian, A., Atherton, G.T., Taylor, J.D. and Malik, A.N. (1991) Molecular genetics of *Pseudomonas syringae* pathovar *pisi*: plasmid involvement in cultivar-specific incompatibility. *Journal of General Microbiology* 137, 2231–2239.

Bodker, L., Lewis, B.G. and Coddington, A. (1993) The occurrence of a new genetic variant of *Fusarium oxysporum* f. sp. *pisi*. *Plant Pathology* 42 (in press).

Bonas, U., Stall, R.E. and Staskawicz, B.J. (1989) Genetic and structural characterization of the avirulence gene *avrBs3* from *Xanthomonas campestris* pv. *vesicatoria*. *Molecular and General Genetics* 218, 127–136.

Breiteneder, H., Pettenburger, K., Bito, A., Valenta, R., Kraft, D., Rumpold, H., Scheiner, O. and Breitenbach, M. (1989) The gene coding for the major birch pollen allergen *BetvI*, is highly homologous to a pea disease resistance response gene. *EMBO Journal* 8, 1935–1938.

Carrington, J.C. and Herndon, K.L. (1992) Characterization of the potyviral HC-Pro autoproteolytic cleavage site. *Virology* 187, 308–315.

Carrington, J.C., Cary, S.M., Parks, T.D. and Dougherty, W.G. (1989) A second proteinase encoded by a plant potyvirus genome. *EMBO Journal* 8, 365–370.

Chiang, C.C. and Hadwiger, L.A. (1990) Cloning and characterization of a disease resistance response gene in pea inducible by *Fusarium solani*. *Molecular Plant-Microbe Interactions* 3, 78–85.

Crowell, D.N., John, M.E., Russell, D. and Amasino, R.M. (1992) Characterization of a stress-induced, developmentally regulated gene family from soybean. *Plant Molecular Biology* 18, 459–466.

Dangl, J.L., Ritter, C., Gibbon, M.J., Mur, L.A.J., Wood, J.R., Goss, S., Mansfield, J., Taylor, J.D. and Vivian, A. (1992a) Functional homologs of the Arabidopsis *RPMI* disease resistance gene in bean and pea. *Plant Cell* 4, 1359–1369.

Dangl, J.L, Debener, T., Gerwin, M., Kiedrowski, S., Ritter, C. Bendahmane, A., Liedgens, H. and Lewald, J. (1992b) Genetic approaches to an understanding of specific resistance responses of *Arabidopsis thaliana* against phytopathogenic pseudomonads. In: Nester, E.W. and Verma, D.P.S. (eds), *Advances in Molecular Genetics of Plant–Microbe Interactions*, vol. 2. Kluwer Academic Publishers, Dordrecht, Netherlands, pp. 405–415.

Daniels, C.H., Fristensky, B., Wagoner, W. and Hadwiger, L.A. (1987) Pea genes associated with non-host disease resistance to *Fusarium* are also active in race-specific disease resistance to *Pseudomonas*. *Plant Molecular Biology* 8, 309–316.

Daniels, C.H., Cody, Y.S. and Hadwiger, L.A. (1988) Host responses in peas to challenge by wall components of *Pseudomonas syringae* pv. *pisi* races 1, 2, and 3. *Phytopathology* 78, 1451–1453.

Deverall, B.J. (1982) Introduction. In: Bailey, J.A. and Mansfield, J.W. (eds), *Phytoalexins*. Blackie, Glasgow, pp. 1–20.

Dewey, F.M. (1992) Detection of plant-invading fungi by monoclonal antibodies. In: Duncan, J.M. and Torrance, L. (eds), *Techniques for the Rapid Detection of Plant Pathogens*. Blackwell Scientific Publications, Oxford, pp. 47–62.

Dickmann, M.B., Podila, G.K. and Kolattukudy, P.E. (1989) Insertion of cutinase gene into a wound pathogen enables it to infect intact host. *Nature, London* 342, 446–448.

Dixon, R.A., Dey, P.M. and Lamb, C.J. (1983) Phytoalexins: enzymology and molecular biology. *Advances in Enzymology and Related Areas of Molecular Biology* 55, 1–136.

Dougherty, W.G. and Carrington, J.C. (1988) Expression and function of potyviral gene products. *Annual Review of Phytopathology* 26, 123–143.

Dougherty, W.G. and Parks, T.D. (1991) Post-translational processing of the tobacco etch virus 49-kDa small nuclear inclusion polyprotein: identification of an internal cleavage site and delimitation of VPg and proteinase domains. *Virology* 183, 449–456.

Ellingboe, A.E. (1984) Genetics of host–parasite relations: an essay. In: Ingrams, D.S. and Williams, P.H. (eds), *Advances in Plant Pathology*, vol. 2. Academic Press, New York, pp. 131–151.

Fillingham, A.J., Wood, J., Bevan, J.R., Crute, I.R., Mansfield, J.W., Taylor, J.D. and Vivian, A. (1992) Avirulence genes from *Pseudomonas syringae* pathovars *phaseolicola* and *pisi* confer specificity towards both host and non-host species. *Physiological and Molecular Plant Pathology* 40, 1–15.

Fischer, D., Ebenau-Jehle, C. and Grisebach, H. (1990) Phytoalexin synthesis in soybean: purification and characterization of NADPH:2'-hydroxydaidzein oxidoreductase from elicitor-challenged soybean cell cultures. *Archives of Biochemistry and Biophysics* 276, 390–395.

Flor, H.H. (1956) The complementary genetic systems in flax and flax rust. *Advances in Genetics* 8, 29–54.

Fraser, R.S.S. (1990) Recognition and response in plant–virus interactions: some underlying concepts. In: Fraser, R.S.S. (ed.), *Recognition and Response in Plant–Virus Interactions*, NATO ASI Series H: vol. 41. Springer-Verlag, Berlin, pp. 1–15.

Fraser, R.S.S. and Gerwitz, A. (1987) The genetics of resistance and virulence in plant virus disease. In: Day, P.R. and Jellis, G.J. (eds), *Genetics and Plant Pathogenesis*. Blackwell Scientific Publications, Oxford, pp. 33–44.

Fristensky, B., Riggleman, R.C., Wagoner, W. and Hadwiger, L.A. (1985) Gene expression in susceptible and disease resistant interactions of peas induced with *Fusarium solani* pathogens and chitosan. *Physiological Plant Pathology* 27, 15–28.

Fristensky, B., Horovitz, D and Hadwiger, L.A. (1988) cDNA sequences for pea disease resistance response genes. *Plant Molecular Biology* 11, 713–715.

Goulden, M.G., Lomonossoff, G.P., Davies, J.W. and Wood, K.R. (1990) The complete nucleotide sequence of PEBV RNA2 reveals the presence of a novel open reading frame and provides insights into the structure of tobraviral subgenomic promoters. *Nucleic Acids Research* 18, 4507–4512.

Hadwiger, L.A. and Webster, D.M. (1984) Phytoalexin production in five cultivars of peas differentially resistant to three races of *Pseudomonas syringae* pv. *pisi*. *Phytopathology* 74, 1312–1314.

Hagedorn, D.J. (1984) Pea diseases. In: Hagedorn, D.J. (ed.), *Compendium of Pea Diseases*. American Phytopathological Society, St Paul, Minnesota, pp. 2–4.

Haglund, W.A. and Kraft, J. (1979) *Fusarium oxysporum* f. sp. *pisi*, race 6: occurrence and distribution. *Phytopathology* 69, 818–820.

Hampton, R.O. (1984a) Diseases caused by viruses. In: Hagedorn, D.J. (ed.), *Compendium of Pea Diseases*. American Phytopathological Society, St Paul, Minnesota, pp. 31–32.

Hampton, R.O. (1984b) Pea seedborne mosaic. In: Hagedorn, D.J. (ed.), *Compendium of Pea Diseases*. American Phytopathological Society, St Paul, Minnesota, pp. 34–35.

Hampton, R.O., Mink, G.I., Bos, L., Inouye, T., Musil, M. and Hagedorn, D. (1981) Host differentiation and serological homology of pea seed-borne mosaic virus isolates. *Netherlands Journal of Plant Pathology* 87, 1–10.

Harker, C.L., Ellis, T.H.N. and Coen, E.S. (1990) Identification and genetic regulation of the chalcone synthase multigene family in pea. *Plant Cell* 2, 185–194.

Harrison, B.D. and Robinson, D.J. (1986) Tobraviruses. In: van Regenmortel, M.H.V. and Frankel-Conrat, H. (eds), *The Plant Viruses*, vol. 2. Plenum, New York, pp. 339–369.

Holloway, B.W. (1992) *Pseudomonas* in the late twentieth century. In: Galli, E., Silver, S. and Witholt, B. (eds), *Pseudomonas: Molecular Biology and Biotechnology*. American Society for Microbiology, Washington DC, pp. 1–8.

Ichinose, Y., Kawamata, S., Yamada, T., An, C. and Kajiwara, T. (1992) Molecular cloning of chalcone synthase cDNAs from *Pisum sativum*. *Plant Molecular Biology* 18, 1009–1018.

Ingham, J.L. (1982) Phytoalexins from the Leguminosae. In: Bailey, J.A. and Mansfield, J.W. (eds), *Phytoalexins*. Blackie, Glasgow, pp. 21–132.

Johansen, E., Rasmussen, O.F., Heide, M. and Borkhardt, B. (1991) The complete nucleotide sequence of pea seed-borne mosaic virus RNA. *Journal of General Virology* 72, 2625–2632.

Jordan, R. and Hammond, J. (1991) Comparison and differentiation of potyvirus isolates and identification of strain-, virus-, subgroup-specific and potyvirus group-common epitopes using monoclonal antibodies. *Journal of General Virology* 72, 25–36.

Keen, N.T. (1991) Characterization and function of bacterial avirulence genes. In: Patil, S.S., Ouchi, S., Mills, D. and Vance, C. (eds), *Molecular Strategies of Pathogens and Host Plants*. Springer-Verlag, Berlin, pp. 59–67.

Khetarpal, R.K. and Maury, Y. (1987) Pea seed-borne mosaic virus: a review. *Agronomie* 7, 215–224.

Klement, Z. (1982) Hypersensitivity. In: Mount, M.S. and Lacey, G.H. (eds), *Phytopathogenic Procaryotes*, vol. 2. Academic Press, New York, pp. 149-177.

Kolattukudy, P.E. (1980) Biopolyester membranes of plants: cutin and suberin. *Science* 208, 990-1000.

Lange, L., Jomantor, A. and Heide, M. (1989) Testing seeds for viruses by dot immuno binding (DIB) directly on plain paper. *Tidsskrift for Planteavl* 93, 93-96.

Lindow, S.E., Arny, D.C. and Upper, C.D. (1978) Distribution of ice nucleation-active bacteria on plants in nature. *Applied and Environmental Microbiology* 36, 831-838.

MacFarlane, S.A. and Davies, J.W. (1992) Plants transformed with a region of the 201-kilodalton replicase gene from pea early browning virus RNA1 are resistant to virus infection. *Proceedings of the National Academy of Sciences USA* 89, 5829-5833.

MacFarlane, S.A., Taylor, S.C., King, D.I., Hughes, G. and Davies, J.W. (1989) Pea early browning virus RNA1 encodes four polypeptides including a putative zinc-finger protein. *Nucleic Acids Research* 17, 2245-2260.

MacFarlane, S.A., Wallis, C.V., Taylor, S.C., Goulden, M.G., Wood, K.R. and Davies, J.W. (1991) Construction and analysis of infectious transcripts synthesized from full-length cDNA clones of both genomic RNAs of pea early browning virus. *Virology* 182, 124-129.

Malik, A.N., Vivian, A. and Taylor, J.D. (1987) Isolation and partial characterization of three classes of mutant in *Pseudomonas syringae* pathovar *pisi* with altered behaviour towards their host, *Pisum sativum*. *Journal of General Microbiology* 133, 2393-2399.

Mazarei, M. and Kerr, A. (1991) Plasmids in *Pseudomonas syringae* pv. *pisi* carry genes for pathogenicity. *Plant Pathology* 40, 408-414.

Moulton, P.J. (1991) The molecular genetics of *Pseudomonas syringae* pv. *pisi*. PhD. thesis, CNAA, Bristol Polytechnic.

Murphy, J.F., Rhoads, R.E., Hunt, A.G. and Shaw, J.G. (1990) The VPg of tobacco etch virus RNA is the 49-kDa proteinase or the N-terminal 24-kDa part of the proteinase. *Virology* 178, 285-288.

Provvidenti, R. (1990) Inheritance of resistance to pea mosaic virus in *Pisum sativum*. *Journal of Heredity* 81, 143-145.

Provvidenti, R. (1991) Inheritance of resistance to the NL-8 strain of bean common mosaic virus in *Pisum sativum*. *Journal of Heredity* 82, 353-355.

Rainey, P.B., Brodey, C.L. and Johnstone, K. (1993) Identification of a gene cluster encoding 3 high-molecular-weight proteins, which is required for synthesis of tolaasin by the mushroom pathogen *Pseudomonas tolaasii*. *Molecular Microbiology* 8, 643-652.

Riggleman, R.C., Fristensky, B. and Hadwiger, L.A. (1985) The disease resistance response in pea is associated with increased levels of specific mRNAs. *Plant Molecular Biology* 4, 81-86.

Romling, U., Grothues, D., Bautsch, W. and Tummler, B. (1989) A physical genome map of *Pseudomonas aeruginosa* PAO. *EMBO Journal* 8, 4081-4089.

Ryder, T.B., Cramer, C.L., Bell, J.N., Robbins, M.P., Dixon, R.A. and Lamb, C.J. (1984) Elicitor rapidly induces chalcone synthase mRNA in *Phaseolus vulgaris* cells at the onset of the phytoalexin defense response. *Proceedings of the National Academy of Sciences USA* 81, 5724-5728.

Sackett, W.G. (1916) A bacterial stem blight of field and garden peas. *Bulletin of the Colorado Agricultural Experimental Station* 218, 3-43.
Salazar, L.F. and Querci, M. (1992) Detection of viroids and viruses by nucleic acid probes. In: Duncan, J.M. and Torrance, L. (eds), *Techniques for the Rapid Detection of Plant Pathogens*. Blackwell Scientific Publications, Oxford, pp. 129-144.
Schaefer, W., Straney, D., Ciuffetti, L., Van Etten, H.D. and Yoder, O.C. (1989) One enzyme makes a fungal pathogen, but not a saprophyte, virulent on a new host plant. *Science* 246, 247-249.
Sequeira, L. (1992) Research on plant-microbe interactions: making it relevant. In: Nester, E.W. and Verma, D.P.S. (eds), *Advances in Molecular Genetics of Plant-Microbe Interactions*, vol. 2. Kluwer Academic Publishers, Dordrecht, Netherlands, pp. 3-14.
Stahl, D.J. and Schafer, W. (1992) Cutinase is not required for fungal pathogenicity on pea. *Plant Cell* 4, 621-629.
Staskawicz, B.J., Dahlbeck, D. and Keen, N.T. (1984) Cloned avirulence gene of *Pseudomonas syringae* pv. *glycinea* determines race-specific incompatibility on *Glycine max* (L.) Merr. *Proceedings of the National Academy of Sciences USA* 81, 6024-6028.
Staskawicz, B., Dahlbeck, D., Keen, N. and Napoli, C. (1987) Molecular characterization of cloned avirulence genes from race 0 and race 1 of *Pseudomonas syringae* pv. *glycinea*. *Journal of Bacteriology* 169, 5789-5794.
Stead, D.E. (1992) Techniques for detecting and identifying plant pathogenic bacteria. In: Duncan, J.M. and Torrance, L. (eds), *Techniques for the Rapid Detection of Plant Pathogens*. Blackwell Scientific Publications, Oxford, pp. 76-111.
Sun, Y., Wu, Q., Van Etten, H.D. and Hdrazina, G. (1991) Stereoisomerism in plant disease resistance: induction and isolation of the 7,2'-dihydroxy-4',5'-methylenedioxyisoflavone oxidoreductase, an enzyme introducing chirality during synthesis of isoflavonoid phytoalexins in pea (*Pisum sativum* L). *Archives of Biochemistry and Biophysics* 284, 167-173.
Taylor, J.D. (1972) Races of *Pseudomonas pisi* and sources of resistance in field and garden peas. *New Zealand Journal of Agricultural Research* 15, 441-447.
Taylor, J.D., Bevan, J.R., Crute, I.R. and Reader, S.L. (1989a) Genetic relationship between races of *Pseudomonas syringae* pv. *pisi* and cultivars of *Pisum sativum*. *Plant Pathology* 38, 364-375.
Taylor, P.N., Lewis, B.G. and Matthews, P. (1989b) Pathotypes of *Peronospora viciae* in Britain. *Journal of Phytopathology* 127, 100-106.
Thornbury, D.W., Hellman, G.M., Rhoads, R.E. and Pirone, T.P. (1985) Purification and characterization of potyvirus helper component. *Virology* 144, 260-267.
Tiemann, K., Hinderer, W. and Barz, W. (1987) Isolation of NADPH:isoflavone oxidoreductase, a new enzyme of pterocarpan phytoalexin biosynthesis in cell suspension cultures of *Cicer arietinum*. *FEBS Letters* 213, 324-328.
Timmerman, G.M., Calder, V.L. and Bolger, L.E.A. (1990) Nucleotide sequence of the coat protein gene of pea seed-borne mosaic potyvirus. *Journal of General Virology* 71, 1869-1872.
Torrance, L. (1992) Serological methods to detect plant viruses: production and use of monoclonal antibodies. In: Duncan, J.M. and Torrance, L. (eds), *Techniques for the Rapid Detection of Plant Pathogens*. Blackwell Scientific Publications, Oxford, pp. 7-33.

Van Etten, H.D., Matthews, D.E. and Matthews, P.S. (1989) Phytoalexin detoxification: importance for pathogenicity and practical implications. *Annual Review of Plant Pathology* 27, 143–164.

van Loon, L.C. (1985) Pathogenesis-related proteins. *Plant Molecular Biology* 4, 111–116.

Vivian, A. (1992) Identification of plant pathogenic bacteria using nucleic acid technology. In: Duncan, J.M. and Torrance, L. (eds), *Techniques for the Rapid Detection of Plant Pathogens*. Blackwell Scientific Publications, Oxford, pp. 145–161.

Vivian, A. and Mansfield, J.W. (1993) A proposal for a uniform genetic nomenclature for avirulence genes in phytopathogenic pseudomonads. *Molecular Plant-Microbe Interactions* 6, 9–10.

Vivian, A., Atherton, G.T., Bevan, J.R., Crute, I.R., Mur, L.A.J. and Taylor, J.D. (1989) Isolation and characterization of cloned DNA conferring specific avirulence in *Pseudomonas syringae* pv. *pisi* to pea (*Pisum sativum*) cultivars, which possess the resistance allele, R2. *Physiological and Molecular Plant Pathology* 34, 335–344.

Walter, M.H., Liu, J.-W., Grand, C., Lamb, C.J. and Hess, D. (1990) Bean pathogenesis-related (PR) proteins deduced from elicitor-induced transcripts are members of a ubiquitous new class of conserved PR proteins including pollen allergens. *Molecular and General Genetics* 222, 353–360.

Wang, D. and Maule, A.J. (1992) Early embryo invasion as a determinant in pea of the seed transmission of pea seed-borne mosaic virus. *Journal of General Virology* 73, 1615–1620.

Wang, R., Woods, D., Cockbain, A.J., Maule, A.J. and Biddle, A.J. (1993) The susceptibility of pea cultivars to pea seed-borne mosaic virus infection and virus seed transmission in the UK. *Plant Pathology* 42, 42–47.

Weltring, K.-M., Turgeon, B.G., Yoder, O.C. and Van Etten, H.D. (1988) Isolation of a phytoalexin-detoxification gene from the plant pathogenic fungus *Nectria haematococca* by detecting its expression in *Aspergillus nidulans*. *Gene* 68, 335–344.

Whalen, M.C., Stall, R.E. and Staskawicz, B.J. (1988) Characterization of a gene from a tomato pathogen determining hypersensitive resistance in non-host species and genetic analysis of this resistance in bean. *Proceedings of the National Academy of Sciences USA* 85, 6743–6747.

Whitehead, D.S., Coddington, A. and Lewis, B.G. (1992) Classification of races by DNA polymorphism analysis and vegetative compatibility grouping in *Fusarium oxysporum* f. sp. *pisi*. *Physiological and Molecular Plant Pathology* 41, 295–305.

Whitlock Jr., J.P. (1986) The regulation of cytochrome P-450 gene expression. *Annual Review of Pharmacology and Toxicology* 26, 333–369.

Woloshuk, C.P. and Kolattukudy, P.E. (1986) Mechanism by which contact with plant cuticle triggers cutinase gene expression in the spores of *Fusarium solani* f. sp. *pisi*. *Proceedings of the National Academy of Sciences USA* 83, 1704–1708.

Root Nodules, *Rhizobium* and Nitrogen Fixation

N.J. Brewin, M.J. Ambrose and J.A. Downie
John Innes Institute, Colney Lane, Norwich, NR4 7UH, UK

Introduction

Biology

The nitrogen-fixing root nodules of peas are a fairly typical example of the *Rhizobium*–legume symbiosis (Allen and Allen, 1981; Stacey *et al.*, 1992). *Rhizobium* is a genus of Gram-negative nitrogen-fixing bacteria (Young and Johnston, 1989) and individual bacterial species and strains of *Rhizobium* interact successfully only with a particular set of legume host plants (Dénarié *et al.*, 1992; Table 9.1). *Pisum* belongs to a very narrow 'cross-inoculation group': nodulation is induced specifically by *Rhizobium leguminosarum* biovar *viciae*, which also nodulates effectively with other members of the tribe Viciaceae, namely *Vicia*, *Lathyrus* and *Lens* spp. Host range in *R. leguminosarum* is determined by a set of nodulation (*nod*) genes encoded on a large plasmid, the symbiotic plasmid (P-*sym*). Biovars *trifolii* and *phaseoli* are different variants of *R. leguminosarum* that harbour symbiotic plasmids with different host specificities (for *Trifolium* and *Phaseolus* spp. respectively). Genetic replacement of one symbiotic plasmid with another plasmid conferring a different host range effectively changes the 'biovar' of the *R. leguminosarum* strain (Johnston *et al.*, 1978).

The characteristic anatomical features of a legume root nodule have been well described for pea (Bond, 1948; Newcomb, 1976). An uninfected apical meristem gives rise to an outer uninfected zone containing peripheral vasculature and a central tissue mass which becomes infected (Fig. 9.1). Within this central tissue, bacteria become internalized by host plant cells and subsequently develop the capacity for nitrogen fixation (Fig. 9.2). At the ultrastructural

Table 9.1. Host range of *Rhizobium leguminosarum* and other species of *Rhizobium*.

Species	Biovar*	Host legumes	Genus
R. leguminosarum	viciae	peas	Pisum
		vetches	Vicia
			Lathyrus
		lentils	Lens
	trifolii	clovers	Trifolium
	phaseoli	French beans	Phaseolus
R. meliloti	-	alfalfa	Medicago
			Melilotus
R. loti	-	-	Lotus
R. fredii	-	soyabean	Glycine
R. galegae	-	-	Galega
R. tropici	-	French beans	Phaseolus

*Note: Host range among biovars of *R. leguminosarum* is encoded on a symbiotic megaplasmid (Johnston et al., 1978).

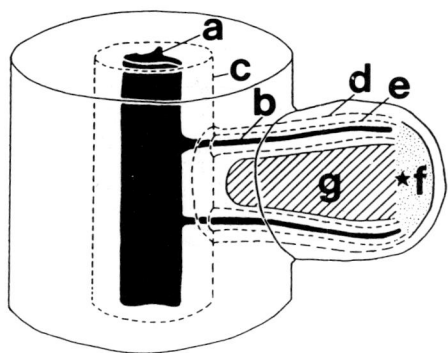

Fig. 9.1. Diagram illustrating the anatomy of a pea root nodule showing vascular connections (adapted from Bond, 1948). Key: (a) root vascular tissue; (b) open-ended nodule vascular tissue; (c) root endodermis; (d) nodule endodermis; (e) nodule vascular endodermal sheath; (f) nodule apical meristem (uninfected); (g) the central infected region, within which there is progressive differentiation of the symbiotic interaction within host cells along the developmental axis from nodule apex to base. The invasion zone (asterisk) marks the starting-point for tissue and cell invasion by *Rhizobium*.

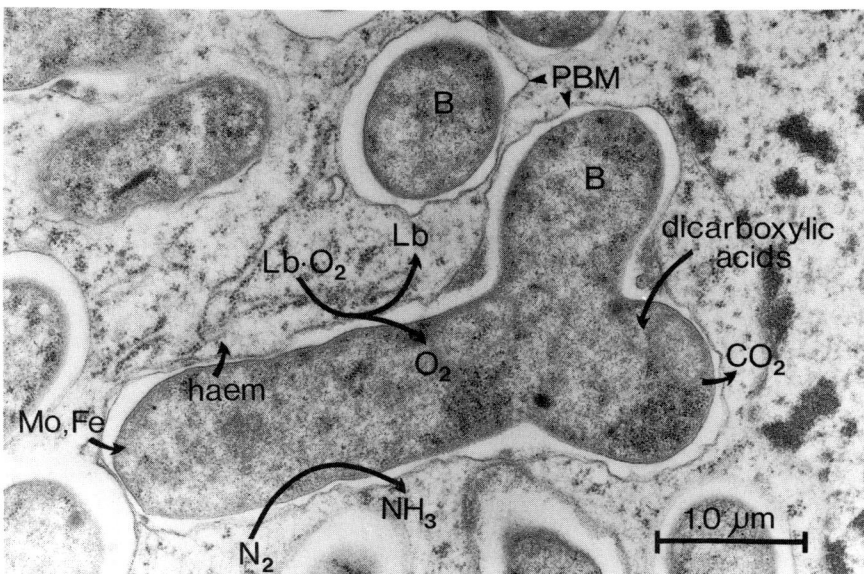

Fig. 9.2. Electron micrograph showing a section of the cytoplasm from an infected pea nodule cell (courtesy of K.A. VandenBosch). Each nitrogen-fixing bacteroid (B) is individually enclosed within a plant-derived peribacteroid membrane (PBM), forming an organelle-like structure, termed the 'symbiosome'. Bacteroids 'fix' nitrogen using the nitrogenase enzyme system, which is only functional under very low oxygen concentrations. Nitrogenase requires ions of iron (Fe) and molybdenum (Mo), which are assembled into a complex inorganic cofactor (FeMoco). Ammonia (NH_3), which is the immediate product of nitrogen fixation, is excreted by bacteroids and assimilated through the activity of glutamine synthase in the host plant cytoplasm. Dicarboxylic acids, supplied through the peribacteroid membrane, are the principal substrates for oxidative phosphorylation by *Rhizobium* bacteroids. Leghaemoglobin (Lb) present in the cytoplasm of host plant cells serves as a facilitated carrier for oxygen.

level, symbiosis involves the creation of a novel intracellular compartment occupied by the bacteria: this compartment can be considered as a quasi-autonomous organelle, which has recently been termed the 'symbiosome' (Mellor, 1989; Roth and Stacey, 1989).

The physiological adaptations in legume nodules have been studied in a wide variety of legume species, particularly soyabeans, *Phaseolus* beans, lupins, alfalfa and peas. The biochemical characterization is most extensive for soyabean nodules, which are larger and easier to harvest than pea nodules. There are, however, quite specific differences between the nodules of these two legumes, particularly with regard to nodule structure and the mechanism of nitrogen assimilation. In this review, we will emphasize the studies with pea nodules but, where appropriate, referring to relevant work on other legume species.

Several discrete stages have been recognized during the progressive development of an effective symbiosis (Vincent, 1980; Bauer, 1981; Newcomb, 1981). These stages are the consequence of the active communication between the plant and microbial partners, which results in the induction of a nodule meristem and subsequently leads to the progressive differentiation of specialized cells and tissues within the developing nodule structure. Concurrent with these processes, there is cell and tissue invasion by *Rhizobium* bacteria, leading ultimately to the differentiation of the intracellular symbiosome as the functional unit of symbiosis. During nodule development, a specialized set of tissue-specific host genes is expressed. These gene products have been termed nodulins (Legocki and Verma, 1980; Verma and Delauney, 1988; Nap and Bisseling, 1990).

Agronomy

The ability of peas to fix atmospheric nitrogen has given them an important role in agricultural systems and crop rotation cycles for many centuries. The pea crop has no agronomic requirement for nitrogenous fertilizers and, moreover, it results in an enrichment of the soil for following crops, especially autumn-sown cereals. The high transportation costs of fertilizers in less developed countries of the world have led to a much greater reliance on leguminous crops in these areas. In more developed regions, legume production may also be set to increase through the current move towards low-input farming systems and current concern over the environmental impact of nitrogen applications, which, inevitably, introduce nitrates into aquifers and water tables.

A review of nitrogen economy of the pea crop and the benefit to succeeding crops is given by Jensen (1989). The recommendations for growing peas state that no N-fertilizer is required. Applications of nitrogen have only rarely been demonstrated to result in increases in yield, and then only very small increases were observed. Indeed, a subsequent decrease in the amount of nitrogen fixed in the nodules is often observed. Nitrogen applied early in the season to stimulate early growth of the crop prior to nodules becoming active has been found to suppress and delay the formation of nodules.

The total amount of nitrogen fixed by the pea crop is of the order of 100–250 kg N ha^{-1}, accounting for between 45 and 80% of the total nitrogen accumulated, with the remainder coming from mineral soil N (Jensen, 1986a). Fluctuations in the levels of fixed nitrogen have been recorded as being due to climatic conditions, the level of mineral nitrogen in the soil and the cultivar used (Jensen, 1986b, 1987). Through the use of ^{15}N, the amount of nitrogen fixed has been found to roughly balance the nitrogen that is removed in the seeds. The cultivation of peas, therefore, does not increase soil nitrogen *per se*. Residual nitrogen following a pea crop is in the form of nitrates and ammonium and comes from the roots and other haulm material left after combining. Levels of up to 25 kg N ha^{-1} have been recorded (Jensen, 1989). These levels are at their highest immediately after harvest. If left fallow, this residual nitrogen is almost entirely lost by the following spring. In order to make full use of the residual nitrogen, it is therefore essential that crops be sown as soon as possible after harvesting the peas. This can be achieved through the sowing of either winter cereals or a nitrogen catch crop such as white mustard. The benefit to winter wheat has been shown to be in the range 0–20 kg N ha^{-1} following peas, while winter oil-seed rape has shown even bigger benefits of between 50 and 80 kg N ha^{-1}. The difference is presumed to be due to the higher nitrogen requirement early in establishment (Plancquaert and Desbureaux, 1984).

Nodule Structure and Function

Nodule structure

The anatomy of the pea nodule shows considerable adaptation to the specialized physiological requirements of this organ. The outer tissues of the nodule remain uninfected, whereas many (but not all) of the central cells become invaded by rhizobia. The nodule endodermis is a single layer of cells with suberized (corky) cell walls, which effectively divides the outer cortex from the central nodule parenchyma (Figs 9.1 and 9.3). It forms an impermeable barrier to the lateral diffusion of solutes (Miao et al., 1991). Because it is established very early in pea nodule development (Rae et al., 1991), its presence could influence the differentiation of neighbouring cells and tissues. A second endodermal sheath also surrounds each vascular bundle that develops in the uninfected tissue just beneath the nodule endodermis. Both the vascular endodermis and the nodule endodermis contain a cell-wall glycoprotein, which has been localized immunologically in pea nodule sections (Rae et al., 1991) and provides an early molecular marker for the differentiation of this tissue.

Underneath the nodule endodermis are several layers of uninfected cells, termed either the inner cortex or the uninfected nodule parenchyma (van de Wiel et al., 1990b). Unlike the diploid outer cortex, the inner cortex of pea nodules is composed of tetraploid cells (Mitchell, 1965). Apart from the apical meristem, the inner cortex is the only nodule tissue in which cells are tightly packed without intercellular air spaces. It may therefore act as a barrier to oxygen diffusion into pea nodules, as in other legumes (Parsons and Day, 1990).

In the central tissues of the nodule, relatively small uninfected cells are interspersed with the enlarged cells that harbour rhizobia. The pattern of gene expression is clearly differentiated in the uninfected interstitial cells, relative to their infected counterparts (Scheres et al., 1990b). Interstitial cells

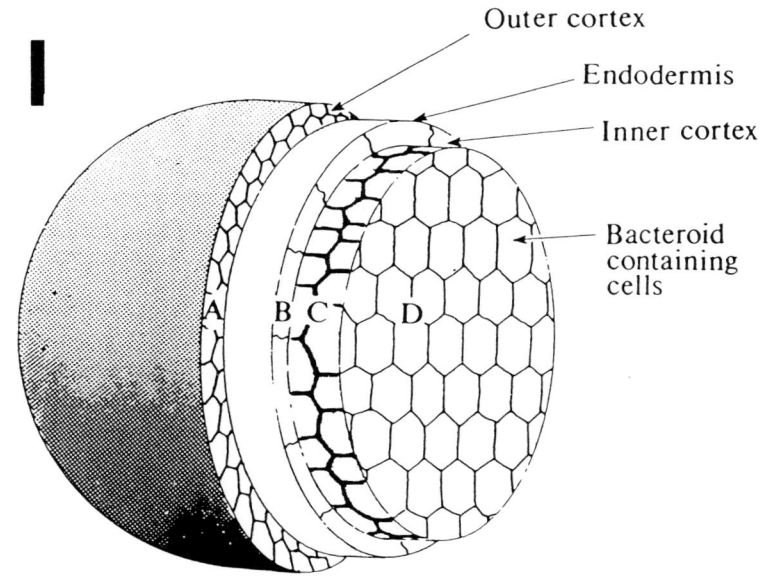
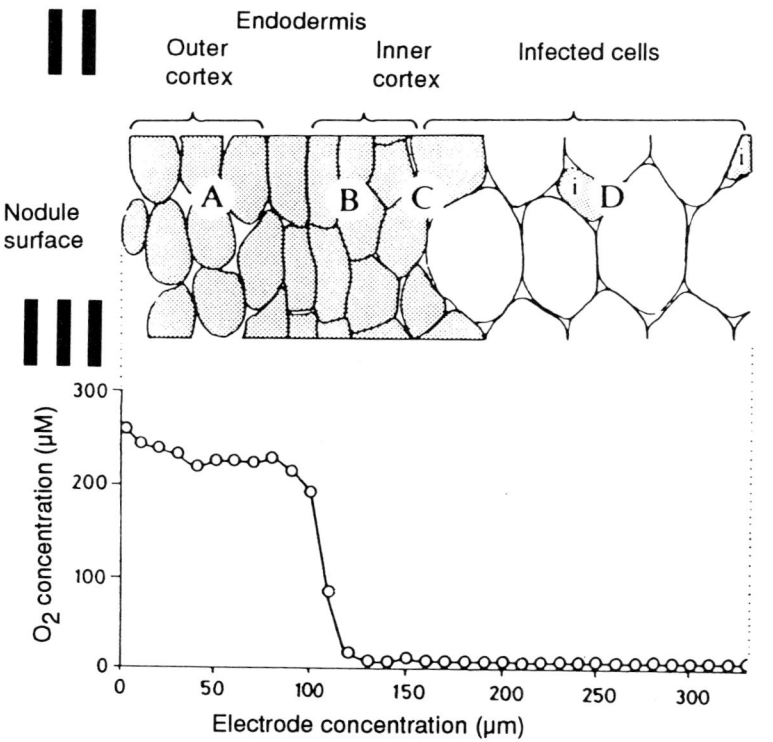

may represent a distribution network between the vascular tissue of the inner cortex and the infected cells, conducting carbon substrates towards the nitrogen-fixing cells and organic nitrogen compounds away from this area. It is also important to point out that the central infected tissues are essentially aerenchymatous, the large intercellular air spaces providing pathways for the diffusion of oxygen. Interestingly, most of the mitochondria in the infected cells are to be found adjacent to these air spaces (Newcomb, 1976).

Nodule physiology

The major constraint on nodule structure and function is the oxygen-sensitivity of the nitrogen fixation reaction. It appears that it is primarily to overcome the problems of oxygen supply that leguminous plants have evolved an intracellular symbiosis within a highly specialized root nodule, rather than, for example, to develop a nitrogen-fixing symbiosis with bacteria that grow in intercellular spaces. The legume nodule provides a highly specialized microaerobic environment that optimizes the process of nitrogen fixation (Sprent and Raven, 1992). Paradoxically this low-oxygen environment must be maintained despite the fact that the many nitrogen-fixing bacteria (bacteroids) within each nodule cell require a rapid flux of oxygen to form the ATP needed in the energy-costly process of reducing nitrogen. Additional features of the nodule include a requirement for a constant supply of photosynthate and a mechanism for the rapid removal of the NH_4^+ produced.

Nitrogen

N_2 is remarkably unreactive and is, in effect, an inert gas for eukaryotic cells, which have no enzymes that react with it. Only prokaryotes have developed the enzymes to reduce N_2. Although there are some slight differences in their nitrogenase enzymes, the nitrogenase complex is remarkably similar and conserved throughout a very wide range of bacteria, including cyanobacteria. In addition to *Rhizobium* spp. (which interact almost exclusively with legumes), other nitrogen-fixing microorganisms have evolved symbioses with higher plants. Examples include *Frankia* spp., a filamentous actinomycete that forms

---

Fig. 9.3. Generalized features of root nodule anatomy involved in the control of O_2 supply to the bacteroids (adapted from Witty et al., 1986). I: Tissue layers showing interconnected air spaces between cells as dark lines; II: cellular arrangement of the outer layers of a pea nodule: (A) outer cortex, (B) inner cortex (tightly packed), (C) inner cortex with air spaces or intercellular matrix material, (D) infected cells with air spaces, (i) interstitial (uninfected cells - all uninfected cells are stippled.); III: oxygen concentrations across the above section of nodule tissue, as measured by insertion of an oxygen microelectrode.

root nodules on woody shrubs, e.g. *Casuarina* and *Alnus*, and *Nostoc*, a cyanobacterium that forms nodules on cycads.

The nitrogenase complex was first defined using the enzyme complex from the free-living nitrogen-fixing bacteria *Klebsiella pneumoniae* and *Azotobacter vinelandii*. Its general features also hold true for the *Rhizobium* enzyme (Dean and Jacobson, 1992). The complex consists of two components, which were called component I and component II based on their elution positions during purification on column chromatography. Component I contains the site of N_2 reduction and is also referred to as dinitrogenase. It is an iron–sulphur (4Fe–4S) enzyme that contains four subunits ($\alpha_2\beta_2$) and two identical cofactors containing Fe and Mo (often referred to as FeMoco). These cofactors constitute at least part of the active centre of the enzyme and are highly sensitive to oxygen damage. In some bacteria, it is evident that forms of dinitrogenases can exist in which Mo may be replaced by vanadium (V) or even iron (Fe), but no such enzyme has yet been identified among symbiotic nitrogen-fixing bacteria.

Component II provides electrons that reduce component I and is therefore called dinitrogenase reductase. It is a homodimeric protein that also contains a [4Fe–4S] cluster. During nitrogen fixation, the dinitrogenase reductase is reduced by single electron additions. It then associates with dinitrogenase and donates a single electron to it in a reaction that is coupled to ATP hydrolysis. The two complexes then dissociate and multiple cycles of these reactions are coupled to successive reductions of N_2 by electrons. The overall nitrogen-fixing reaction is summarized by the equation:

$$N_2 + 8\ e^- + 8\ H^+ + 16\ MgATP \rightarrow 2\ NH_3 + H_2 + 16\ MgADP + 16\ Pi$$

Thus, eight electrons, eight protons and 16 ATP molecules are consumed per N_2 reduced and a necessary by-product of the catalysis is the production of one molecule of H_2 per N_2 reduced. The enzyme is not very specific for N_2 as a substrate and can also reduce C_2H_2 to C_2H_4 (Dilworth, 1966). Indeed, the reduction of acetylene to ethylene is frequently used as a measure of nitrogenase activity, although this method is prone to some inaccuracies (Minchin *et al.*, 1986). Several *Rhizobium* genes are required to make a functional nitrogenase complex. The α and β subunits of dinitrogenase are encoded by *nifD* and *nifK* respectively and the nitrogenase reductase is encoded by *nifH*. In addition, there are several genes required for the synthesis and assembly of the Fe–Mo cofactor (reviewed in Dean and Jacobsen, 1992) and it is likely that still more *Rhizobium* genes remain to be identified.

The nitrogenase reductase in *K. pneumoniae* is itself reduced by a flavodoxin (NifF), which is in turn reduced by a pyruvate-flavodoxin oxidoreductase (NifJ). In rhizobia it appears that dinitrogenase reductase is reduced via a different mechanism that probably involves the products of the *fixABCX* genes (Schetgens *et al.*, 1985; Gubler and Hennecke, 1986; Earl *et al.*, 1987; Gronger *et al.*, 1987), analogues of which are found in other aerobic and microaerobic diazotrophs (Dean and Jacobsen, 1992).

The primary control of expression of rhizobial nitrogen-fixing genes appears to be via an oxygen-sensing system (David et al., 1988; Batut et al., 1989). Given the controlled O_2 environment within legume nodules (see below), this makes a great deal of sense. The normal regulatory control system for most free-living nitrogen-fixing bacteria is at the level of N availability, an option unsuited to bacteroids, which must produce an excess of fixed nitrogen. It is now evident that the protein encoded by *fixL* is a haem protein (Gilles-Gonzales et al., 1991) that senses the O_2 environment and phosphorylates the regulatory protein FixJ, which then activates *nifA* expression. The NifA protein then activates the expression of other genes required for nitrogen fixation. It has been proposed that the level of *nifA* expression is governed by the *fixK* gene product, which represses *nifA* expression (Batut et al., 1989).

Several plant genes are highly expressed in nodules to facilitate the assimilation of the NH_4^+ produced by the bacteroids (Schubert, 1986; Lea et al., 1990; Coruzzi, 1991; Cullimore and Bennet, 1992). The primary enzyme of ammonia assimilation in legume nodules is glutamine synthetase, which catalyses the formation of glutamine from glutamate and ammonia (Forde and Cullimore, 1989). Peas contain at least three genes that encode glutamine synthetases, and one of these genes is strongly enhanced (10–20-fold) in its expression in pea nodules (Tingey et al., 1987). This gene is also expressed at low levels in roots and leaves (Brears et al., 1991) and therefore the situation is slightly different from that in *Phaseolus* beans or soyabeans, where nodule-specific forms of glutamine synthetase are found (Sengupta-Gopalan and Pitas, 1986; Forde et al., 1989). In all three legumes, the induction of the nodule-enhanced form of glutamine synthetase(s) is apparently not under ammonium control, because there is still induction of glutamine synthetase in nodules induced by mutant rhizobia unable to fix nitrogen (Sengupta-Gopalan and Pitas, 1986; Walker and Coruzzi, 1989; Cock et al., 1990). Thus, expression must be developmentally regulated, which is in contrast to the case for root and leaf glutamine synthetases (Walker and Coruzzi, 1989).

Given the high rates of NH_4^+ assimilation, it is important that there is a constant supply of glutamate and this is provided by the plastidic enzyme glutamate synthase (commonly called GOGAT: glutamine-oxoglutarate aminotransferase). A nodule-specific glutamate synthase has been identified in beans and, like glutamine synthetase, it is strongly induced in nodules (Chen and Cullimore, 1988, 1989). Glutamate cycles via the two reactions catalysed by glutamine synthetase and glutamate synthase:

2 glutamate + 2 NH_3 + 2 ATP → 2 glutamine + 2 ADP + 2 Pi
glutamine + α-ketoglutarate + NADH → 2 glutamate + $NADP^+$ + H_2O

to give the net reaction

α-ketoglutarate + 2 NH_3 + 2 ATP + $NADH^+$ → glutamine + 2 ADP + 2 Pi + NADP + H_2O

Subsequent steps of nitrogen assimilation involve the transfer (by aspartate aminotransferase) of the amino group from glutamate to oxaloacetate, forming aspartate. Subsequently, asparagine is formed by the action of asparagine synthetase, which forms asparagine from the aspartate using glutamine as a nitrogen donor. Two asparagine synthetase genes have been identified in peas and both are strongly induced in nodule tissue (Tsai and Coruzzi, 1990). It is mostly asparagine and to a lesser extent glutamine that is transferred via the xylem from pea nodules to other plant tissues (van Beuschem, 1983; Peoples *et al.*, 1987).

Oxygen

It is most probably oxygen supply that is the rate-limiting factor in the symbiosis (Layzell and Hunt, 1990; Layzell *et al.*, 1990). The flow of oxygen to the bacteroids is regulated by the plant on at least two levels – a physical barrier to oxygen diffusion through the nodule cortex and an oxygen carrier system mediated via leghaemoglobin (Appleby, 1984; Minchin *et al.*, 1985; Sheehy *et al.*, 1985; Witty *et al.*, 1986).

Using an O_2-selective microelectrode, Witty *et al.* (1987) demonstrated that there is a variable barrier to O_2 diffusion in the cortical cell region just outside the region of infected pea nodule cells. This zone was estimated to be one to five cell layers thick in a region where the cells are tightly packed with relatively few intercellular spaces through which gas exchange can occur (Fig. 9.3). When the oxygen concentration outside the nodule was increased, this was paralleled by an increase in resistance of the oxygen barrier (Witty *et al.*, 1987). It has been suggested that legumes may be able to alter the rates of O_2 exchange through air-spaces by filling the spaces with gel-like extracellular glycoproteins (James *et al.*, 1991). One potential component has been identified in nodule intracellular spaces using a highly specific monoclonal antibody that recognizes an extracellular glycoprotein (VandenBosch *et al.*, 1989; Rae *et al.*, 1991).

The second level of control of oxygen flux is located within the infected nodule cells. The high rate of O_2 consumption by bacteroids within nodules, coupled to the physical oxygen barrier, tends to keep the oxygen concentration low within the nodule. Within this environment, the oxygen-carrier leghaemoglobin acts as an O_2 buffer, maintaining a very low free oxygen concentration (Monroe *et al.*, 1989). Leghaemoglobins constitute the major class of proteins within the infected host cells (Govers *et al.*, 1985), and this can maintain a high rate of oxygen flux whilst the free (unbound) oxygen concentration is kept in the 3–10 nM range (Appleby, 1984). Using immunolocalization techniques, Robertson *et al.* (1984) demonstrated that the leghaemoglobin is localized in the cytoplasm of infected pea nodule cells.

There is some evidence to suggest that the haem moiety of leghaemoglobin is synthesized by the bacteroids since there is a parallel increase in total nodule haem and the activity of bacteroid haem synthetic enzymes (Nadler and Avissar, 1977). Furthermore, *Rhizobium* mutants blocked in haem synthesis induce nodules lacking leghaemoglobin (Leong et al., 1982; Stanley et al., 1988). It does appear that in some legumes the bacteroids can take up the haem precursor δ-aminolaevulinic acid from the plant cell cytoplasm and use it for haem biosynthesis (Sangwan and O'Brian, 1991).

Peas contain a small gene family that encodes at least four different apoproteins of leghaemoglobin (Uheda and Syono, 1982a,b; Govers et al., 1985). The expression of these genes is synchronized with nodule development and they are not under oxygen control (Govers et al., 1986b), although in nodules formed by mutants of *R. leguminosarum* unable to fix nitrogen the amount of leghaemoglobin found in nodules is greatly reduced. However, since leghaemoglobin mRNA was present in these ineffective nodules, it was concluded that there is a post-transcriptional control of leghaemoglobin biosynthesis (Govers et al., 1986a).

As a result of the high affinity of leghaemoglobins for O_2 (Appleby, 1984), it has been argued that the nitrogen-fixing bacteroids must have an electron transport pathway that terminates with an unusually high-affinity oxidase to efficiently capture the low free oxygen concentration (Bergersen, 1982; Appleby, 1984). Strains of *R. leguminosarum* are known to possess three cytochrome oxidases under free living conditions: cytochromes d, aa_3 and o (Soberon et al., 1989; Barquera et al., 1991). By analogy with the *E. coli* cytochrome d complex, only cytochrome d is likely to have a high enough affinity for oxygen (Hill et al., 1990). However, there is good genetic evidence that the pathway of electrons to the terminal oxidase is via the cytochrome bc_1 complex (Thony-Meyer et al., 1989, C. Vargas and J.A. Downie, unpublished), which does not normally transfer electrons to pathways terminated with cytochrome o or cytochrome d. Therefore, it is thought that bacteroids may have a fourth terminal oxidase that accepts electrons (indirectly) from cytochromes bc_1 (Bott et al., 1990; Hennecke et al., 1993).

Carbon

Nodules can take up an estimated 16–33% of the net photosynthate of the pea plant (Minchin et al., 1981). Sucrose is the primary source of carbon that is translocated to the nodules and the first enzyme involved in its degradation is sucrose synthase (Schubert, 1986), the activity of which is high in nodules (Morell and Copeland, 1985).

A significant proportion of the carbon in sugars entering glycolysis within pea nodules is destined to be re-exported in the form of asparagine or glutamine (Schubert, 1986). Therefore there is a high demand for malate and oxaloacetate for the biosynthesis of amino acids. Malate is also the major source of carbon used by the bacteroids for respiration. It has been argued (de Vries et al., 1980; Vance and Heichel, 1991) that the low O_2 tension within nodules induces high levels of phosphoenolpyruvate (PEP) carboxylase, which catalyses the first step in a 'malic' fermentation, and indeed there are reports of higher levels of PEP carboxylase in nodule tissue than in root tissue (de Vries et al., 1980; Deroche et al., 1983; Gadal, 1983). However, Smith (1985) clearly demonstrated that, when expressed on a protein basis, the activity of PEP carboxylase and malate dehydrogenase in pea nodules was high, but not significantly different from levels found in pea root tips. This indicated that the pathways of fermentation of nodules and roots are essentially similar and therefore a malic fermentation pathway in nodules is unlikely. Rosendahl et al. (1990) measured [^{14}C]-CO_2 incorporation into actively metabolizing nodules and showed a rapid and extensive incorporation of label, principally into malate and aspartate. This is consistent with the idea that PEP carboxylase and malate dehydrogenase direct much of the photosynthate into amino acid metabolism.

The availability of malate in nodules correlates well with the ability of isolated R. leguminosarum bacteroids to utilize dicarboxylic acids but not sugars (Glenn et al., 1980, 1984; Ronson et al., 1981; Arwas et al., 1985; McKay et al., 1988). Mutants of R. leguminosarum blocked in dicarboxylic acid transport (dct) are unable to fix nitrogen (Ronson et al., 1981; Finan et al., 1983) and the bacteroids in nodules senesce at an early stage (Watson et al., 1988; Engelke et al., 1989). Therefore there is clear evidence for dicarboxylic acids such as malate and succinate being the major carbon source in bacteroids. In addition, there is good evidence for the presence of a dicarboxylate transporter in the peribacteroid membrane in soyabean (Udvardi et al., 1988), and Rosendahl et al. (1990) showed that there was a rapid appearance of [^{14}C]-malate and succinate in bacteroids following the incorporation of [^{14}C]-CO_2 into pea nodules via PEP carboxylase.

Although dicarboxylic acids are the principal carbon source for bacteroids, it is evident that bacteria within nodules must also have access to other carbon sources, because a number of different auxotrophic mutants of R. leguminosarum can form nitrogen fixing nodules on peas (Beringer et al., 1980), as can gluconeogenic mutants (McKay et al., 1985).

Hydrogen

H_2 is an unavoidable by-product of the process of biological N_2 fixation (Evans et al., 1987). Some strains of R. leguminosarum have the capacity to recycle this potentially useful energy source by means of an uptake hydrogenase system (hup). The genes encoding the hydrogenase complex are only present in some strains of R. leguminosarum and, where characterized, have been found to be on the symbiotic plasmid (Brewin et al., 1980). If nodule activity is carbon-limited, the additional yield of ATP generated by bacteroids that can respire H_2 should theoretically lead to a more carbon-efficient nitrogen-fixing symbiosis. However, a comparison of pea plants inoculated with Hup^+ and Hup^- isogenic strains did not give a clear increase in nitrogen yield (Cunningham et al., 1985; Kagan and Brewin, 1985). On the other hand, if nodule activity is limited by the supply of oxygen rather than by the supply of carbon, the presence or absence of a hydrogen recycling activity might not be expected to make any significant difference to the overall efficiency of symbiosis and resulting plant yield.

Other nutrients

It is evident that there must be supplies of many other nutrients delivered to rhizobia within nodules. In particular, since the bacteroids are cytochrome-rich, they must have access to iron and in addition there are obvious requirements for phosphate, sulphate, K^+, molybdate, etc. However, relatively little is known about their transport into bacteroids.

Nodule Initiation and Development

The initiation of nodule morphogenesis in leguminous plants is based on a two-way communication process. Initially, flavonoid or isoflavonoid molecules normally secreted from legume roots stimulate the expression of rhizobial nodulation genes. Several of these '*nod*' genes are involved in the biosynthesis of lipo-oligosaccharide signalling molecules that can induce nodule structures in some legumes. These characteristic plant responses are induced at astonishingly low ($< 10^{-11}$ M) concentrations. Much of the elucidation of this work was based on an understanding of the *nod* genes from different rhizobia – in particular work on *R. meliloti* and on *R. leguminosarum*. This work has been comprehensively reviewed recently (Downie, 1991; Fisher and Long, 1992; Long, 1992; Schlaman et al., 1992b; Spaink, 1992) and in this section we will focus principally on work carried out on the pea nodulating strain *R. leguminosarum* biovar *viciae*, which is commonly referred to as *R. l. viciae*.

Characterization of Rhizobium nod *genes*

At an early stage in *R. leguminosarum* genetics, it was recognized that the determinants of host specificity were located on a plasmid rather than on the chromosome. This was established by the isolation of a transposon-induced mutant unable to make one of the *R. l. viciae* bacteriocins (Beringer et al., 1978). It was found that the antibiotic resistance marker on the transposon used to make the mutation was highly transmissible between *Rhizobium* strains;

most importantly, when the plasmid (carrying the transposon) was transferred to *R. l. phaseoli*, the new derivative of *R. l. phaseoli* acquired the ability to nodulate peas rather than *Phaseolus* beans, the normal host of that strain (Johnston *et al.*, 1978). The so-called *Sym* (symbiotic) plasmid identified in these genetic studies was called pRL1JI and has been the subject of intensive study over the past 12 years (see Downie and Johnston, 1988; Downie, 1991). It carries the nodulation (*nod*) genes, which are the principal determinants of nodulation specificity (i.e. the range of legumes nodulated).

Several *R. l. viciae* nodulation-deficient mutants were isolated (Downie *et al.*, 1983a,b, 1985; Wijffelman *et al.*, 1985) and a region of about 20 kb was identified as carrying a cluster of *nod* genes. Plasmids carrying this cloned region of DNA were found to confer on strains such as *R. l. phaseoli*, *R. l. trifolii* and even *Agrobacterium tumefaciens* the ability to nodulate legumes such as peas and vetch (*Vicia hirsuta*) (Downie *et al.*, 1983a; Govers *et al.*, 1986a). Therefore it was concluded that the key determinants of host recognition were contained within a relatively short region of DNA. In a series of publications involving isolation of mutants and DNA sequencing of the *nod* gene region of pRL1JI (Downie *et al.*, 1983a; Rossen *et al.*, 1984; Downie *et al.*, 1985; Wijffelman *et al.*, 1985; Evans and Downie, 1986; Shearman *et al.*, 1986; Surin and Downie, 1988, 1989; Canter-Cremers *et al.*, 1989; de Maagd *et al.*, 1989; Economou *et al.*, 1989, 1990; Surin *et al.*, 1990), 13 *nod* genes were identified and their arrangement is outlined in Fig. 9.4 (see also Table 9.2). These genes are arranged in five operons – *nodABCIJ*, *nodD*, *nodFEL*, *nodMNT* and *nodO*.

Legume recognition and nod gene regulation

When different *nod* genes such as *nodC*, *nodF* and *nodD* were fused to the *lacZ* gene (β-galactosidase from *E. coli*), it was found that the *nodD* gene was expressed constitutively in *R. l. viciae* strains grown in laboratory media, but the *nodABC* and *nodFEL* operons were not (Rossen *et al.*, 1985; Shearman *et al.*, 1986; Spaink *et al.*, 1987a). However, the observation that the *nodABC* genes and *nodFEL* genes were strongly induced in media that contained root exudate from axenically grown peas (Rossen *et al.*, 1985) stimulated a search for the factors required for their induction. It was established (Firmin *et al.*, 1986; Zaat *et al.*, 1987) that a number of different flavonoids secreted from legume roots could induce *nod* gene expression in *R. l. viciae* strains. The structures of some of these compounds are shown in Fig. 9.5.

The induction of *nod* gene expression was found to be dependent on the presence of the *nodD* gene product (Rossen *et al.*, 1985) which is a transcriptional activator protein that belongs to the LysR family of bacterial regulatory proteins (Schlaman *et al.*, 1992a). NodD has been shown to be a DNA-binding protein (Hong *et al.*, 1987; Schlaman *et al.*, 1992a) and it is also thought to

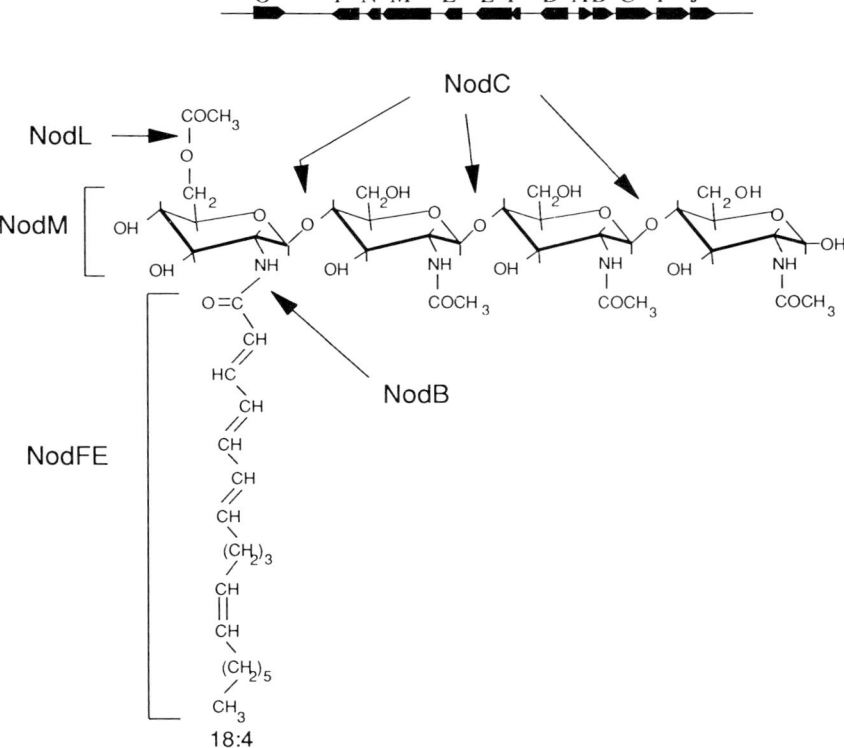

Fig. 9.4. Diagram showing the nodulation genes of *R. leguminosarum* bv. *viciae* that have been identified. The black arrows in the upper part of the figure indicate the positions and orientations of the nodulation genes. All the genes indicated have been localized on the symbiotic plasmid pRL1JI. The *nodX* gene, which extends the range of pea cultivars nodulated, was identified downstream of *nodJ*, and is only present in a minority of strains of *R. leguminosarum* bv. *viciae*. Potential functions of the *nod* genes are listed in Table 9.2. The lower part of the diagram illustrates the chemical structure of a lipo-oligosaccharide, termed NodRlv-IV (Ac, C18:4), which is secreted by *Rhizobium leguminosarum* bv. *viciae* and induces root hair deformation and cortical cell division in seedling roots of pea and *Vicia*. This structure is synthesized by the combined activity of several nodulation (*nod*) genes. Their probable roles in the biosynthesis of the lipo-oligosaccharide nodulation factor are indicated in the diagram.

Table 9.2. Nodulation (*nod*) genes of *Rhizobium leguminosarum* bv. *viciae* and their probable functions.

nod gene product	Possible role or homologue	Reference
NodA	Unknown	
NodB	Deacetylation of N-acetylglucosamine polymer to allow N-acylation	John *et al.*, 1993
NodC	Chitin synthase; synthesis of poly-N-acetylglucosamine oligomer	Bulawa and Wasco, 1991 Atkinson and Long, 1992 Debelle *et al.*, 1992
NodD	Positive regulator of *nod* gene expression	Rossen *et al.* 1985 Shearman *et al.*, 1986
NodE	β ketoacyl synthase; synthesis of host-specific C18:4 acyl group	Bibb *et al.*, 1989 Spaink *et al.*, 1991
NodF	Acyl carrier protein; synthesis of host-specific C18:4 acyl group	Shearman *et al.*, 1986 Geiger *et al.*, 1991 Spaink *et al.*, 1991
NodI	ATP-binding membrane-associated protein involved in transport	Evans and Downie, 1986 Schlaman *et al.*, 1990
NodJ	Integral membrane protein Transport function	Evans and Downie, 1986 Surin *et al.*, 1990
NodL	O-acetyltransferase; acetylates acylated glucosamine in mature nodulation factor	Downie, 1989 Spaink *et al.*, 1991
NodM	Glucosamine-6-P synthase, makes glucosamine precursors of nodulation factor	Marie *et al.*, 1992
NodN	Unknown	
NodO	Secreted nodulation-signalling protein; forms membrane pores	de Maagd *et al.*, 1989 Economou *et al.*, 1990 Sutton *et al.*, 1993
NodT	Outer membrane protein, homologous to proteins from bacterial transport complexes; transport function	Surin *et al.*, 1990 R. Rivilla, personal communication
NodX	O-acetyltransferase; acetylates the 6-C of the reducing glucosamine on the nodulation factor	J. Firmin, personal communication

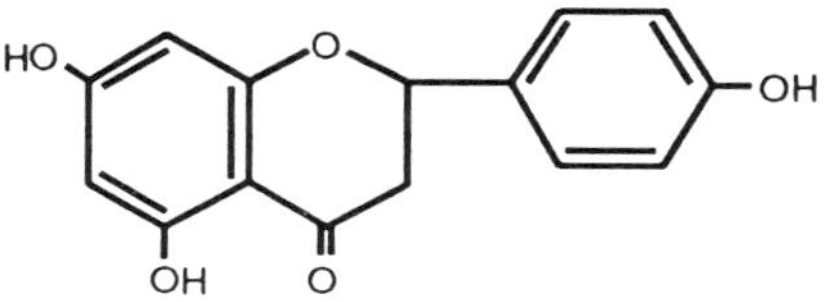

Fig. 9.5. Examples of flavonoid compounds secreted by *Vicia* roots which act as inducers of *nod* gene transcription for *R.l. viciae* (adapted from Recourt *et al.*, 1992). 1: 7,4'-dihydroxyflavanone (liguiritigenin); 2: 7,4'-dihydroxy-3'-methoxyflavanone; 3: 7,3'-dihydroxy-4'-methoxyflavanone; 4: 5,7,4'-trihydroxyflavanone (naringenin).

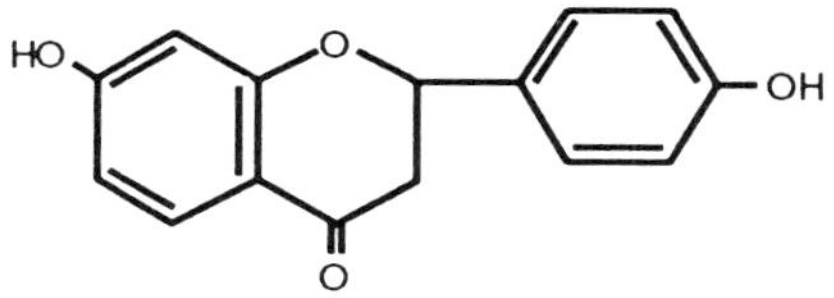

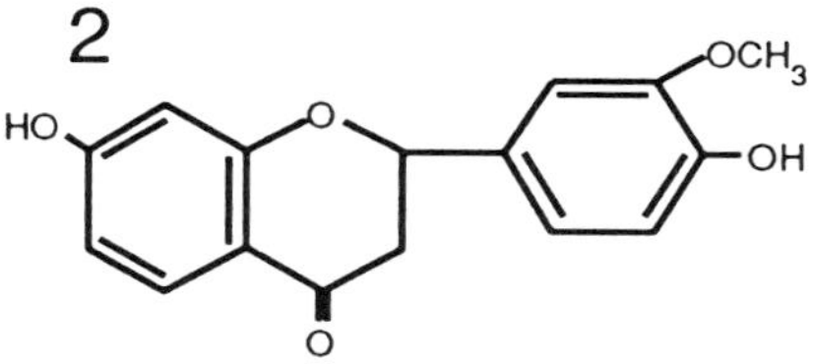

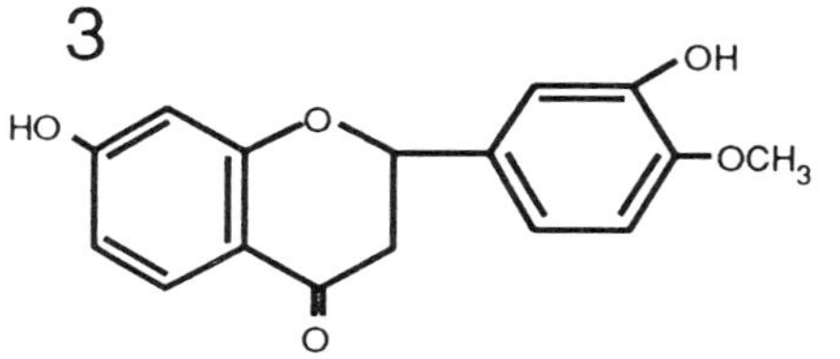

interact directly with flavonoid molecules (Burn *et al.*, 1987; Spaink *et al.*, 1989a). The NodD protein is found to be membrane-associated (Schlaman *et al.*, 1989) and it is thought that the flavonoids partition across the bacterial inner membrane in response to the pH gradient, and are subsequently bound by NodD. However, it appears that NodD binding to DNA is not flavonoid-dependent (Hong *et al.*, 1987; Burn *et al.*, 1989) and therefore the flavonoids may induce some kind of conformational change in *nodD* that results in activation of *nod* gene transcription.

Although the *R. l. viciae* and *R. l. trifolii* strains studied contain only one *nodD* gene, it is evident that different rhizobial strains can carry multiple *nodD* genes. The various *nodD* gene products appear to have relatively different specificities for the various flavonoids secreted from legume roots and therefore the NodD protein acts as a determinant of specificity by recognizing the flavonoids secreted by the appropriate legume (Spaink *et al.*, 1989a). In some cases it has been possible to alter the range of legumes nodulated (or increase the efficiency of nodulation) by transferring a *nodD* gene from one *Rhizobium* strain to another (Spaink *et al.*, 1987b; Györgypal *et al.*, 1988; Surin and Downie, 1989).

The NodD protein binds to a highly conserved DNA sequence in the promoter region of the *nodABCIJ*, *nodFEL*, *nodMNT* and *nodO* operons (Hong *et al.*, 1987; Schlaman *et al.*, 1992a). Using a DNA probe homologous to this region, it was established that there were only four hybridizing bands found in the *R. l. viciae* genomic DNA (Rodriguez-Quinones *et al.*, 1987; Economou et al., 1989), indicating that all of the operons under NodD control have been identified in this strain.

Nodulation signalling molecules

It had been known for some time, that filter-sterilized medium taken from the roots of legumes inoculated with rhizobia could induce marked root hair deformation on axenically grown legumes (Yao and Vincent, 1969) and that a low-molecular-weight compound ($M_r \approx 1200$) was responsible for this effect (Bhuvaneswari and Solheim, 1985). Subsequently, Van Brussel et al. (1986) demonstrated that R. l. viciae grown axenically in the presence of nod-gene-inducing flavonoids secreted signals that could induce strong root hair deformation. Spaink et al. (1991) purified the active fractions and showed that the major active components were oligomers of four or five glucosamine residues that carried an unusual N-linked acyl group (C18:4) and an O-linked acetyl group (Fig. 9.4). These nodulation factors made by R. l. viciae were similar to those made by R. meliloti, except that the R. meliloti factor contained a different acyl group (C16:2) and contained a sulphate group on the terminal glucosamine residue (Roche et al., 1991a,b; Schultze et al., 1992).

These two different classes of nodulation factors made by R. l. viciae and R. meliloti constitute a new class of signalling molecules in plant biology. When added at concentrations as low as 10^{-11} M, they induce pronounced root hair deformation and the induction of plant nodule-specific genes (Lerouge et al., 1990; Spaink et al., 1991; Truchet et al., 1991). At higher concentrations (10^{-8} M) the R. meliloti factor can induce complete nodule organogenesis in the absence of added bacteria (Truchet et al., 1991) and at a similar concentration the R. l. viciae factor can induce the early stages of organogenesis in V. sativa roots (Spaink et al., 1991). However, the nodulation factor made by R. meliloti is only very weakly (if at all) active on the roots of peas or Vicia spp. and the R. l. viciae factor is not active on the roots of Medicago spp. (Faucher et al., 1989).

On the basis of genetic work, it is now evident that the sulphate group is an important determinant for specificity in those legumes nodulated by meliloti, which has additional nod genes (compared with R. l. viciae) that encode the sulphate transfer enzymes (Schwedock and Long, 1990; Roche et al., 1991a). Other important determinants of specificity are the nodF and nodE genes, since transfer of these genes from R. l. viciae to R. l. trifolii enables the latter strain to nodulate peas and vetch (Spaink et al., 1989b; Surin and Downie, 1989). A nodE mutant of R. l. viciae did make nodulation-signalling molecules but the signals lacked the C18:4 acyl group and had instead a C18:1 acyl group. It had already been evident that the NodF and NodE proteins might be involved in some way in fatty acid synthesis on the basis that NodF showed homology to acyl carrier proteins (Shearman et al., 1986) and NodE was homologous to the β-ketoacyl synthase (condensing enzyme) activity of the fatty acid complex (Bibb et al., 1989). Therefore it is highly likely that NodF and NodE participate in fatty acid biosynthesis to form a novel fatty acid that is attached to the nodulation signal molecule: in the absence of these gene functions, a C18:1 acyl group (the major fatty acid found in R. l. viciae) can be incorporated instead (Spaink, 1992).

The nodA, nodB and nodC gene products are all essential for the biosynthesis of the signal molecules (Spaink et al., 1991) and mutants of R. l. viciae lacking these genes evoke no obvious plant response (Rossen et al., 1984). It is evident from protein homologies that NodC is similar to chitin synthase (Bulawa and Wasco, 1991) and cellulose synthase (Atkinson and Long, 1992) and is therefore likely to be involved in the polymerization of the N-acetylglucosamine backbone. It appears that NodB removes an N-acetyl group from one of the sugar residues prior to the N-acetylation of this residue (John et al., 1993). This is consistent with the observations that nodB and nodC do not confer any legume-specific nodulation to rhizobial strains (Fisher et al., 1985). Thus the gene products of nodA, nodB and nodC synthesize a 'core' nodulation signal that can be modified by a number of other gene products. One of these other gene products is NodL, which is strongly homologous to a group of O-acetyl transferases (Downie, 1989): it is required for the attachment of the O-acetyl group (Fig. 9.4) on to the acylated glucosamine residue (Spaink et al., 1991).

Another gene that modifies the signal molecule is the nodX gene. This gene is absent from most strains of R. l. viciae but was first identified in a strain of R. l. viciae isolated from Turkey (Brewin et al., 1980; Hombrecher et al., 1984; Götz et al., 1985; Davis et al., 1988). The strain from which nodX was isolated has the ability to nodulate peas of the cultivar Afghanistan, which are normally resistant to nodulation by a wide variety of R. l. viciae strains isolated from Western European or North American soils (Lie, 1978). However, when nodX is transferred to such isolates, they then acquire the ability to nodulate cv. Afghanistan (Götz et al., 1985). This is correlated with an O-acetylation of the 6-carbon on the reducing glucosamine of the nodulation factor (J.L. Firmin, personal communication).

The NodM protein is a glucosamine synthase that is involved in the formation of glucosamine precursors required for the synthesis of the sugar backbone of the nodulation factor (Marie et al., 1992). The observation that mutations in nodM do not significantly affect nodulation (Surin and Downie, 1988) is explained by the presence of two glucosamine synthases; NodM is encoded on the symbiotic plasmid and GlmS encoded on the chromosome. Thus mutations in nodM are compensated by expression of the chromosomal gene (Marie et al., 1992). Of the other nod genes in R. l. viciae, only nodN appears to be involved in the synthesis of the signal; mutations of the nodN homologue in R. meliloti decreased the production of the nodulation-signalling factor (Baev et al., 1991).

The other *nod* gene products, NodI, NodJ, NodT and NodO, are not involved in the biosynthesis of the lipo-oligosaccharide nodulation signalling molecules. NodI and NodJ are thought to be involved in transport of components across the bacterial inner membrane; NodI is associated with the bacterial inner membrane (Schlaman *et al.*, 1990) and is homologous to a large group of bacterial proteins involved in ATP-dependent transport across bacterial membranes (Evans and Downie, 1986; Higgins *et al.*, 1986, 1990). NodJ is very hydrophobic (Evans and Downie, 1986) and probably spans the bacterial inner membrane (Surin *et al.*, 1990). Interestingly, *nodT* encodes a protein that is located in the outer membrane of *R. l. viciae* (Surin *et al.*, 1990; R. Rivilla and J.A. Downie, unpublished observations) and it is possible that NodT, NodI and NodJ may function together to form a complex involved in the secretion of components across the inner and outer membranes of *R. l. viciae*.

The *nodO* gene encodes a Ca^{2+}-binding protein that is secreted into the growth medium of *R. l. viciae* (de Maagd *et al.*, 1989; Economou *et al.*, 1990) and plays a role in plant signalling (Downie and Surin, 1990). In the absence of the *nodFE* genes, inefficient nodulation occurs, presumably as a result of the wrong (i.e. C18:1-containing) nodulation factor being produced (Spaink *et al.*, 1991). However, the residual pea nodulation (about 30–50% of normal) induced by *nodFE* mutants depends on the *nodO* gene product (Downie and Surin, 1990). It is evident that NodO is not involved in the modification of the lipo-oligosaccharide nodulation factor because a normal signal is made in the absence of *nodO* (Spaink *et al.*, 1991). Therefore it is possible that NodO may in some way facilitate the presentation of the nodulation signals to the plant. It is apparent that NodO makes pores in membranes (Sutton *et al.*, 1993) and it is possible that such pores could facilitate the uptake of bacterially made nodulation signals.

In addition to the *nod* genes that have been described, it is quite likely that there may be other, as yet unidentified, characteristics that improve the efficiency of the *R. l. viciae* symbiosis with its compatible legumes. Such genes may improve nodulation on different lines or species within the range of legumes nodulated by *R. l. viciae*. Adjacent to the *nod* gene cluster of *R. l. viciae*, a cluster of genes has been identified which are expressed in the rhizosphere and during free-living growth of the bacteria, but not within nodules (Dibb *et al.*, 1984; Economou *et al.*, 1989). The role of these genes is unclear, but it does appear that they may be important in the nodulation process (Cubo *et al.*, 1992), although they are under different genetic control from the *nod* genes. These genes may be specific for the *R. l. viciae* symbiosis since they have not been found in other rhizobial strains.

Pea nodules are rich in the amino acid homoserine and it appears that *R. l. viciae* strains, but not other rhizobial strains tested, have the ability to utilize homoserine as a source of both carbon and nitrogen for growth (van Egeraat, 1975). The genes encoding this ability are present on the *R. l. viciae* symbiotic plasmid pRL1JI (Johnston et al., 1988). This property may confer upon R. l. viciae strains the ability to increase their numbers in the environment outside nodules.

Plant responses to bacterial Nod-factors

The *Rhizobium*-derived lipo-oligosaccharides (Nod-factors) are examples of an increasing group of oligosaccharides which have morphogenetic effects on plant development (Ryan and Farmer, 1991). It is now important to determine how they function. The possibility that nodule initiation proceeds as a consequence of signal transduction into one or more secondary messengers is suggested by recent experiments with alfalfa root hairs, which show a rapid depolarization in membrane potential (within 5-20 min) following application of a purified Nod-factor (Ehrhardt et al., 1992). Moreover, in the pea-nodulating strain R. *leguminosarum* bv. *viciae* the NodO gene product (which apparently potentiates the plant response to lipo-oligosaccharide) appears to function as a calcium-dependent ion-channel protein when introduced into artificial membrane bilayers (Sutton et al., 1993).

When purified Nod-factors are applied to seedling roots, several apparently unconnected biological responses are set in motion. The earliest observed effect is on growing root hairs, where cell wall deformation and curling is visible within 3-4 h (Dart, 1975): this perhaps implies an effect of Nod-factors on the coordination of cell growth through modification of the structure or functioning of the plasma membrane, or of the underlying cytoskeleton. The application of purified Nod-factor also causes the reactivation of cortical cells, which develop centralized nuclei and transcellular cytoplasmic strands (Van Brussel et al., 1992). In the inner cortex of *Vicia* and pea, these premitotic structures predetermine the orientation of cell plate formation during cell division in the nodule primordium: in the outer cortex, similar cytoplasmic strands predetermine the pathway for transcellular infection thread growth without prior cell division (Bakhuizen, 1988; Rae et al., 1992). Another curious phenomenon that has been observed is that the application of a relatively high concentration of Nod-factor to seedling roots of *Vicia* results in the development of structures resembling root hairs from the subepidermal cortical cell layer (van Brussel et al., 1992), which again indicates that a perturbation of the plant cytoskeleton may be involved. The mitotic activity that is initiated in the inner cortex frequently develops into a nodule primordium. In the absence of rhizobia, this leads to the development of empty uninvaded nodule structures (Truchet et al., 1991).

Another plant response that is observed following the application of Nod-factors to *Vicia* roots is a change in the spectrum of flavonoids synthesized and secreted by root cells (Van Brussel et al., 1990; Recourt et al., 1992). This up-regulation of flavonoid biosynthesis could simply serve to raise the level of bacterial *nod* gene activity. However, certain flavonoids have been reported to function as auxin transport inhibitors (Faulkner and Rubery, 1992), and inhibitors of polar auxin transport have been shown to cause the induction of nodule formation (Hirsch *et al.*, 1989). Thus it is possible to imagine that the local application of Nod-factors might induce local flavonoid biosynthesis and thereby alter the local balance of auxin and other phytohormones, leading to the initiation of nodule development in response to a strain-specific signal from *Rhizobium* (Schmidt *et al.*, 1992). Such a model is, however, completely speculative at the present time.

Flavonoids require chalcone synthase (CHS) for their production, and CHS is strongly expressed in cells of nodule and lateral root meristems (Yang *et al.*, 1992). CHS is encoded by a multigene family in pea, and no mutants are known that completely lack CHS activity. Mutations at the *a* and *a2* loci of *Pisum* block CHS activity in flowers, but gene expression still occurs in roots (Harker *et al.*, 1990), perhaps reflecting the potential importance of this biosynthetic pathway for the *Rhizobium*-legume symbiosis. Flavonoid compounds, particularly isoflavonoids, also figure prominently in the Leguminosae as broad-spectrum antimicrobial compounds known as phytoalexins (Harborne, 1971). However, phytoalexins are not normally induced within nodules (Werner *et al.*, 1985; Schmidt *et al.*, 1992).

In the absence of rhizobia, empty nodules can sometimes originate 'spontaneously' from the root of legume plants. This phenomenon occurs at high frequency in selected plant genotypes of alfalfa (Truchet *et al.*, 1989a). Similar nodule-like structures can also be induced experimentally in pea and other legumes by altering the hormonal balance of the root system following the exogenous application of phytohormones or inhibitors of the polar transport of auxin (Hirsch *et al.*, 1989). Transcription of some of the so-called 'nodule-specific' genes (early nodulins) is induced within these structures (van de Wiel *et al.*, 1990a), and, like *Rhizobium*-filled nodules, 'empty' nodules arise by the proliferation of root cortical cells (Truchet *et al.*, 1989b). These observations indicate that the nodule is essentially a plant-derived organ (Caetano-Anolles and Gresshoff, 1991), although rhizobia seem to have found the molecular keys with which to unlock this developmental pathway (Fig. 9.6).

Tissue and cell invasion

Although the soluble bacterial Nod-factor is capable of inducing branching and deformation of the root hair, the physical presence of live rhizobia is required to induce the growing root hair tip to curl to its maximum extent. Direct physical contact between the cell surfaces of the symbionts thus appears to play a prominent role in the process of tissue and cell invasion by *Rhizobium*. Root attachment may involve a 14,000 M_r Ca^{2+}-binding bacterial cell-surface protein termed rhicadhesin (Smit *et al.*, 1989), which appears to

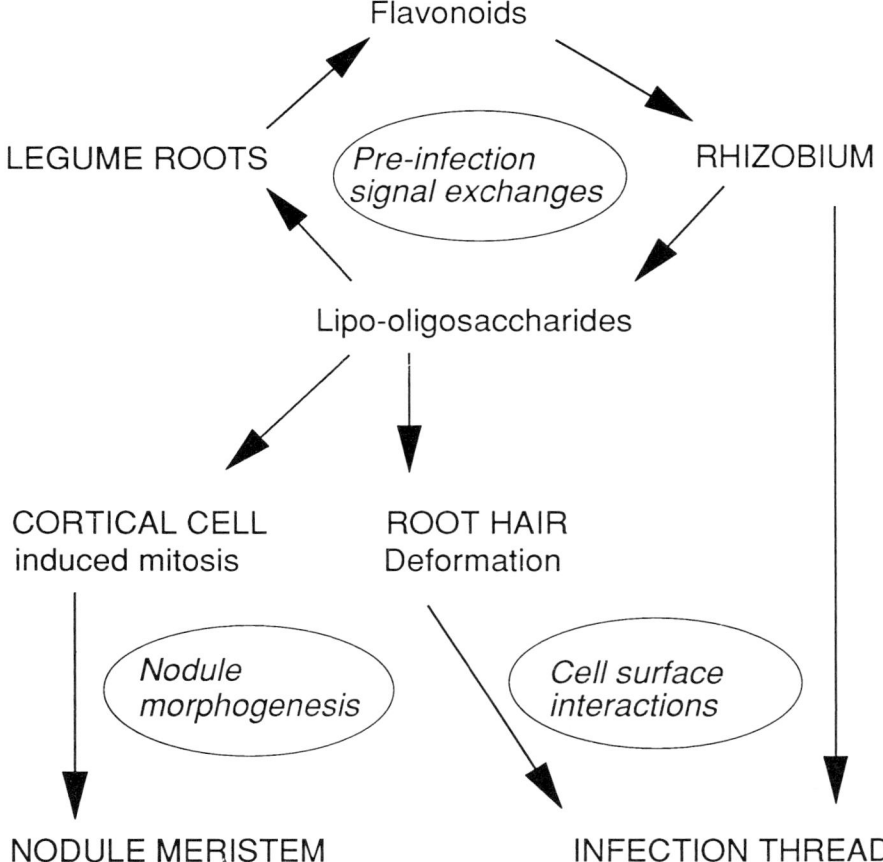

Fig. 9.6. Diagram illustrating signal exchanges in legume nodule initation and development.

be a common component of all strains of *Rhizobium* and *Agrobacterium*. However the relevant bacterial gene has not yet been cloned, nor has the plant receptor been identified that mediates rhicadhesin-dependent attachment to root hair cells.

Under appropriate conditions, the accumulation of rhizobia on the root hair tip is accelerated by pea lectin molecules (Kijne *et al.*, 1988). Native pea lectin is a 49,000 M_r protein with sugar-binding specificity for glucose and mannose residues. It has been shown by immunofluorescence microscopy to accumulate at the tips of growing root hairs (Diaz *et al.*, 1990). Genetic transformation of white clover roots with a functional pea lectin gene (Gatehouse *et al.*, 1987) was achieved using an *Agrobacterium rhizogenes* vector. The transgenic roots acquired the ability to be nodulated by

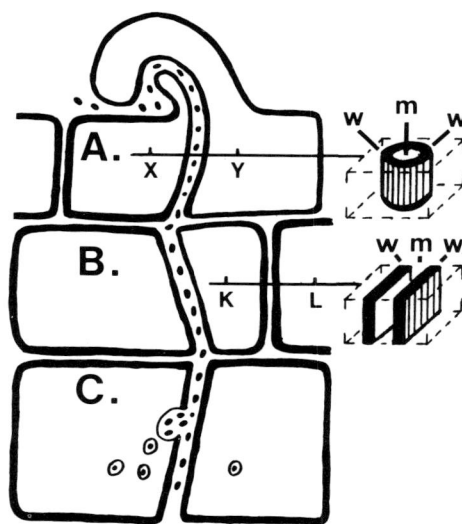

Fig. 9.7. Diagram illustrating the structure and topology of an infection thread at three successive stages of tissue and cell invasion by *Rhizobium* (adapted from Rae *et al.*, 1992). (A) Root hair curling and infection thread initiation; (B) Propagation of the infection thread as a transcellular tunnel through the root cortex and also in the subapical region of the developing nodule; (C) Release of rhizobia by endocytosis into the intracytoplasmic domain from unwalled infection droplets. The transects X-Y and K-L compare and contrast the topology of a transcellular infection thread (which is a cylinder) with a transcellular cell plate or cell wall (which is planar): w = wall; m = extracellular matrix material for the infection thread lumen and for the contents of intercellular spaces and m = middle lamella for the cell plate.

R. leguminosarum bv. *viciae*, although nodulation was limited and delayed in time (Diaz *et al.*, 1989). Because *R. leguminosarum* bv. *viciae* normally induces root hair curling on white clover but does not normally infect the cells, these results suggest that the presence of the pea lectin gene product somehow promotes cell invasion and infection-thread formation in this hairy root system of clover. However, the mechanism and specificity of this effect are still unclear (Kijne, 1992).

Within the pocket formed by the curled root hair (the so-called shepherd's crook), the plant cell wall is weakened and the bacteria intrude into the cytoplasmic space as an invagination of the plant cell membrane. This intrusion grows as a result of the localized deposition of new wall material. It develops into a tubular structure, termed the infection thread (Fig. 9.7), which is surrounded by a wall that is continuous with the plant cell wall and has a similar polysaccharide composition (Kijne, 1992; Rae *et al.*, 1992). The lumen of the infection thread is filled with an amorphous matrix material similar to that of intercellular spaces and containing at least one major glycoprotein of plant origin (VandenBosch *et al.*, 1989). Proline-rich proteins (PRP) and arabinogalactan proteins (AGP) have also been implicated in infection-thread growth, and tissue-specific genes (PsENOD12 and PsENOD5) that probably encode such products have been identified as transcripts in root cortical cells shortly after infection by *Rhizobium* (Franssen et al., 1992). In pea, infection threads are relatively broad, with a diameter of 3–4 μm, and micrographs often show several bacteria in a single cross-section. These bacteria are encapsulated (Latchford et al., 1991). Perhaps significantly, mutants of *Rhizobium leguminosarum* bv. *viciae* that are unable to make extracellular polysaccharide are unable to promote infection-thread development (Borthakur et al., 1986).

The mechanisms that allow the formation of a tubular intracellular infection structure are not known, but they probably result from a combination of the following: (i) secretion of plant matrix glycoproteins into the lumen; (ii) expansion and division of bacterial cells within the lumen, providing an inward force against the turgor pressure of the plant cell; (iii) forces of cytoplasmic streaming, coupled through the cytoskeleton and the plasma membrane. The infection thread normally grows down into the root hair cell, following closely the migration of the plant nucleus towards the base of the cell (Lloyd et al., 1987; Brewin, 1990). Having reached the base of the root hair, the thread somehow exits by fusion with the mother cell wall, thereby releasing rhizobia into the intercellular space. Normally a new infection thread is immediately initiated in the adjacent cell and thus the thread continues its growth into the underlying layers of the root cortex. However, because threads are interrupted at the middle lamella, the passage from one cell to the next must involve a new penetration of the adjacent plant cell wall. The fact that rhizobia do not produce detectable amounts of hydrolytic enzymes has made it difficult to sustain the hypothesis of penetration by the localized degradation of the cell wall (Dart, 1975). More probably, the orientation and local organization of wall growth and wall dissolution are organized by the cytoskeleton, and the growing infection thread follows the path laid down by polarized cytoplasmic bridges, which are observed in the cortical cells ahead of it (Van Brussel et al., 1992; Rae et al., 1992).

Several classes of bacterial surface polysaccharides seem to play important roles during the invasion of host tissues and cells. This has been investigated mainly by using mutants defective in the biosynthesis of a particular species of surface polysaccharide (Noel, 1991; Carlson et al., 1992; Leigh and Coplin, 1992). However, in no case has a clear function been ascribed, nor is it clear to what extent any of these functions might be strain- or species-specific. It seems probable that general properties such as surface charge or hydrophobicity may be the essential requirement provided by these surface polysaccharides. For example, the acidic extracellular polysaccharide of *Rhizobium* is known to form gels in the presence of Ca^{2+} ions (Morris et al., 1989), and this property could provide the physical force needed in the initial stages of root hair cell invasion by *Rhizobium* (Brewin, 1991). In addition, rhizobia within infection threads are encapsulated by acidic polysaccharide, which may contribute to the masking of invading bacteria in such a way as to prevent eliciting host defence responses.

Even in the 'correct' host legume, tissue and cell invasion by *Rhizobium* is frequently arrested. The premature abortion of a large percentage of infection threads has recently been suggested to depend on a plant defence response induced at early stages of bacterial invasion (Vasse *et al.*, 1993). These observations point to the fact that the distinction between pathogenic and symbiotic interaction is finely balanced, and depends on forms of plant–microbial signal exchange which have yet to be elucidated. It has also been observed that bacterial mutants with defective lipopolysaccharides frequently induce a host defence response within infection threads (Yang *et al.*, 1992), suggesting that complete LPS structures either directly or indirectly suppress the release of a potential 'elicitor' function (Kannenberg *et al.*, 1992).

Bacteria are released from the infection threads and colonize the plant cytoplasm surrounded by a plant-derived membrane (Fig. 9.7). In pea and clover, the infection threads give rise to droplet-like structures which contain matrix material but lack the wall that surrounds the infection thread (Newcomb, 1981), possibly as a result of the uncoupling of membrane biosynthesis from cell wall biosynthesis (Rae *et al.*, 1992). The bacteria contained in the droplet are individually released into the cytoplasm by a process resembling endocytosis which is apparently driven by very close contact between plant plasma membrane and the bacterial surface (Robertson *et al.*, 1985). The closeness between these two surfaces is not only suggested by morphological observations, but also from the fact that the plant matrix material contained in the infection droplet is apparently excluded from the space between the bacteria and the plant membrane during this process of endocytosis (Vanden-Bosch *et al.*, 1989; Rae *et al.*, 1992). Although a physical association between the plant membrane and the bacterial surface has been shown *in vitro* (Bradley *et al.*, 1986) and many of the surface components of the plant membrane glycocalyx have been characterized (Bradley *et al.*, 1988; Pennell *et al.*, 1989; Perotto *et al.*, 1991), the mechanism of endocytosis of rhizobia from the infection droplet remains obscure (Robertson *et al.*, 1978).

After their release into the plant cell, the bacteria continue to divide actively (Fig. 9.8). The term bacteroid is normally used to describe all forms of released bacteria, although it should be remembered that there are major differences between early released bacteroids and fully differentiated nitrogen-fixing bacteroids. During differentiation of the bacteroids, the *nod* genes are switched off (Sharma and Signer, 1990, Schlaman *et al.*, 1991, 1992b). It is evident that this control is not due to inhibition of *nodD* expression because this gene does appear to be expressed in nodules (Schlaman *et al.*, 1991, 1992b). There is evidence to suggest that there is a bacteroid-specific inhibition of *nod*-gene expression, possibly via a DNA-binding repressor (Schlamann *et al.*, 1991, 1992b) The reasons for the increase in size of bacteroids is not clear. It is possible that the bacterial genes *fts z*, which, in *E. coli*, are involved

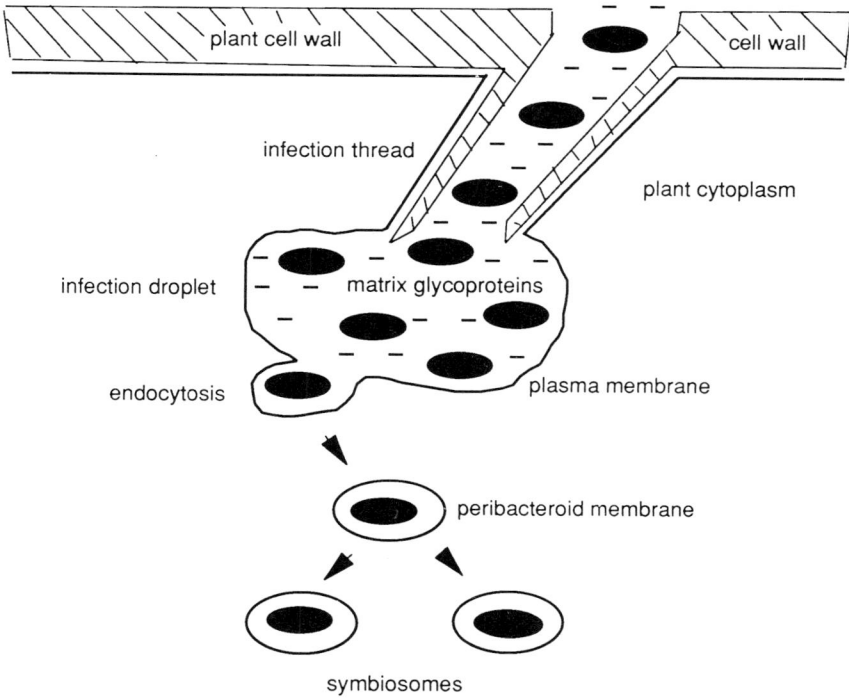

Fig. 9.8. Diagram illustrating successive stages in endocytosis of bacteria and the development of symbiosomes. Stage 1, cell penetration through an infection thread; stage 2, endocytosis from an unwalled infection droplet; stage 3, division of intracytoplasmic bacteria (bacteroids) with concomitant division of the enveloping peribacteroid membrane, which is of plant origin. Following release of bacteria into the host cytoplasm, there is progressive differentiation of the peribacteroid membrane and of enclosed bacteroids, leading to the development of nitrogen-fixing symbiosomes.

in septum formation during cell division, may, in *Rhizobium*, play a role in the inhibition of bacteroid cell division (Margolin *et al.*, 1991).

In pea, the mature peribacteroid membrane normally contains a single pleiomorphic (Y-shaped) bacteroid (Robertson and Lyttleton, 1984), except in the case of a symbiotically defective (non-nitrogen-fixing) pea mutant (Borisov *et al.*, 1993), where 8–12 bacteroids are frequently packaged together. The bacteroid and the plant-derived peribacteroid membrane grow, divide and differentiate together as the symbiosome compartment. Elements in the process of differentiation of the peribacteroid membrane include the loss of some components specific for the plasma membrane (e.g. cellulose synthase) and the insertion of proteins involved in symbiotic interactions and exchange of metabolites with the endosymbiotic bacterium. Specific information for the correct targeting of plant components directed to this subcellular compartment

Table 9.3. (a) Genetic loci (b) mutants in *Pisum* involved in nodulation and nitrogen fixation.

Locus/mutant	Linkage group	Phenotype	Name and number of pea variety	Registered stock of pea genotype	Reference
(a) Genetic loci					
brz	4	Few nodules	Sparkle E107	JI 2616	Kneen et al., 1990b
nod-1; nod-2	(1)	High nodule number	Parvus		Gelin and Blixt, 1964
nod-3		Supernodulating	Rondo-nod3		Jacobsen and Feenstra, 1984
Sym-1		Temperature-sensitive > 26°C	Iran PrI H721	WBH 2419	Lie, 1971
sym-2	1	Strain-specific nodulation	Afghanistan	JI 1357	Holl, 1975
sym-3		fix⁻	Afghanistan	JI 1357	Holl, 1975
Sym-4		Strain-specific nodulation	*P. elatius*	JI 261	Lie, 1984
sym-5	1	Non-nodulating, nodulated at 12°C	Sparkle E2	JI 2498	Kneen and LaRue, 1984a
sym-6		Strain-specific nodulation	Afghanistan	JI 1357	Lie et al., 1987
sym-7	3	nod⁻	Sparkle E69		Kneen and LaRue, 1986
sym-8	6	nod⁻	Sparkle R25		Kneen and LaRue, 1986
sym-9		nod⁻	Sparkle R72		Kneen and LaRue, 1986
sym-10	(1)	nod⁻	Sparkle N15		Weeden et al., 1990
sym-11	7	nod⁻	Sparkle N24		Weeden et al., 1990
sym-12		Few nodules	Rondo K5		Jacobsen, 1984
sym-13	7	fix⁻	Sparkle E135F	JI 2497	Kneen et al., 1990a
sym-14	2	nod⁻	Sparkle E135N		Kneen et al., 1990a
sym-15	7	Few nodules	Sparkle E151		Weeden et al., 1990
sym-16	5	Few nodules	Sparkle R50		Weeden et al., 1990
sym-17	6	Few nodules	Sparkle R82	PI 562570	Weeden et al., 1990
sym-18	1	Strain-specific nodulation	Sparkle E54		Weeden et al., 1990

sym-19	1	nod⁻	Rondo K24		Jacobsen, 1984
sym-20		nod⁻	Sparkle R80		LaRue and Weeden, 1992
sym-21		Few nodules	Sparkle E132		LaRue and Weeden, 1992
Sym-22	(2)	Low nodule number	P. humile	JI 1794	LaRue and Weeden, 1992
sym-23*		fix⁻	Frisson P59		Duc and Messager, 1989
sym-24*		fix⁻	Frisson P60		Duc and Messager, 1989
sym-25*		fix⁻	Frisson P61		Duc and Messager, 1989
sym-26*		fix⁻	Frisson P63		Duc and Messager, 1989
sym-27*		fix⁻	Frisson P12		Duc and Messager, 1989
sym-28*		Hypernodulating	Frisson 190F		Duc and Messager, 1989

(b) **Mutants**

P1	nod⁻ myc⁻	Frisson P1 (P2, P3, F4-1)**		Duc et al., 1989
P4	nod⁻ myc⁻	Frisson P4 (F4-141)		Duc et al., 1989
P6	nod⁻ myc⁻	Frisson P6		Duc et al., 1989
F4-122	nod⁻ myc⁺	Frisson F4-122		Duc et al., 1989
F4-58	nod⁻ myc⁺	Frisson F4-58 (P5, P7)		Duc et al., 1989
FN1	fix⁻	Rondo FN1	JI 2314	Postma et al., 1990
Sprint-2fix⁻	fix⁻	Sprint-2-Fix⁻	JI 2613	Borisov et al., 1993
k301	Supernodulating	Ramonskii77 k301		Sidorova and Uzhintseva, 1992
k287	fix⁻	Ramonskii77 k287		Sidorova and Uzhintseva, 1992
k1005m	nod⁻	Ramonskii77 k1005m		Sidorova and Uzhintseva, 1992
KN7	nod⁻	nod-3, KN7		Postma et al., 1988b

*Provisional symbols. Linkage group data as in Weeden et al. (1992). Linkage groups in parentheses have not been ratified.
**Additional mutants in the same complementation group.
Registered stock abbreviations: Pl = USDA Pullman, JI = John Innes Institute, WBH = Nordic Gene Bank (now dealt with at JI). nod⁻, nod-nodulating; myc⁻, does not form mycorrhizal associations; myc⁺, found to form mycorrhizal associations when checked; fix⁻, forms nodules that do not fix nitrogen.

would therefore be expected. Recently a glycosylphosphatidylinositide antigen has been found to be developmentally regulated during the progressive differentiation of the peribacteroid membrane in pea nodules (Brewin et al., 1993).

Host Genes Involved in Symbiosis

The symbiotic interaction between legumes and rhizobia provides a convenient model system for the study of plant morphogenesis and plant–microbe interaction. Recent advances in studying the pea symbiosis have been based on a genetic approach, which has allowed the identification of many plant and bacterial determinants necessary for the establishment of the symbiotic interaction. The corresponding plant and bacterial mutants, which are impaired at different stages during nodule initiation, now provide an important resource for future analysis of plant–microbe interaction and signal exchange.

Genetics

Many symbiotic plant mutants have been obtained through mutagenesis programmes using both chemical and X-ray mutagens. In pea, these programmes have proved very successful in the generation of symbiotic mutants at different points in the process: over 30 genes have been identified that are involved in nodulation and fixation. A compilation of genes involved in nodulation and fixation identified through allelism tests is presented (Table 9.3a), along with mutants of interesting phenotype where the full allelism tests have not been completed (Table 9.3b). More mutants are still in the process of being characterized and include a number of new genes and alleles to several of the registered genes (LaRue and Weeden, 1992; G. Duc, personal communication). Five main phenotypic groups have been identified (Table 9.4).

The non-nodulating mutants are reported as showing no visible response when exposed to rhizobia, except for the genotype carrying the *sym*-9 locus, where root swelling has been observed (Markwei and LaRue, 1992). An important finding with a number of non-nodulating mutants has been the pleiotropic expression of resistance to vesicular-arbuscular (VA) mycorrhizal symbioses (Duc *et al.*, 1989). The fact that given mutations block the establishment of symbioses with such diverse microorganisms suggests similarities in the early steps in the signal transduction systems. The study of the two complementation groups of Nod$^-$, Myc$^-$ mutants defined by Duc and co-workers (Duc *et al.*, 1989; Gianinazzi-Pearson *et al.*, 1991) should enable further investigation of the common aspects of these symbiotic associations.

Strain specificity was first recorded in *Pisum sativum* cv. Afghanistan (JI 1357), which was found to be resistant to strains from northern Europe and

Table 9.4. Phenotypic classes and observations made on symbiotically defective pea mutants. Developmental stages: RHC = root hair curling, INF = infection thread, NOD = nodulation, FIX = fixation.

Phenotype class	Locus/mutant	RHC	INF	NOD	FIX	Phenotype comments	Reference
Non-nodulating	sym-8	–	–	–	–	no root deformation	Markwei and LaRue, 1992
	sym-9	–	–	–	–	root swelling	Markwei and LaRue, 1992
	k1005m	–	–	–	–	resistant to strain 250a	Sidorova and Uzhintseva, 1992
	sym-19	+	–	–	–	occasionally single nodules	Postma et al., 1988a
	P1, P4, P6	–	–	–	–	resistant to mycorrhiza	Duc et al., 1989
							Postma et al., 1988b
Strain specificity	sym-2	+	+/–	+/–	+/–	nodulation only with some strains of R. leguminosarum	LeGal and Hobbs, 1989
	Sym-4	+	+/–	+/–	+/–	as above	LaRue and Weeden, 1992
	sym-4	+	+/–	+/–	+/–	as above	LaRue and Weeden, 1992
	sym-18	+	+/–	+/–	+/–	as above	Weeden et al., 1990
Low frequency of nodule formation	sym-5	+	+	+	+	increased by a period at 12°C	Fearn and LaRue, 1991
	sym-12	+	+	+	+	delayed nodulation	Postma et al., 1988a
	sym-17	+	+	+	+	thick short roots	Lee and LaRue, 1992
Non-fixing	sym-3	+	+	+	–	fixation only 10–20% of normal	LeGal and Hobbs, 1989
	sym-13	+	+	+	–	nodules remain white, small, senesce early	Kneen et al., 1990a
	sym-15	+	+	+	–	nodules remain white, short lateral roots	Kneen and LaRue, 1986
	sym-16	+	+	+	–	nodules remain white, short lateral roots	Kneen and LaRue, 1986
	k287	+	+	+	–	white nodules	Sidorova and Uzhintseva, 1992
	Sprint-2fix⁻	+	+	+	–	symbiosomes contain multiple bacteroids	Borisov et al., 1993
High number of nodules	nod-3	+	+	++	+	controlled by root, nitrate-resistant	Postma et al., 1990
	sym-28	+	+	++	+	controlled by shoot, nitrate-resistant	Duc and Messager, 1989
	FN1	+	+	++	–	spherical nodules	Postma et al., 1990
	k301	+	+	++	+	nitrate-resistant	Sidorova and Uzhintseva, 1992

North America but was nodulated by particular strains from the Middle East, notably by the TOM strain (Lie, 1971). The *sym-2* gene involved has been found to be widespread in a range of ecotypes and land races of *Pisum* from the eastern Mediterranean, Turkey and Iran (Young and Matthews, 1982; Kneen and LaRue, 1984b). Studies on other wild accessions of peas and a range of strains from this secondary centre of diversity have revealed further variation in the efficiency of nodulation with specific strains (Lie, 1978). This has led to the concept of co-evolution of the legume–*Rhizobium* association (Lie *et al.*, 1987).

Hypernodulating mutants have been isolated, and two genes identified (*nod-3* and *sym-28*). The number of nodules produced is in excess of ten times that of the parental line (Postma *et al.*, 1990) but individual nodules tend to be very small. The whole nodule mass of the *nod-3* mutant of Rondo was only twice that of the parent line (Rosendahl *et al.*, 1989). Grafting experiments have shown that the hypernodulating character of *nod-3* is dependent only on the root genotype (Postma *et al.*, 1990). This, interestingly, is the reverse for mutant 190F, where nodulation phenotype is dependent on the genotype of the shoot (Duc and Messager, 1989).

One interesting feature is the discovery of a cluster of four *sym* genes (*sym-2*, *sym-5*, *sym-18*, *sym-19*) located on chromosome 1 (Weeden *et al.*, 1990). The region of *sym-5* has been referred to as a mutational hot spot on the basis of the number of mutants obtained (Kneen and LaRue, 1988). Recent linkage data also show *nod-1* to be associated with this cluster (T.A. LaRue, personal communication). This clustering within a small 10 cM region is unusual for functionally related genes in higher plants.

Some of the mutants are temperature sensitive for nodulation. *Sym-1*, reported as occurring naturally in a primitive line of pea from Iran, was the first temperature-sensitive nodulation mutant to be described in peas; nodules were formed when roots were kept at 26°C but failed to form below 20°C (Lie, 1971). Fearn and LaRue (1991) reported another temperature-sensitive mutant (*sym-5*), which failed to nodulate when grown at a constant 20°C but nodulated successfully when given a short period (6 h) at 12°C early in the infection process. The effects of temperature were to some extent mediated by the moisture content of the soil. Under dry conditions no nodules were observed. The authors postulated that the observed effects may be due to an increased sensitivity to ethylene, because inhibitors of ethylene production, such as Ag^+ ions or aminoethoxyvinylglycine (AVG), increased the nodulation frequency of *sym-5* mutants grown at higher temperatures (Fearn and LaRue, 1991; Guinel and LaRue, 1992). As more of the nodulation mutants are studied, so further temperature specificity is being observed, especially in the low-nodule-forming mutants (T.A. LaRue, personal communication).

Tissue-specific gene expression

As an alternative point of entry into the molecular genetics of the host plant, nodulin genes have been isolated by differential screening of a pea nodule cDNA library (Table 9.5). The nodulins were arbitrarily divided into 'early' and 'late' nodulins, reflecting the temporal sequence of induction during nodule development (Franssen et al., 1992; Verma, 1992; Verma et al., 1992). As originally defined, nodulins were 'nodule-specific', but several nodulins have subsequently been shown to be expressed in tissues other than the nodule (Nap and Bisseling, 1990; Scheres et al., 1990a). Hence, the term 'nodulin' now refers to any gene product that is up-regulated in response to Rhizobium or Rhizobium-derived signals (Franssen et al., 1992).

Early nodulins are more directly involved in the initiation of the infection process. Presumably they function in nodule morphogenesis, i.e. the organization of plant cell growth and development (Gloudemans et al., 1989), or they participate in the host plant response to microbial invasion. The nodulin gene PsENOD40 is expressed in meristematic cells induced in the nodule primordium (Bisseling et al., 1993). Two gene products identified on two-dimensional protein gels, RH-42 and RH-44, are up-regulated in pea root hairs following interaction with R. leguminosarum bv. viciae (Gloudemans et al., 1989). Nodulin PsENOD12 is also expressed in root hair cells (Pichon et al., 1992). The PsENOD12 protein sequence contains a putative signal peptide at the N-terminus and the rest of the protein is composed of two repeating penta-peptides, each containing two prolines: therefore the nodulin is likely to be a proline- or hydroxyproline-rich cell-wall protein. In situ hybridization studies show that this gene is expressed in cells already containing, and in cells preparing for, the growing infection thread (Scheres et al., 1990a). The PsENOD12 protein may be part of the infection thread and/or the modified cell wall of the root cortical cell.

The PsENOD5 protein is also a proline-rich early nodulin, but it does not show a repetitive structure like that of PsENOD12. PsENOD5 is rich in Pro, Ala, Gly and Ser residues and may belong to a class of arabinogalactan proteins (Scheres et al., 1990b). Unlike PsENOD12, the PsENOD5 transcript is only expressed in root cortical cells containing the infection tip and therefore its function may be directly associated with infection thread growth. Also in contrast to PsENOD12, the PsENOD5 transcript is detected in the infected cells of the early symbiotic zone in maturing pea nodules, suggesting that this nodulin continues to function during the phase of intracellular bacterial proliferation when infection thread growth has ceased.

The infected cells of the early symbiotic zone also show transient expression of genes encoding PsENOD3 and PsENOD14, two small polypeptides with 55% sequence homology, each having a putative N-terminal transit peptide and M_r 6000. Each protein contains four cysteine residues with a conserved spatial distribution resembling that of metal-binding proteins (Scheres et al.,

Table 9.5. cDNA clones corresponding to nodule-enhanced gene products in *P. sativum*.

Designation	Description	Reference
Early nodulins		
PsENOD2	Cell-wall protein; proline- or hydroxyproline-rich; localized in nodule parenchyma	van de Wiel et al., 1990a
PsENOD12	Cell-wall protein; proline- or hydroxyproline-rich; localized in infected and in uninfected cells in the invasion zone	Scheres et al., 1990a
PsENOD5	Possibly an arabinogalactan protein; localized in infected cells	Scheres et al., 1990b
PsENOD40	Localized in meristematic cells; down-regulated after cell invasion by *Rhizobium*	Bisseling et al., 1993
PsENOD3, PsENOD14	Small cysteine-rich proteins; transient expression in infected cells	Scheres et al., 1990b
Late nodulins		
pPsLb102	Leghaemoglobin	Govers et al., 1987
PsNOD6	Similar to ENOD3, but expressed at a later stage in nodule development	Govers et al., 1987; I.V. Kardailsky, personal communication
GS_{N1}, GS_{N2}	Glutamine synthetases; apparently not nodule-specific in peas	Tingey et al., 1987; Walker and Coruzzi, 1989
Asparagine synthetase		Tsai and Coruzzi, 1990
Aspartate aminotransferase		Tsai and Coruzzi, 1990

1990b). Later in nodule development, these two nodulins are apparently replaced by a third protein from the same gene family, PsNOD6 (I.V. Kardailsky, personal communication). At the same time, leghaemoglobin and other late nodulin genes are also expressed.

Late nodulin genes (Table 9.5) are expressed concomitantly with the release of bacteria from the infection thread, but prior to the induction of nitrogenase and the commencement of nitrogen fixation. These gene products may contribute to nodule physiology, i.e. carbon, oxygen and nitrogen metabolism (Verma, 1992; Verma *et al.*, 1992). Nodulins often represent minor structural variants of proteins that are common in other parts of the plant (Nap and Bisseling, 1990). For example, Nodulin 26, a component of the peribacteroid membrane in soyabean (Fortin *et al.*, 1987) shows homology to JM7a, a turgor-regulated protein present in pea roots (Guerrero *et al.*, 1990).

The uninfected cells of the inner cortex also express a nodule-specific transcript, PsENOD2 (van de Wiel *et al.*, 1990b), which encodes a proline-or hydroxyproline-rich protein with two repeating pentapeptide motifs, Pro–Pro–Glu–Tyr–Gln and Pro–Pro–His–Glu–Lys (Franssen *et al.*, 1992). These motifs are characteristic of a general class of proline-rich plant cell wall glycoproteins. The function of these cell-wall glycoproteins is unknown, but perhaps the presence of PsENOD2 (glyco)protein limits cell wall expansion, resulting in a layer of relatively small, tightly packed cells with few intercellular air spaces.

Cytological analysis reveals that nodulins are expressed at many different times and in many different tissues during nodule development (Scheres *et al.*, 1990b). It remains to be seen whether all or most nodulins share a common component of gene promoter structure that is associated with nodule-enhanced expression. An analysis of promoter structure has been undertaken for leghaemoglobin (Szabados *et al.*, 1990) and comparisons made with other nodulin genes (Stougaard *et al.*, 1990). These studies have been conducted in transgenic *Lotus* and tobacco plants using modified nodulin promoters fused to a reporter gene such as chloramphenicol acetyltransferase (de Bruijn *et al.*, 1990). The results suggest a hierarchy of control systems governing tissue specificity. Transcriptional regulation occurs through organ-specific elements (OSE) and negative elements (NE) located upstream of the TATAA box and is further controlled by strong (SPE) or weak positive elements (WPE), which act as other binding sites for *trans*-acting factors (Jensen, 1990).

Biotechnology

Artificial inocula have been developed to ensure the presence of sufficient levels of rhizobia for legume crops grown in soils lacking appropriate strains. Such commercial inocula have been of particular benefit where new legume crops have been introduced, for example, into Australia, USA and Europe. For pea, the almost ubiquitous availability of suitable naturally occurring strains of rhizobia has resulted in this particular legume crop having one of the lowest uses of artificial inoculants. Only on land where peas have not been grown for over ten years is there any measurable increase in yield in response to inoculation (Bingefors *et al.*, 1986). Rhizobia are considered relatively fast-growing bacteria, but under adverse soil conditions, especially low pH, growth is slow and the bacteria show poor persistence. Symbiotic nitrogen fixation can itself contribute to the further acidification of the rhizosphere through the excretion of H^+ ions (Sprent *et al.*, 1988). Acidification can also lead to an increase in the solubilization of potentially toxic ions such as aluminium and manganese.

As pea plants do not flourish in low pH conditions, it is more common to try to correct the pH of the land rather than to try to raise the levels of rhizobia. The market for inoculants has therefore been developed principally for other legumes such as soyabean, *Phaseolus*, peanut and lucerne. Preparations of commercial inoculants are currently made available in forms that are easy for the farmer to use. They have a long shelf-life of up to 2 years, and deliver approximately 10^9 rhizobia per gram of carrier. The main carrier today is gamma-irradiated peat, applied as a slurry dressing immediately prior to sowing. The use of organic polymers as carriers is being studied, but none has yet been found to match the properties of peat (Clarke *et al.*, 1988).

Under normal agronomic conditions, *Rhizobium* populations are less affected by environmental stresses than is the host plant. Soil moisture levels which are still too low to sustain plant growth generally contain sufficient moist microsites to ensure rhizobial survival (Sprent *et al.*, 1988). There are, however, large areas of land marginal to cultivation which suffer from conditions such as low fertility, accumulating acidity and salinity. The improvement of nitrogen fixation in such regions could make a significant contribution to the agricultural economy, as well as contributing to the improvement of soil structure and fertility. The ability to select *Rhizobium* strains tolerant to environmental stresses, such as salinity and acidity, has been demonstrated in a number of cases in other legumes (Phillips, 1991). This indicates that the natural genetic variation in *Rhizobium* is sufficiently large to be used as the basis for selecting strains suitable for particular types of soil.

One of the most difficult problems faced when trying to introduce a selected strain of *Rhizobium* into the rhizosphere is that the newly introduced strain is unable to compete with the indigenous soil bacteria (Triplett and Sadowsky, 1992). A possible procedure for overcoming this problem has been outlined by Fobert *et al.* (1991). The strategy is to make use of the specific interaction between the pea *sym-2* gene (which encodes resistance to nodulation) and the bacterial *nodX* gene (which overcomes this resistance). Introduction of the *nodX* gene into an isolate of *R. leguminosarum* that was prevalent in a particular region was found to overcome earlier problems of poor competitiveness of foreign strains carrying the *nodX* gene. Greenhouse experiments were successful in establishing nodulation with the *nodX*-carrying strain, even where the levels of the original isolate (without *nodX*) were high. However, no field trials have yet been reported using this method.

The large number of nodulation and fixation mutants that have been characterized in the pea will become a powerful resource for further study of the physiology and biochemistry of these processes. Moreover, the development of isogenic series will greatly assist the study of individual genes and gene combinations. As well as providing the opportunity for further scientific investigation, the use of these alleles by breeders should also be encouraged (Phillips, 1991). For example, it may be possible to select more efficient genetic combinations by the introduction of characters into different backgrounds where the penetrance of genes may vary. More specifically, work on the mechanisms of temperature specificity may help to overcome the present inability of nodules to fix nitrogen early in crop development, which remains one of the main areas where selection of both plant and symbiont would be profitable.

In many ways the kit of parts is now available to modify this system successfully and efficiently. Undoubtedly more studies will be required on nodule development and physiology before we have a complete understanding of the symbiosis. What is also needed are more multidisciplinary programmes whereby expertise from a number of fields are brought together in specific projects.

References

Allen, O.N. and Allen, E.K. (1981) *The Leguminosae*. Macmillan Publishing Company, London.
Appleby, C.A. (1984) Leghemoglobin and *Rhizobium* respiration. *Annual Review of Plant Physiology* 35, 443–478.
Arwas, R., McKay, I.A., Rowney, F.R.P., Dilworth, M.J. and Glenn, A.R. (1985) Properties of organic acid utilization mutants of *Rhizobium leguminosarum* strain 300. *Journal of General Microbiology* 131, 2059–2066.
Atkinson, E.M. and Long, S.R. (1992) Homology of *Rhizobium meliloti* NodC to polysaccharide polymerizing enzymes. *Molecular Plant–Microbe Interactions* 5, 439–442.
Baev, N., Endre, G., Petrovics, G., Banfalvi, Z. and Kondorosi, A. (1991) Six nodulation genes of *nod* box locus 4 in *Rhizobium meliloti* are involved in nodulation signal production: *nodM* codes for D-glucosamine synthetase. *Molecular and General Genetics* 228, 113–124.
Bakhuizen, R. (1988) The plant cytoskeleton in the *Rhizobium*–legume symbiosis. PhD thesis, Leiden University, The Netherlands.
Barquera, B., Garcia-Horsman, A. and Escamilla, J. (1991) Cytochrome d expression and regulation pattern in free-living *R. phaseoli*. *Archives of Microbiology* 155, 114–119.
Batut, J., Daveran-Mingot, M.L., David, M., Jacobs, J., Garnerone, A.M. and Kahn, D. (1989) *fixK*, a gene homologous with *fnr* and *crp* from *Escherichia coli*, regulates nitrogen fixation genes both positively and negatively in *Rhizobium meliloti*. *EMBO Journal* 8, 1279–1286.
Bauer, W.D. (1981) Infection of legumes by rhizobia. *Annual Review of Plant Physiology* 32, 407–449.
Bergersen, F.J. (1982) *Root Nodules of Legumes: Structure and Functions*. Wiley & Sons Ltd., Chichester.
Beringer, J.E., Beynon, J.L., Buchanan-Wollaston, A.V. and Johnston, A.W.B. (1978) Transfer of the drug resistance transposon Tn5 to Rhizobium. *Nature* 276, 633–634.
Beringer, J.E., Brewin, N.J. and Johnston, A.W.B. (1980) The genetic analysis of *Rhizobium* in relation to symbiotic nitrogen fixation. *Heredity* 45, 161–186.

Bhuvaneswari, T.V. and Solheim, B. (1985) Root hair deformation in the white clover/ *Rhizobium trifolii* symbiosis. *Physiologia Plantarum* 63, 25-34.

Bibb, M.J., Biro, S., Motamedi, H., Collins, J.F. and Hutchinson, C.R. (1989) Analysis of the nucleotide sequence of the *Streptomyces glaucescens tcm1* genes provides information about the enzymology of polyketide antibiotic biosynthesis. *EMBO Journal* 8, 2727-2736.

Bingefors, S., Johansson, N. and Rydberg, I. (1986) Improved field pea cultivars for modern farming. In: Olsson, G. (ed.), *Research and Results in Plant Breeding*. Svalof, Stockholm, pp. 185-195.

Bisseling, T., Franssen, H., Heidstra, R., Horvath, B., Katinakis, P., Moerman, M., Spaink, H., van Brussel, T. and Vijn, I. (1993) *Rhizobium* Nod metabolites and early nodulin gene expression. In: Nester, E.W. and Verma, D.P.S. (eds), *Advances in Molecular Genetics of Plant-Microbe Interactions*. Kluwer Academic Publishers, Dordrecht, pp. 365-368.

Bond, L. (1948) Origin and developmental morphology of root nodules of *Pisum sativum*. *Botanical Gazette* 109, 411-434.

Borisov, A.Y., Morzina, E.V., Kulikova, O.A., Tchetkova, S.A., Lebsky, V.K. and Tikhonovich, I.A. (1993) New symbiotic mutants of pea (*Pisum sativum* L) affecting either nodule initiation or symbiosome development. *Symbiosis* (in press).

Borthakur, D., Barber, C.E., Lamb, J.W., Daniels, M.J., Downie, J.A. and Johnston, A.W.B. (1986) A mutation that blocks exopolysaccharide synthesis prevents nodulation of peas by *Rhizobium leguminosarum* but not of beans by *Rhizobium phaseoli* and is corrected by cloned DNA from *Rhizobium* or the phytopathogen *Xanthomonas*. *Molecular and General Genetics* 203, 320-323.

Bott, M., Bollinger, M. and Hennecke, H. (1990) Genetic analysis of the cytochrome c-aa$_3$ branch of the *B. japonicum* respiratory chain. *Molecular Microbiology* 4, 2147-2157.

Bradley, D.J., Butcher, G.W., Galfre, G., Wood, E.A. and Brewin, N.J. (1986) Physical association between the peribacteroid membrane and lipopolysaccharide from the bacteroid outer membrane in *Rhizobium*-infected pea root nodule cells. *Journal of Cell Science* 85, 47-61.

Bradley, D.J., Wood, E.A., Larkins, A.P., Galfre, G., Butcher, G.W. and Brewin, N.J. (1988) Isolation of monoclonal antibodies reacting with peribacteroid membranes and other components of pea root nodules containing *Rhizobium leguminosarum*. *Planta* 173, 149-160.

Brears, T., Walker, E.L. and Coruzzi, G.M. (1991) A promoter sequence involved in cell-specific expression of the pea glutamine synthetase *GS3A* gene in organs of transgenic tobacco and alfalfa. *Plant Journal* 1, 235-244.

Brewin, N.J. (1990) The role of the plant plasma membrane in symbiosis. In: Larsson, C. and Moller, I.M. (eds), *The Plant Plasma Membrane: Structure, Function and Molecular Biology*. Springer-Verlag, Berlin, pp. 351-375.

Brewin, N.J. (1991) Development of the legume root nodule. *Annual Review of Cell Biology* 7, 191-226.

Brewin, N.J., Beringer, J.E., Buchanan-Wollaston, A.V., Johnston, A.W.B. and Hirsch, P.R. (1980) Transfer of symbiotic genes with bacteriocinogenic plasmids in *Rhizobium leguminosarum*. *Journal of General Microbiology* 116, 261-270.

Brewin, N.J., Perotto, S., Kannenberg, E.L., Rae, A.L., Rathbun, E.A., Lucas, M.M., Kardailsky, I., Donovan, N. and Drobak, B.K. (1993) Mechanism of cell and tissue invasion by *Rhizobium leguminosarum*: the role of cell surface interactions. In: Nester, E.W. and Verma, D.P.S. (eds), *Advances in Molecular Genetics of Plant-Microbe Interactions*. Kluwer Academic Publishers, Dordrecht, pp. 369-380.

Bulawa, C.E. and Wasco, W. (1991) Chitin and nodulation. *Nature* 353, 710.

Burn, J.E., Rossen, L. and Johnston, A.W.B. (1987) Four classes of mutations in the *nodD* gene of *Rhizobium leguminosarum* biovar *viciae* that affect its ability to autoregulate and/or activate other *nod* genes in the presence of flavonoid inducers. *Genes and Development* 1, 456-464.

Burn, J.E., Hamilton, W.D.O., Wootton, J.C. and Johnston, A.W.B. (1989) Single and multiple mutations affecting properties of the regulatory gene nodD of Rhizobium. *Molecular Microbiology* 3, 1567-1577.

Caetano-Anolles, G. and Gresshoff, P.M. (1991) Plant genetic control of nodulation. *Annual Reviews of Microbiology* 45, 345-382.

Canter-Cremers, H.C.J., Spaink, H.P., Wijfjes, H.M., Pees, E., Wijffelman, C.A., Okker, R.J.H. and Lugtenberg, B.J.J. (1989) Additional nodulation genes on the Sym plasmid of *Rhizobium leguminosarum* biovar *viciae*. *Plant Molecular Biology* 13, 163-174.

Carlson, R.W., Bhat, U.R. and Reuhs, B. (1992) *Rhizobium* lipopolysaccharides: their structure and evidence for their importance in the nitrogen-fixing symbiotic infection of their host legumes. In: Gresshoff, P.M. (ed.), *Plant Biotechnology and Development*. CRC Press, Boca Raton, pp. 33-44.

Chen, F.-L. and Cullimore, J.V. (1988) Two isoenzymes of NADH-dependent glutamate synthase in root nodules in *Phaseolus vulgaris* L. *Plant Physiology* 88, 1411-1417.

Chen, F.-L. and Cullimore, J.V. (1989) Location of two isoenzymes of NADH-dependent glutamate synthase in root nodules of *Phaseolus vulgaris* L. *Planta* 179, 441-447.

Clarke, K.W., Brockwell, J. and Thompson, J.A. (1988) Role of inoculants in improving nitrogen fixation in legumes. In: Summerfield, R.J. (ed.), *World Crops: Cool Season Food Legumes*. Kluwer, Dordrecht, pp. 731-743.

Cock, J.M., Mould, R.M., Bennett, M.J. and Cullimore, J.V. (1990) Expression of glutamine synthase genes in roots and nodules of *Phaseolus vulgaris* following changes in the ammonium supply and infection with various *Rhizobium* mutants. *Plant Molecular Biology* 14, 549-560.

Coruzzi, G.M. (1991) Molecular approaches to the study of amino acid biosynthesis in plants. *Plant Science* 74, 145-155.

Cubo, T., Economou, A., Murphy, G., Johnston, A.W.B. and Downie, J.A. (1992) Molecular characterization and regulation of the rhizosphere-expressed genes *rhiABCR* that can influence nodulation by *Rhizobium leguminosarum* biovar *viciae*. *Journal of Bacteriology* 174, 4026-4035.

Cullimore, J.V. and Bennet, M.J. (1992) Nitrogen assimiliation in the legume root nodule: current status of the molecular biology of the plant enzymes. *Canadian Journal of Microbiology* 38, 461-466.

Cunningham, S.D., Kapulnik, Y., Brewin, N.J. and Phillips, D.A. (1985) Uptake hydrogenase activity determined by plasmid pRL6JI in *Rhizobium leguminosarum* does not increase symbiotic nitrogen fixation. *Applied and Environmental Microbiology* 50, 791-794.

Dart, P.J. (1975) Legume root nodule initiation and development. In: Torrey, J.G. and

Clarkson, D.T. (eds), *The Development and Function of Roots*. Academic Press, London, New York, pp. 467–506.

David, M., Daveran, M.-L., Batut, J., Dedieu, A., Domergue, O., Ghai, J., Hertig, C., Boistard, P. and Kahn, D. (1988) Cascade regulation of *nif* gene expression in *Rhizobium meliloti*. *Cell* 54, 671–683.

Davis, E.O., Evans, I.J. and Johnston, A.W.B. (1988) Identification of nodX, a gene that allows *Rhizobium leguminosarum* biovar *viciae* strain TOM to nodulate Afghanistan peas. *Molecular and General Genetics* 212, 531–535.

Dean, D.R. and Jacobsen, M.R. (1992) Biochemical genetics of nitrogenase. In: Stacey, G., Burris, R.H. and Evans, H.J. (eds), *Biological Nitrogen Fixation*. Chapman and Hall, New York, pp. 763–834.

Debelle, F., Rosenberg, C. and Denarie, J. (1992) The *Rhizobium*, *Bradyrhizobium*, and *Azorhizobium* NodC proteins are homologous to yeast chitin synthases. *Molecular Plant–Microbe Interactions* 5, 443–446.

de Bruijn, F.J., Szabados, L. and Schell, J. (1990) Chimeric genes and transgenic plants are used to study the regulation of genes involved in symbiotic plant–microbe interactions (nodulin genes). *Developmental Genetics* 11, 182–196.

de Maagd, R.A., Wijfjes, H.M., Spaink, H.P., Ruiz-Sainz, J.E., Wijffelman, C.A., Okker, R.J.H. and Lugtenberg, B.J.J. (1989) nodO, a new nod gene of the *Rhizobium leguminosarum* biovar *viciae* Sym plasmid encodes a secreted protein. *Journal of Bacteriology* 171, 6764–6770.

Dénarié, J., Debelle, F. and Rosenberg, C. (1992) Signalling and host range variation in nodulation. *Annual Reviews of Microbiology* 46, 497–531.

Deroche, M.-E., Carrayol, E. and Jolivet, E. (1983) Phosphoenolpyruvate carboxylase in legume nodule. *Physiologie Végétale* 21, 1075–1081.

De Vries, G.E., In'T Veld, P. and Kijne, J.W. (1980) Production of organic acids in *Pisum sativum* root nodules as a result of oxygen stress. *Plant Science Letters* 20, 115–123.

Diaz, C.L., Melchers, L.S., Hooykaas, P.J.J., Lugtenberg, B.J.J. and Kijne, J.W. (1989) Root lectin as a determinant of host-plant specificity in the *Rhizobium*-legume symbiosis. *Nature* 338, 579–581.

Diaz, C.L., Hosselet, M., Logman, T.J.J., van Driessche, E., Lugtenberg, B.J.J. and Kijne, J.W. (1990) Distribution of glucose/mannose-specific isolectins in pea (*Pisum sativum* L.) seedlings. *Planta* 181, 451–461.

Dibb, N.J., Downie, J.A. and Brewin, N.J. (1984) Identification of a rhizosphere protein encoded by the symbiotic plasmid of *Rhizobium leguminosarum*. *Journal of Bacteriology* 158, 621–627.

Dilworth, M.J. (1966) Acetylene reduction by nitrogen-fixing preparations from *Clostridium pasteuranium*. *Biochimica Biophysica Acta* 127, 283–294.

Downie, J.A. (1989) The *nodL* gene from *Rhizobium leguminosarum* is homologous to the acetyl transferases encoded by *lacA* and *cysE*. *Molecular Microbiology* 3, 1649–1651.

Downie, J.A (1991) A *nod* of recognition. *Current Biology* 1, 382–384.

Downie, J.A. and Johnston, A.W.B. (1988) Nodulation of legumes by *Rhizobium*. *Plant, Cell and Environment* 11, 403–412.

Downie, J.A. and Surin, B.P. (1990) Either of two *nod* gene loci can complement the nodulation defect of a *nod* deletion mutant of *Rhizobium leguminosarum* bv. *viciae*. *Molecular and General Genetics* 222, 81–86.

Downie, J.A., Hombrecher, G., Ma, Q.-S., Knight, C.D., Wells, B. and Johnston, A.W.B. (1983a) Cloned nodulation genes of *Rhizobium leguminosarum* determine host-range specificity. *Molecular and General Genetics* 190, 359-365.

Downie, J.A., Ma, Q.-S., Knight, C.D., Hombrecher, G. and Johnston, A.W.B. (1983b) Cloning of the symbiotic region of *Rhizobium leguminosarum*: the nodulation genes are between the nitrogenase genes and the *nifA*-like gene. *EMBO Journal* 2, 947-952.

Downie, J.A., Knight, C.D. and Johnston, A.W.B. (1985) Identification of genes and gene products involved in nodulation of peas by *Rhizobium leguminosarum*. *Molecular and General Genetics* 198, 255-262.

Duc, G. and Messager, A. (1989) Mutagenesis of pea (*Pisum sativum*) and the isolation of mutants for nodulation and nitrogen fixation. *Plant Science* 60, 207-213.

Duc, G., Trouvelot, A., Gianinazzi-Pearson, V. and Gianinazzi, S. (1989) First report of non-mycorrhizal plant mutants (Myc$^-$) obtained in pea (*Pisum sativum* L.) and fababean (*Vicia faba* L.). *Plant Science* 60, 215-222.

Earl, C.D., Ronson, C.W. and Ausubel, F.M. (1987) Genetic and structural analysis of the *Rhizobium meliloti fixA*, *fixB*, *fixC* and *fixX* genes. *Journal of Bacteriology* 169, 1127-1136.

Economou, A., Hawkins, F.K.L., Downie, J.A. and Johnston, A.W.B. (1989) Transcription of *rhiA*, a gene on a *Rhizobium leguminosarum* bv. *viciae* Sym plasmid, requires *rhiR* and is repressed by flavonoids that induce *nod* genes. *Molecular Microbiology* 3, 87-93.

Economou, A., Hamilton, W.D.O., Johnston, A.W.B. and Downie, J.A. (1990) The *Rhizobium* nodulation gene *nodO* encodes a Ca^{2+}-binding protein that is exported without N-terminal cleavage and is homologous to haemolysin and related proteins. *EMBO Journal* 9, 349-354.

Ehrhardt, D.W., Atkinson, E.M. and Long, S.R. (1992) Depolarization of alfalfa root hair membrane potential by *Rhizobium meliloti* Nod factors. *Science* 256, 998-1000.

Engelke, T., Jording, D., Kapp, D. and Pühler, A. (1989) Identification and sequence analysis of the *Rhizobium meliloti dctA* gene encoding the C_4-dicarboxylate carrier. *Journal of Bacteriology* 171, 5551-5560.

Evans, H.J., Harker, A.R., Papen, H., Russell, S.A., Hanus, F.J. and Zuber, M. (1987) Physiology, biochemistry and genetics of the uptake hydrogenase in *Rhizobium*. *Annual Reviews of Microbiology* 41, 335-361.

Evans, I.J. and Downie, J.A. (1986) The *nodI* gene product of *Rhizobium leguminosarum* is closely related to ATP-binding bacterial transport proteins: nucleotide sequence analysis of the *nodI* and *nodJ* genes. *Gene* 43, 95-101.

Faucher, C., Camut, S., Dénarié, J. and Truchet, G. (1989) The *nodH* and *nodQ* host range genes of *Rhizobium meliloti* behave as avirulence genes in *R. leguminosarum* bv. *viciae* and determine changes in the production of plant-specific extracellular signals. *Molecular Plant-Microbe Interactions* 2, 291-300.

Faulkner, I.J. and Rubery, P.H. (1992) Flavonoids and flavonoid sulphates as probes of auxin-transport regulation in *Cucurbita pepo* hypocotyl segments and vesicles. *Planta* 186, 618-625.

Fearn, J.C. and LaRue, T.A. (1991) A temperature-sensitive nodulation mutant (*sym5*) of *Pisum sativum* L. *Plant Cell and Environment* 14, 221-227.

Finan, T.M., Wood, J.M. and Jordan, D.C. (1983) Symbiotic properties of

C$_4$-dicarboxylic acid transport mutants of *Rhizobium leguminosarum*. *Journal of Bacteriology* 154, 1403–1413.

Firmin, J.L., Wilson, K.E., Rossen, L. and Johnston, A.W.B. (1986) Flavonoid activation of nodulation genes in *Rhizobium* reversed by other compounds present in plants. *Nature* 324, 90–92.

Fisher, R.F. and Long, S.R. (1992) *Rhizobium*–plant signal exchange. *Nature* 357, 655–660.

Fisher, R.F., Tu, J.K. and Long, S.R. (1985) Conserved nodulation genes in *Rhizobium meliloti* and *Rhizobium trifolii*. *Applied and Environmental Microbiology* 49, 1439–1445.

Fobert, P.R., Roy, N., Nash, J.H.E. and Iyer, V.N. (1991) Procedure for obtaining efficient root nodulation of a pea cultivar by a desired *Rhizobium* strain and preempting nodulation by other strains. *Microbiology* 57, 1590–1594.

Forde, B.G. and Cullimore, J.V. (1989) The molecular biology of glutamine synthetase in higher plants. *Oxford Survey of Plant Molecular and Cellular Biology* 6, 247–296.

Forde, B.G., Day, H.M., Turton, J.F., Shen, W.J., Cullimore, J.V. and Oliver, J.E. (1989) Two glutamine synthetase genes from *Phaseolus vulgaris* L. display contrasting developmental spatial patterns of expression in transgenic *Lotus corniculatus* plants. *Plant Cell* 1, 391–401.

Fortin, M.G., Morrison, N.A. and Verma, D.P.S. (1987) Nodulin-26, a peribacteroid membrane nodulin is expressed independently of the development of the peribacteroid compartment. *Nucleic Acids Research* 15, 813–824.

Franssen, H.J., Vijn, I., Yang, W.C. and Bisseling, T. (1992) Developmental aspects of the *Rhizobium*–legume symbiosis. *Plant Molecular Biology* 19, 89–107.

Gadal, P. (1983) Phosphoenolpyruvate carboxylase and nitrogen fixation. *Physiologie Végétale* 21, 1069–1074.

Gatehouse, J.A., Brown, D., Evans, I.M., Gatehouse, L.N., Jobes, D., Preston, P. and Croy, R.R.D. (1987) Sequence of the seed lectin gene from pea (*Pisum sativum* L.). *Nucleic Acids Research* 15, 7642–7643.

Geiger, O., Spaink, H.P. and Kennedy, E.P (1991) Isolation of the *Rhizobium leguminosarum* NodF nodulation protein: NodF carries a 4'-phosphopantetheine prosthetic group. *Journal of Bacteriology* 173, 2872–2878.

Gelin, O. and Blixt, S. (1964) Root nodulation in peas. *Agri Hortique Genetica* 22, 149–152.

Gianinazzi-Pearson, V., Gianinazzi, S., Guillemin, J.P., Trouvelot, A. and Duc, G. (1991) Genetic and cellular analysis of resistance to vesicular arbuscular (VA) mycorrhizal fungi in pea mutants. In: Hennecke, H. and Verma, D.P.S. (eds), *Advances in Molecular Genetics of Plant–Microbe Interactions*, vol. 1. Kluwer Academic Publishers, Dordrecht, pp. 336–342.

Gilles-Gonzales, M.A., Ditta, G. and Helinski, D.R. (1991) A haemoprotein with kinase activity encoded by the oxygen-sensor of *Rhizobium meliloti*. *Nature* 350, 170–172.

Glenn, A.R., Poole, P.S. and Hudman, J.F. (1980) Succinate uptake by free-living and bacteroid forms of *Rhizobium leguminosarum*. *Journal of General Microbiology* 119, 267–271.

Glenn, A.R., McKay, I.A., Arwas, R. and Dilworth, M.J. (1984) Sugar metabolism and the symbiotic properties of carbohydrate mutants of *Rhizobium leguminosarum*. *Journal of General Microbiology* 130, 239-245.

Gloudemans, T., Bhuvaneswari, T.V., Moerman, M., Van Brussel, A.A.N., Van Kammen, A. and Bisseling, T. (1989) Involvement of *Rhizobium leguminosarum* nodulation genes in gene expression in pea root hairs. *Plant Molecular Biology* 12, 157-167.

Götz, R., Evans, I.J., Downie, J.A. and Johnston, A.W.B. (1985) Identification of the host-range DNA which allows *Rhizobium leguminosarum* strain TOM to nodulate cv. Afghanistan peas. *Molecular and General Genetics* 201, 296-300.

Govers, F., Gloudemans, T., Moerman, M., Van Kammen, A. and Bisseling, T. (1985) Expression of plant genes during the development of pea root nodules. *EMBO Journal* 4, 861-867.

Govers, F., Moerman, M., Downie, J.A., Hooykaas, P.J.J., Franssen, H.J., Louwerse, J., Van Kammen, A. and Bisseling, T. (1986a) *Rhizobium nod* genes are involved in inducing an early nodulin gene. *Nature* 323, 564-566.

Govers, F., Moerman, M., Hooymans, J., Van Kammen, A. and Bisseling, T. (1986b) Microaerobiosis is not involved in the induction of pea nodulin gene expression. *Planta* 169, 513-517.

Govers, F., Nap, J.-P., Moerman, M., Franssen, H.J., Van Kammen, A. and Bisseling, T. (1987) cDNA cloning and developmental expression of pea nodulin genes. *Plant Molecular Biology* 8, 425-435.

Gronger, P., Manian, S., Reilander, H., O'Connell, M., Priefer, U.B. and Puhler, A. (1987) Organisation and partial sequence of a DNA region of the *Rhizobium leguminosarum* symbiotic plasmid pRL6JI containing the genes *fixABC*, *nifA*, *nifB* and a novel open reading frame. *Nucleic Acids Research* 15, 31-49.

Gubler, M. and Hennecke, H. (1986) *fixA*, *B* and *C* genes are essential for symbiotic and free-living microaerobic nitrogen fixation. *FEBS Letters* 220, 186-192.

Guerrero, F.D., Jones, J.T. and Muller, J.E. (1990) Turgor-responsive gene transcription and RNA levels increase rapidly when pea shoots are wilted. Sequence and expression of three inducible genes. *Plant Molecular Biology* 15, 11-26.

Guinel, F.C. and LaRue, T.A. (1992) Ethylene inhibitors partly restore nodulation to pea mutant E107 (Brz). *Plant Physiology* 99, 515-518.

Györgypal, Z., Iyer, N. and Kondorosi, A. (1988) Three regulatory *nodD* alleles of diverged flavonoid-specificity are involved in host-dependent nodulation by *Rhizobium meliloti*. *Molecular and General Genetics* 212, 85-92.

Harborne, J.B. (1971) Distribution of flavonoids in the Leguminosae. In: Harborne, J.B., Boulter, D. and Turner, B.L. (eds), *Chemotaxonomy of the Leguminosae*, Academic Press, London, pp. 31-71.

Harker, C.L., Ellis, T.H.N. and Coen, E.S. (1990) Identification and genetic regulation of the chalcone synthase multigene family in pea. *Plant Cell* 2, 185-194.

Hennecke, H., Anhamatten, D., Babst, M., Bott, M., Fischer, H.M., Kaspar, T., Kullik, I., Loferer, H., Preisig, O., Ritz, D. and Weidenhaupt, M. (1993) Genetic and physiologic requirements for optimal bacteroid function in the *Bradyrhizobium japonicum* soybean symbiosis. In: Nester, E.W. and Verma, D.P.S. (eds), *Advances in the Genetics of Plant-Microbe Interactios*, Kluwer, Dordrecht, pp. 199-207.

Higgins, C.F., Hiles, I.D., Salmond, G.P.C., Gill, D.R., Downie, J.A., Evans, I.J.,

Holland, I.B., Gray, L., Bucke, S.D., Bell, A.W. and Hermodson, M.A. (1986) A family of related ATP-binding subunits coupled to many distinct biological processes in bacteria. *Nature* 323, 448–450.

Higgins C.F., Hyde, S.C., Mimmack, M.M., Gileadi, U., Gill, D.R. and Gallagher, M.P. (1990) Binding protein-dependent transport systems. *Journal of Bioenergetics and Biomembranes* 22, 571–592.

Hill, S., Viollet, S., Smith, A.J. and Anthony, C. (1990) Roles for enteric δ-type cytochrome oxidase in N_2 fixation and microaerobiosis. *Journal of Bacteriology* 172, 2071–2078.

Hirsch, A.M., Bhuvaneswari, T.V., Torrey, J.G. and Bisseling, T. (1989) Early nodulin genes are induced in alfalfa root outgrowths elicited by auxin transport inhibitors. *Proceedings of the National Academy of Sciences USA* 86, 1244–1248.

Holl, F.B. (1975) Host plant control of the inheritance of dinitrogen fixation in the *Pisum–Rhizobium* symbiosis. *Euphytica* 24, 767–770.

Hombrecher, G., Götz, R., Dibb, N.J., Downie, J.A., Johnston, A.W.B. and Brewin, N.J. (1984) Cloning and mutagenesis of nodulation genes from *Rhizobium leguminosarum* TOM, a strain with extended host range. *Molecular and General Genetics* 184, 293–298.

Hong, G.-F., Burn, J.E. and Johnston, A.W.B. (1987) Evidence that DNA involved in the expression of nodulation nod genes in *Rhizobium* binds to the product of the regulatory gene *nodD*. *Nucleic Acids Research* 15, 9677–9690.

Jacobsen, E. (1984) Modification of symbiotic interaction of pea (*Pisum sativum*) and *Rhizobium leguminosarum* by induced mutation. *Plant and Soil* 82, 155–167.

Jacobsen, E. and Feenstra, W.J. (1984) A new pea mutant with efficient nodulation in the presence of nitrate. *Plant Science Letters* 33, 337–344.

James, E.K., Sprent, J.I., Minchin, F.R. and Brewin, N.J. (1991) Intercellular location of glycoprotein in soybean nodules: effect of altered rhizosphere oxygen concentration. *Plant, Cell and Environment* 14, 467–476.

Jensen, E.O. (1990) Regulation of nodule expressed soybean genes. In: Henneke, H. (ed.), *Proceedings of 5th International Symposium on the Molecular Genetics of Plant–Microbe Interactions*. Kluwer Academic, Dordrecht, pp. 310–316.

Jensen, E.S. (1986a) Symbiotic N_2 fixation in pea and field bean estimated by ^{15}N fertilizer dilution in field experiments with barley as a reference crop. *Plant and Soil* 92, 3–13.

Jensen, E.S. (1986b) The influence of rate and time of nitrate supply on nitrogen fixation and yield in pea (*Pisum sativum*). *Fertility Research* 10, 193–202.

Jensen, E.S. (1987) Seasonal patterns of growth and nitrogen fixation in field-grown pea. *Plant Soil* 101, 29–37.

Jensen, E.S. (1989) The role of pea cultivation in the nitrogen economy of soils and succeeding crops. In: Plancquaert, P. and Hagger, R. (eds), *Legumes in Farming Systems*, Kluwer, Dordrecht, pp. 3–15.

John, M., Röhrig, H., Schmidt, J., Wieneke, U. and Schell, J. (1993) *Rhizobium* NodB protein involved in nodulation signal synthesis is a chitooligosaccharide deacetylase. *Proceedings of the National Academy of Sciences USA* 90, 625–629.

Johnston, A.W.B., Beynon, J.L., Buchanan-Wollaston, A.V., Setchell, S.M., Hirsh, P. and Beringer, J.E. (1978) High frequency transfer of nodulating ability between strains and species of *Rhizobium*. *Nature* 276, 634–636.

Johnston, A.W.B., Burn, J.E., Economou, A., Davis, E.O., Hawkins, F.K.L. and Bibb,

M.J. (1988) Genetic factors affecting host range in *Rhizobium leguminosarum*. In: Palacios, R. and Verma, D.P.S. (eds), *Molecular Genetics of Plant–Microbe Interactions*, APS Press, St Paul, MN, pp. 378–384.

Kagan, S.A. and Brewin, N.J. (1985) Mutagenesis of a *Rhizobium* plasmid carrying hydrogenase determinants. *Journal of General Microbiology* 131, 1141–1147.

Kannenberg, E.L., Rathbun, E.A. and Brewin, N.J. (1992) Molecular dissection of structure and function in the lipopolysaccharide of *Rhizobium leguminosarum* strain 3841 using monoclonal antibodies and genetic analysis. *Molecular Microbiology* 6, 2477–2487.

Kijne, J.W. (1992) The *Rhizobium* infection process. In: Stacey, G., Burris, R.H. and Evans, H.J. (eds), *Biological Nitrogen Fixation*. Chapman and Hall, New York, London, pp. 349–398.

Kijne, J.W., Smit, G., Diaz, C.L. and Lugtenberg, B.J.J. (1988) Lectin-enhanced accumulation of manganese-limited *Rhizobium leguminosarum* cells on pea root hair tips. *Journal of Bacteriology* 170, 2994–3000.

Kneen, B.E. and LaRue, T.A. (1984a) Nodulation resistant mutant of *Pisum sativum*. *Journal of Heredity* 75, 238–240.

Kneen, B.E. and LaRue, T.A. (1984b) Peas (*Pisum sativum*) with strain specificity for *Rhizobium leguminosarum*. *Heredity* 52, 383–389.

Kneen, B.E. and LaRue, T.A. (1986) Additional mutants defective in nodulation. *Pisum Newsletter* 8, 33.

Kneen, B.E. and LaRue, T.A. (1988) Induced symbiosis mutants of pea (*Pisum sativum*) and sweetclover (*Melilotus alba annua*). *Plant Science* 58, 177–182.

Kneen, B.E., LaRue, T.A., Welch, R.M. and Weeden, N.F. (1990a) Pleiotropic effects of *brz*: a mutation in *Pisum sativum* (L.) cv 'Sparkle' conditioning decreased nodulation and increased iron uptake and leaf necrosis. *Plant Physiology* 93, 717–722.

Kneen, B.E., LaRue, T.A., Hirsch, A.M., Smith, C.A. and Weeden, N.F. (1990b) *sym*13 – a gene conditioning ineffective nodulation in *Pisum sativum*. *Plant Physiology* 94, 899–905.

LaRue, T.A. and Weeden, N.F. (1992) The symbiosis genes of pea. *Pisum Genetics* 24, 5–12.

Latchford, J.W., Borthakur, D. and Johnston, A.W.B. (1991) The products of *Rhizobium* genes *psi* and *pss*, which affect exopolysaccharide production, are associated with the bacterial cell surface. *Molecular Microbiology* 5, 2107–2114.

Layzell, D.B. and Hunt, S. (1990) Oxygen and regulation of N fixation in legume nodules. *Physiologia Plantarum* 802, 322–327.

Layzell, D.B., Hunt, S. and Palmer, G.R. (1990) Mechanism of nitrogenase inhibition in soybean nodules. Pulse modulated spectroscopy indicates nitrogenase activity is limited by O_2. *Plant Physiology* 92, 1102–1107.

Lea, P.J., Robinson, S.A. and Steward, G.R. (1990) The enzymology and metabolism of glutamine, glutamate and asparagine. In: Miflin, B.J. and Lea, P.J. (eds), *The Biochemistry of Plants. Intermediary Nitrogen Metabolism*, Academic Press, San Diego, pp. 121–159.

Lee, K.H. and LaRue, T.A. (1992) Pleiotropic effects of *sym-17* a mutation in *Pisum sativum* L. cv Sparkle causes decreased nodulation, altered root and shoot growth and increases ethylene production. *Plant Physiology* 100, 1326–1333.

LeGal, M.F. and Hobbs, S.L.A. (1989) Cytological studies of the infection process in nodulating and non-nodulating pea genotypes. *Canadian Journal of Botany* 67, 2435-2443.

Legocki, R.P. and Verma, D.P.S. (1980) Identification of 'nodule-specific' host proteins (nodulins) involved in the development of *Rhizobium*-legume symbiosis. *Cell* 20, 153-163.

Leigh, J.A. and Coplin, D.L. (1992) Exopolysaccharides in plant–bacterial interactions. *Annual Reviews of Microbiology* 46, 307-346.

Leong, S.A., Ditta, G.S. and Helinski, D.R. (1982) Heme biosynthesis in *Rhizobium* identification of a cloned gene coding for delta amino levulinic-acid synthetase from *Rhizobium meliloti*. *Journal of Biological Chemistry* 257, 8724-8730.

Lerouge, P., Roche, P., Faucher, C., Maillet, F., Truchet, G., Promé, J.-C. and Dénarié, J. (1990) Symbiotic host-specificity of *Rhizobium meliloti* is determined by a sulphated and acylated glucosamine oligosaccharide signal. *Nature* 344, 781-784.

Lie, T.A. (1971) Symbiotic nitrogen fixation under stress conditions. *Plant and Soil* 31, 117-127.

Lie, T.A. (1978) Symbiotic specialisation in pea plants: the requirement of specific *Rhizobium* strains for peas from Afghanistan. *Annals of Applied Biology* 88, 462-465.

Lie, T.A. (1984) Host genes in *Pisum sativum* L. conferring resistance to European *Rhizobium leguminosarum* strains. *Plant and Soil* 82, 415-425.

Lie, T.A., Goktan, D., Engin, M., Pijnenborg, J. and Anlarsal, E. (1987) Co-evolution of the legume–*Rhizobium* association. *Plant and Soil* 100, 171-181.

Lloyd, C.W., Pearce, K.J., Rawlins, D.J., Ridge, R.W. and Shaw, P.J. (1987) Endoplasmatic microtubules connect the advancing nucleus to the tip of legume root hairs, but F-actin is involved in basipetal migration. *Cell Motility and the Cytoskeleton* 8, 27-36.

Long, S.R. (1992) Genetic analysis of *Rhizobium* nodulation. In: Stacey, G., Burris, R.H. and Evans, H.J. (eds), *Biological Nitrogen Fixation*. Chapman and Hall, New York, London, pp. 560-597.

McKay, I.A., Glenn, A.R. and Dilworth, M.J. (1985) Gluconeogenesis in *Rhizobium leguminosarum* MNF3841. *Journal of General Microbiology* 131, 2067-2073.

McKay, I.A., Dilworth, M.J. and Glenn, A.R. (1988) C_4-Dicarboxylate metabolism in free-living and bacteroid forms of *Rhizobium leguminosarum* MNF3841. *Journal of General Microbiology* 134, 1433-1440.

Margolin, W., Corbo, C. and Long, S.R. (1991) Cloning and characterisation of a *Rhizobium meliloti* homolog of the *Escherichia coli* cell division gene *ftsZ*. *Journal of Bacteriology* 173, 5822-5830.

Marie, C., Barny, M.-A. and Downie, J.A (1992) *Rhizobium leguminosarum* has two glucosamine synthases, GlmS and NodM, required for nodulation and development of nitrogen-fixing nodules. *Molecular Microbiology* 6, 843-851.

Markwei, C.M. and LaRue, T.A. (1992) Phenotypic characterisation of *sym8* and *sym9*, two genes conditioning non-nodulation in *Pisum sativum* 'Sparkle'. *Canadian Journal of Microbiology* 38, 548-554.

Mellor, R.B. (1989) Bacteroids in the *Rhizobium*-legume symbiosis inhabit a plant internal lytic compartment: implications for other microbial endosymbioses. *Journal of Experimental Botany* 40, 831-839.

Miao, G.-H., Hirel, B., Marsolier, M.C., Ridge, R.W. and Verma, D.P.S. (1991)

Ammonia-regulated expression of a soybean gene encoding cytosolic glutamine synthetase in transgenic *Lotus corniculatus*. *Plant Cell* 3, 11–22.

Minchin, F.R., Summerfield, R.J., Hadley, P., Roberts, E.H. and Rawsthorne, S. (1981) Carbon and nitrogen nutrition of nodulated roots of grain legumes. *Plant Cell and Environment* 4, 5–26.

Minchin, F.R., Sheehy, J.E., Ines-Minguez, M. and Witty, J.F. (1985) Characterization of the resistance to oxygen diffusion in legume nodules. *Annals of Botany* 55, 53–60.

Minchin, F.R., Sheehy, J.E. and Witty, J.F. (1986) Further errors in the acetylene-reduction assay: effects of plant disturbance. *Journal of Experimental Botany* 37, 1581–1591.

Mitchell, J.P. (1965) The DNA content of nuclei in pea root nodules. *Annals of Botany* 29, 371–377.

Monroe, J.D., Owens, T.G. and LaRue, T.A. (1989) Measurement of the fractional oxygenation of leghemoglobin in intact detached pea nodules by reflectance spectroscopy. *Plant Physiology* 91, 598–602.

Morell, M. and Copeland, L. (1985) Sucrose synthase of soybean nodules. *Plant Physiology* 78, 149–154.

Morris, V.J., Brownsey, G.J., Harris, J.E., Gunning, A.P. and Stevens, B.J.H. (1989) Cation-dependent gelation of the acidic extracellular polysaccharides of *Rhizobium leguminosarum*: a non-specific mechanism for the attachment of bacteria to plant roots. *Carbohydrate Research* 191, 315–320.

Nadler, K.D. and Avissar, Y.J. (1977) Heme synthesis in soybean root nodules. I. On the role of bacteroid δ-amino laevulinic acid synthetase and δ-amino laevulinic acid dehydratase on the synthesis of heme in leghemoglobin. *Plant Physiology* 60, 433–436.

Nap, J.-P. and Bisseling, T. (1990) The roots of nodulins. *Physiologia Plantarum* 79, 407–414.

Newcomb, W.E. (1976) A correlated light and electron microscopic study of symbiotic growth and differentiation in *Pisum sativum* root nodules. *Canadian Journal of Botany* 54, 2163–2186.

Newcomb, W.E. (1981) Nodule morphogenesis. In: Bourne, G.H. and Danielli, J.F. (eds), *International Review of Cytology*, supplement 13. Academic Press, New York, pp. 246–298.

Noel, K.D. (1991) Rhizobial polysaccharides required in symbioses with legumes. In: Verma, D.P.S. (ed.), *Molecular Signals in Plant–Microbe Interactions*. CRC Press, Boca Raton, pp. 341–357.

Parsons, R. and Day, D.A. (1990) Mechanism of soybean nodule adaptation to different oxygen pressures. *Plant, Cell and Environment* 13, 501–512.

Pennell, R.I., Knox, J.P., Scofield, G.N., Selvendran, R.R. and Roberts, K. (1989) A family of abundant plasma membrane-associated glycoproteins related to the arabinogalactan proteins is unique to flowering plants. *Journal of Cell Biology* 108, 1967–1977.

Peoples, M.B., Sudin, M.-N. and Herridge, D.F. (1987) Translocation of nitrogenase compounds in symbiotic and nitrate-fed amide-exporting legumes. *Journal of Experimental Botany* 38, 567–579.

Perotto, S., VandenBosch, K.A., Butcher, G.W. and Brewin, N.J. (1991) Molecular composition and development of the plant glycocalyx associated with the peribacteroid membrane of pea root nodules. *Development* 112, 763-773.

Phillips, D.A. (1991) Genetic enhancement of nitrogen fixation. In: Dilworth, M.J. and Glenn, A.R. (eds), *Biology and Biochemistry of Nitrogen Fixation*. Elsevier, Amsterdam, pp. 408-428.

Pichon, M., Journet, E.-P., Dedieu, A., de Billy, F., Truchet, G. and Barker, D.G. (1992) *Rhizobium meliloti* elicits transient expression of the early nodulin gene *ENOD*12 in the differentiating root epidermis of transgenic alfalfa. *Plant Cell* 4, 1199-1211.

Plancquaert, P. and Desbureaux, J. (1984) The agronomic effects of peas in rotation with winter wheat and oilseed rape: a progress report. In: Hebbelthwaite, P.D., Heath, M.C. and Dawkins, T.C.K. (eds), *The Pea Crop*. Butterworths, London, pp. 193-202.

Postma, J.G., Jacobsen, E. and Feenstra, W.J. (1988a) Three pea mutants with an altered nodulation studied by genetic analysis and grafting. *Journal of Plant Physiology* 132, 424-430.

Postma, J.G., Jacobsen, E. and Feenstra, W.J. (1988b) Experiments with mutants of pea (*Pisum sativum* L.). In: Bothe, H., de Bruijn, F.J. and Newton, W.E. (eds), *Nitrogen Fixation: Hundred Years After*. Gustav Fischer Verlag, Stuttgart, pp. 629-633.

Postma, J.G., Jager, D., Jacobsen, E. and Feenstra, W.J. (1990) Studies on a non-fixing mutant of pea (*Pisum sativum* L.) I. Phenotypical description and bacteroid activity. *Plant Science* 68, 151-161.

Rae, A.E., Perotto, S., Knox, J.P., Kannenberg, E.L. and Brewin, N.J. (1991) Expression of extracellular glycoproteins in the uninfected cells of developing pea nodule tissue. *Molecular Plant-Microbe Interactions* 4, 563-570.

Rae, A.L., Bonfante-Fasolo, P. and Brewin, N.J. (1992) Structure and growth of infection threads in the legume symbiosis with *Rhizobium leguminosarum*. *Plant Journal* 2, 385-395.

Recourt, K., Verkerke, M., Schripsema, J., Van Brussel, A.A.N., Lugtenberg, B.J.J. and Kijne, J.W. (1992) Major flavonoids in uninoculated and inoculated roots of *Vicia sativa* subsp. *nigra* are four conjugates of the nodulation gene-inhibitor kaempferol. *Plant Molecular Biology* 18, 505-513.

Robertson, J.G. and Lyttleton, P. (1984) Division of peribacteroid membranes in root nodules of white clover. *Journal of Cell Science* 69, 147-157.

Robertson, J.G., Lyttleton, P., Bullivant, S. and Grayston, G.F. (1978) Membranes in lupin root nodules I. The role of Golgi bodies in the biogenesis of infection threads and peribacteroid membranes. *Journal of Cell Science* 30, 129-149.

Robertson, J.G., Wells, B., Bisseling, T., Farnden, K.J.F. and Johnson, A.W.B. (1984) Immuno-gold localization of leghemoglobin in cytoplasm in nitrogen-fixing root nodules of pea (*Pisum sativum*). *Nature* 311, 254-256.

Robertson, J.G., Wells, B., Brewin, N.J., Wood, E.A., Knight, C.D. and Downie, J.A. (1985) The legume-*Rhizobium* symbiosis: a cell surface interaction. *Journal of Cell Science, Supplement* 2, 317-331.

Roche, P., Debellé F., Maillet, F., Lerouge, P., Faucher, C., Truchet, G., Dénarié, J. and Promé, J.-C. (1991a) Molecular basis of symbiotic host specificity in *Rhizobium meliloti: nodH* and *nodPQ* genes encode the sulfation of lipo-oligosaccharide signals. *Cell* 67, 1131-1143.

Roche, P., Lerouge, P., Ponthus, C. and Prome, J.-C. (1991b) Structural determination of bacterial nodulation factors involved in the *Rhizobium meliloti*-alfalfa symbiosis. *Journal of Biological Chemistry* 266, 10933–10940.

Rodriguez-Quinones, F., Banfalvi, Z., Murphy, P.J. and Kondorosi, A. (1987) Interspecies homology of nodulation genes in *Rhizobium*. *Plant Molecular Biology* 8, 61–76.

Ronson, C.W., Lyttleton, P. and Robertson, J.G. (1981) C_4-dicarboxylate transport mutants in *Rhizobium trifolii* form ineffective nodules on *Trifolium repens*. *Proceedings of the National Academy of Sciences USA* 78, 4284–4288.

Rosendahl, L., Vance, C.P., Miller, S.S. and Jacobsen, E. (1989) Nodule physiology of a supernodulating pea mutant. *Physiologia Plantarum* 77, 606–612.

Rosendahl, L., Vance, C.P. and Pederson, W.B. (1990) Products of dark CO_2 fixation in pea root nodules support bacteroid metabolism. *Plant Physiology* 93, 12–19.

Rossen, L., Johnston, A.W.B. and Downie, J.A. (1984) DNA sequence of the *Rhizobium leguminosarum* nodulation genes *nodA, B* and *C* required for root hair curling. *Nucleic Acids Research* 12, 9497–9508.

Rossen, L., Shearman, C.A., Johnston, A.W.B. and Downie, J.A. (1985) The *nodD* gene of *Rhizobium leguminosarum* is autoregulatory and in the presence of plant exudate induces the *nodABC* genes. *EMBO Journal* 4, 3369–3373.

Roth, L.E. and Stacey, G. (1989) Bacterium release into host cells of nitrogen-fixing soybean nodules: the symbiosome membrane comes from three sources. *European Journal of Cell Biology* 49, 13–23.

Ryan, C.A. and Farmer, E.E. (1991) Oligosaccharide signals in plants: a current assessment. *Annual Review of Plant Physiology and Plant Molecular Biology* 42, 651–674.

Sangwan, I. and O'Brian, M.R. (1991) Evidence for an interorganismic heme biosynthetic pathway in symbiotic soybean root nodules. *Science* 251, 1220–1222.

Scheres, B., Van de Weil, C., Zalensky, A., Horvath, B., Spaink, H., Van Eck, H., Zwartkruis, F., Wolters. A.M., Gloudemans, T., Van Kammen, A. and Bisseling, T. (1990a) The ENOD12 gene product is involved in the infection process during the pea-*Rhizobium* interaction. *Cell* 60, 281–294.

Scheres, B., van Engelen, F., van der Knaap, E., van de Wiel, C., van Kammen, A. and Bisseling, T. (1990b) Sequential induction of nodulin gene expression in the developing pea nodule. *Plant Cell* 2, 687–700.

Schetgens, T.M.P., Hontelez, J.G.J., van den Bos, R.C. and Van Kammen, A. (1985) Identification and phenotypical characterisation of a cluster of *fix* genes including a *nif* regulatory gene from *Rhizobium leguminosarum*. *Molecular and General Genetics* 200, 368–374.

Schlaman, H.R.M., Spaink, H.P., Okker, R.J.H. and Lugtenberg, B.J.J. (1989) Subcellular localization of the *nodD* gene product in *Rhizobium leguminosarum*. *Journal of Bacteriology* 171, 4686–4693.

Schlaman, H.R.M., Okker. R.J.H. and Lugtenberg, B.J.J. (1990) Subcellular localization of the *Rhizobium leguminosarum nodI* gene product. *Journal of Bacteriology* 172, 5486–5489.

Schlaman, H.R.M., Horvath, B., Vijgenboom, E., Okker, R.J.H. and Lugtenberg, B.J.J. (1991) Suppression of nodulation gene expression in bacteroids of *Rhizobium leguminosarum* biovar *viciae*. *Journal of Bacteriology* 173, 4277–4287.

Schlaman, H.R.M., Okker, R.J.H. and Lugtenberg, B.J.J. (1992a) Regulation of nodulation gene expression by NodD in rhizobia. *Journal of Bacteriology* 174, 5177-5182.

Schlaman, H.R.M., Lugtenberg, B.J.J. and Okker, R.J.H. (1992b) The NodD protein does not bind to promoters of inducible nodulation genes in extracts of bacteroids of *Rhizobium leguminosarum* Q biovar *viciae*. *Journal of Bacteriology* 174, 6109-6116.

Schmidt, P.E., Parniske, M. and Werner, D. (1992) Production of the phytoalexin glyceollin-1 by soybean roots in response to symbotic and pathogenic infection. *Botanica Acta* 105, 18-28.

Schubert, K.R. (1986) Products of biological N_2 fixation in higher plants: synthesis, transport and metabolism. *Annual Review of Plant Physiology* 37, 539-574.

Schultze, M., Quiclet-Sire, B., Kondorosi, E., Virelizier, H., Glushka, J.N., Endre, G., Gero, S. and Kondorosi, A. (1992) *Rhizobium meliloti* produces a family of sulfated lipo-oligosaccharides exhibiting different degrees of plant host specificity. *Proceedings of the National Academy of Sciences USA* 89, 192-196.

Schwedock, J. and Long, S.R. (1990) ATP sulphurylase activity of the *nodP* and *nodQ* gene products of *Rhizobium meliloti*. *Nature* 348, 644-647.

Sengupta-Gopalan, C. and Pitas, J.W. (1986) Expression of nodule-specific glutamine synthetase genes during nodule development in soybeans. *Plant Molecular Biology* 7, 189-199.

Sharma, S.B. and Signer, E.R. (1990) Temporal and spatial regulation of the symbiotic genes of *Rhizobium meliloti* in planta revealed by transposon Tn5-*gusA*. *Genes and Development* 4, 344-356.

Shearman, C.A., Rossen, L., Johnston, A.W.B. and Downie, J.A. (1986) The *Rhizobium* gene *nodF* encodes a protein similar to acyl carrier protein and is regulated by *nodD* plus a factor in pea root exudate. *EMBO Journal* 5, 647-652.

Sheehy, J.E., Minchin, F.R. and Witty, J.F. (1985) Control of nitrogen fixation in a legume nodule: an analysis of the role of oxygen diffusion in relation to nodule structure. *Annals of Botany* 55, 549-562.

Sidorova, K.K. and Uzhintseva, P. (1992) Use of mutants to detect genes controlling symbiotic characteristics in the pea. *Soviet Genetics* 28, 494-500.

Smit, G., Logman, T.J.J., Boerrigter, M.E.T.I., Kijne, J.W. and Lugtenberg, B.J.J. (1989) Purification and partial characterization of the *Rhizobium leguminosarum* biovar *viciae* Ca^{2+}-dependent adhesin, which mediates the first step in attachment of cells of the family Rhizobiaceae to plant root hair tips. *Journal of Bacteriology* 171, 4054-4062.

Smith, A.M. (1985) Capacity for fermentation in roots and *Rhizobium* nodules of *Pisum sativum* L. *Planta* 166, 264-270.

Soberon, M., Williams, H.D., Poole, R.K. and Escamilla, E. (1989) Isolation of a *Rhizobium phaseoli* cytochrome mutant with enhanced respiration and symbiotic nitrogen fixation. *Journal of Bacteriology* 171, 465-472.

Spaink, H.P. (1992) Rhizobial lipo-oligosaccharides: answers and questions. *Plant Molecular Biology* 20, 977-986.

Spaink, H.P., Okker, R.J.H., Wijffelman, C.A., Pees, E. and Lugtenberg, B.J.J. (1987a) Promoters in the nodulation region of the *Rhizobium leguminosarum* Sym plasmid pRL1JI. *Plant Molecular Biology* 9, 27-39.

Spaink, H.P., Wijffelman, C.A., Pees, E., Okker, R.J.H. and Lugtenberg, B.J.J. (1987b) *Rhizobium* nodulation gene *nodD* is a determinant of host specificity. *Nature* 328, 337-339.

Spaink, H.P., Okker, R.J.H., Wijffelman, C.A., Tak, T., Goosen-de Roo, L., Pees, E., Van Brussel, A.A.N. and Lugtenberg, B.J.J. (1989a) Symbiotic properties of rhizobia containing a flavonoid-independent hybrid *nodD* product. *Journal of Bacteriology* 171, 4045-4053.

Spaink, H.P., Weinman, J., Djordjevic, M.A., Wijffelman, C.A., Okker, R.J.H. and Lugtenberg, B.J.J. (1989b) Genetic analysis and cellular localization of the *Rhizobium* host specificity-determining NodE protein. *EMBO Journal* 8, 2811-2818.

Spaink, H.P., Sheeley, D.M., Van Brussel, A.A.N., Glushka, J., York, W.S., Tak, T., Geiger, O., Kennedy, E.P., Reinhold, V.N. and Lugtenberg, B.J.J. (1991) A novel, highly unsaturated, fatty acid moeity of lipo-oligosaccharide signals determines host specificity of *Rhizobium leguminosarum*. *Nature* 354, 125-130.

Sprent, J.I. and Raven, J.A. (1992) Evolution of nitrogen-fixing symbioses. In: Stacey, G., Burris, R.H. and Evans, H.J. (eds), *Biological Nitrogen Fixation*. Chapman and Hall, New York, pp. 461-496.

Sprent, J.I., Stephens, J.H. and Rupela, O.P. (1988) Environmental effects on nitrogen fixation. In: Summerfield, R.J. (ed.), *World Crops: Cool Season Food Legumes*. Kluwer Academic Publishers, Dordrecht, pp. 801-810.

Stacey, G., Burris, R.H. and Evans, H.J. (1992) *Biological Nitrogen Fixation*. Chapman and Hall, New York.

Stanley, J., Dowling, D.N. and Broughton, W.J. (1988) Cloning of *hemA* from *Rhizobium* sp. NGR234 and symbiotic phenotype of a gene-directed mutant in diverse legume genera. *Molecular and General Genetics* 215, 32-37.

Stougaard, J., Jorgensen, J.-E., Christensen, T., Kuhle, A. and Marcker, K.A. (1990) Interdependence and nodule specificity of *cis*-acting regulatory elements in the soybean leghemoglobin lbc_3 and N23 promoters. *Molecular and General Genetics* 222, 353.

Surin, BP. and Downie, J.A. (1988) Chracterization of the *Rhizobium leguminosarum* genes *nodLMN* involved in efficient host specific nodulation. *Molecular Microbiology* 2, 173-183.

Surin, B.P. and Downie, J.A. (1989) *Rhizobium leguminosarum* genes required for expression and transfer of host specific nodulation. *Plant Molecular Biology* 12, 19-29.

Surin, B.P., Watson, J.M., Hamilton, W.D.O., Economou, A. and Downie, J.A. (1990) Molecular characterization of the nodulation gene *nodT* from two biovars of *Rhizobium leguminosarum*. *Molecular Microbiology* 4, 245-252.

Sutton, M.J., Lea, E.J.A., Crank, S., Rivilla, R., Economou, A., Ghelani, S., Johnston, A.W.B. and Downie, J.A. (1993) NodO: a nodulation protein that forms pores in membranes. In: Nester, E.W. and Verma, D.P.S. (eds), *Advances in Molecular Genetics of Plant–Microbe Interactions*. Kluwer Academic Publishers, Dordrecht, pp. 163-167.

Szabados, L., Ratet, P., Grunenberg, B. and de Bruijn, F.J. (1990) Functional analysis of the *Sesbania rostrata* leghemoglobin *glb3* gene 5'-upstream region in transgenic *Lotus corniculatus* and *Nicotiana tabacum* plants. *Plant Cell* 2, 973-986.

Thony-Meyer, L., Stax, D. and Hennecke, H. (1989) An unusual gene cluster for the cytochrome bc_1 complex in *Bradyrhizobium japonicum* and its requirement for effective root nodule symbiosis. *Cell* 57, 683-697.

Tingey, S.V., Walker, E.L. and Coruzzi, G.M. (1987) Glutamine synthetase genes of pea encode distinct polypeptides which are differentially expressed in leaves, root and nodules. *EMBO Journal* 6, 1-9.

Triplett, E.W. and Sadowsky, M.J. (1992) Genetics of competition for nodulation of legumes. *Annual Review of Microbiology* 46, 399-428.
Truchet, G., Barker, D.G., Camut, S., de Billy, F., Vasse, J. and Huguet, T. (1989a) Alfalfa nodulation in the absence of *Rhizobium*. *Molecular and General Genetics* 219, 65-68.
Truchet, G., Camut, S., de Billy, F., Odorico, R. and Vasse, J. (1989b) The *Rhizobium*-legume symbiosis: two methods to discriminate between nodules and other root-derived structures. *Protoplasma* 149, 82-88.
Truchet, G., Roche, P., Lerouge, P., Vasse, J., Camut, S., de Billy, F., Promé, J-C. and Dénarié, J. (1991) Sulphated lipo-oligosaccharide signals of *Rhizobium meliloti* elicit root nodule organogenisis in alfalfa. *Nature* 351, 670-673.
Tsai, F.-Y. and Coruzzi, G.M. (1990) Dark-induced and organ-specific expression of two asparagine synthetase genes in *Pisum sativum*. *EMBO Journal* 9, 323-332.
Udvardi, M.K., Price, G.D., Gresshoff, P.M. and Day, D.A. (1988) A dicarboxylate transporter on the peribacteroid membrane of soybean nodules. *FEBS Letters* 231, 36-40.
Uheda, E. and Syono, K. (1982a) Physiological role of leghaemoglobin heterogeneity in pea root nodule development. *Plant and Cell Physiology* 23, 75-84.
Uheda, E. and Syono, K. (1982b) Effects of leghaemoglobin components on nitrogen fixation and oxygen consumption. *Plant and Cell Physiology* 23, 85-90.
van Beuschem, M.L. (1983) Xylary charge distribution and nitrogen transport in *Pisum sativum* L. during dinitrogen fixation or nitrate nutrition. *Zeitschrift für Pflanzenphysiologie* 109, 449-458.
Van Brussel, A.A.N., Zaat, S.A.J., Canter-Cremers, H.C.J., Wijffelman, C.A., Pees, E., Tak, T. and Lugtenberg, B.J.J. (1986) Role of plant root exudate and Sym plasmid-localized nodulation genes in the synthesis by *Rhizobium leguminosarum* of Tsr factor, which causes thick and short roots on common vetch. *Journal of Bacteriology* 165, 517-522.
Van Brussel, A.A.N., Recourt, K., Pees, E., Spaink, H.P., Tak, T., Wijffelman, C.A., Kijne, J.W. and Lugtenberg, B.J.J. (1990) A biovar-specific signal of *Rhizobium leguminosarum* bv. *viciae* induces increased nodulation gene-inducing activity in root exudate of *Vicia sativa* subsp. *nigra*. *Journal of Bacteriology* 172, 5394-5401.
Van Brussel, A.A.N., Bakhuizen, R., van Spronsen, P.C., Spaink, H.P., Tak, T., Lugtenberg, B.J.J. and Kijne, J.W. (1992) Induction of pre-infection thread structures in the leguminous host plant by mitogenic lipo-oligosaccharides of *Rhizobium*. *Science* 257, 70-72.
Vance, C.P. and Heichel, G.H. (1991) Carbon in nitrogen fixation: limitation or exquisite adaptation. *Annual Review of Plant Physiology and Plant Molecular Biology* 42, 373-392.
van de Wiel, C., Norris, J.H., Bochenek, B., Dickstein, R., Bisseling, T. and Hirsch, A.M. (1990a) Nodulin gene expression and ENOD2 localization in effective, nitrogen-fixing and ineffective, bacteria-free nodules of alfalfa. *Plant Cell* 2, 1009-1017.
van de Wiel, C., Scheres, B., Franssen, H.J., van Lierop, M.-J., van Lammeren, A., Van Kammen, A. and Bisseling, T. (1990b) The early nodulin transcript ENOD2 is located in the nodule parenchyma (inner cortex) of pea and soybean root nodules. *EMBO Journal* 9, 1-7.

van Egeraat, A.W.S.M. (1975) The possible role of homoserine in the development of *Rhizobium leguminosarum* in the rhizosphere of pea seedlings. *Plant and Soil* 42, 381–386.
VandenBosch, K.A., Bradley. D.J., Knox, J.P., Perotto, S., Butcher, G.W. and Brewin, N.J. (1989) Common components of the infection thread matrix and the intercellular space identified by immunocytochemical analysis of pea nodules and uninfected roots. *EMBO Journal* 8, 335–342.
Vasse, J., de Billy, F. and Truchet, G. (1993) Control of infection in the alfalfa–*Rhizobium meliloti* symbiosis. In: Nester, E.W. and Verma, D.P.S. (eds), *Advances in Molecular Genetics of Plant–Microbe Interactions.* Kluwer Academic Publishers, Dordrecht, pp. 381–384.
Verma, D.P.S. (1992) Signals in root nodule organogenesis and endocytosis of *Rhizobium. Plant Cell* 4, 373–382.
Verma, D.P.S. and Delauney, A.J. (1988) Root nodule symbiosis: nodulins and nodulin genes in plant gene research. In: Verma, D.P.S. and Goldberg, R.B. (eds), *Temporal and Spatial Regulation of Plant Genes.* Springer-Verlag, Berlin, pp. 169–199.
Verma, D.P.S., Hu, C.-A. and Zhang, M. (1992) Root nodule development: origin, function and regulation of nodulin genes. *Physiologia Plantarum* 85, 253–265.
Vincent, J.M. (1980) Factors controlling the legume–*Rhizobium* symbiosis. In: Newton, W.E. and Orme-Johnson, W.H. (eds), *Nitrogen Fixation, Vol. II: Symbiotic Associations and Cyanobacteria*, University Park Press, Baltimore, pp. 103–129.
Walker, E.L. and Coruzzi, G.M. (1989) Developmentally regulated expression of the gene family for cytosolic glutamine synthetase in *Pisum sativum. Plant Physiology* 91, 702–708.
Watson, R.J., Chan, Y.-K., Wheatcroft, R., Yang, A.-F, and Han, S. (1988) *Rhizobium meliloti* genes required for C_4-dicarboxylate transport and symbiotic nitrogen fixation are located on a megaplasmid. *Journal of Bacteriology* 170, 927–934.
Weeden, N.F., Kneen, B.E. and LaRue, T.A. (1990) Genetic analysis of *sym* genes and other nodule-related genes in *Pisum sativum*. In: Gresshoff, P.M., Roth, J., Stacey, G. and Newton, W.E. (eds), *Nitrogen Fixation: Achievements and Objectives.* Chapman and Hall, New York, London, pp. 323–330.
Weeden, N.F., Ambrose, M.J. and Sweicicki, W. (1992) *Pisum sativum*. In: O'Brien, S. (ed.), *Genetic Maps.* Cold Spring Harbor Laboratory, New York.
Werner, D., Mellor, R.B., Hahn, M.G. and Grisebach, H. (1985) Glyceollin 1 accumulation in an ineffective type of soybean nodule with an early loss of peribacteroid membrane. *Zeitschrift für Naturforschung* 40c, 179–181.
Wijffelman, C.A., Pees, E., Van Brussel, A.A.N., Okker, R.J.H. and Lugtenberg, B.J.J. (1985) Genetic and functional analysis of the nodulation region of the *Rhizobium leguminosarum* Sym plasmid pRLlJI. *Archives of Microbiology* 143, 225–232.
Witty, J.F., Minchin, F.R., Skot, L. and Sheehy, J.E. (1986) Nitrogen fixation and oxygen in legume root nodules. *Oxford Surveys of Plant Molecular and Cell Biology* 3, 275–314.
Witty, J.F., Skot, L. and Revsbech, N.P. (1987) Direct evidence for changes in the resistance of legume root nodules to O_2 diffusion. *Journal of Experimental Botany* 38, 1129–1140.

Yang, W.-C., Cremers, H.C.J.C., Hogendijk, P., Katinakis, P., Wijffelman, C.A., Franssen, H.J., Van Kammen, A. and Bisseling, T. (1992) *In-situ* localisation of chalcone synthase mRNA in pea root nodule development. *Plant Journal* 2, 143–151.

Yao, P.Y. and Vincent, J.M. (1969) Host specificity in the root hair 'curling factor' of *Rhizobium* spp. *Australian Journal of Biological Sciences* 22, 413–423.

Young, J.P.W. and Johnston, A.W.B. (1989) The evolution of specificity in the legume–*Rhizobium* symbiosis. *Trends in Ecology and Evolution* 4, 341–349.

Young, J.P.W. and Matthews, P. (1982) A distinct class of peas (*Pisum sativum* L.) from Afghanistan that show strain specificity for symbiotic *Rhizobium*. *Heredity* 48, 203–210.

Zaat, S.A.J., Wijffelman, C.A., Spaink, H.P., Van Brussel, A.A.N., Okker, R.J.H. and Lugtenberg, B.J.J. (1987) Induction of the *nodA* promoter of *Rhizobium leguminosarum* sym plasmid pRL1JI by plant flavanones and flavones. *Journal of Bacteriology* 169, 198–204.

Tissue Culture and Transformation

D.R. DAVIES AND P.M. MULLINEAUX
John Innes Institute, Colney Lane, Norwich, NR4 7UH, UK

Tissue Culture

Pea roots were among the first plant tissues to be cultured *in vitro* (Scheitterer, 1931; Bonner and Addicott, 1937). Given that early advantage, it is somewhat disconcerting that tissue culture of peas remains a challenge, with embryogenesis and organogenesis *in vitro* as well as regeneration from protoplasts still being difficult to achieve on a regular basis. In this respect, peas resemble many of the other grain legumes. It is not within the remit of this chapter to review all aspects of the tissue culture of peas; rather, we will emphasize those that impinge on the area of biotechnology. These include the exploitation of protoplasts for cell fusion and transformation and the use of cell, tissue and organ culture as vehicles for transformation.

Methods of culturing immature pea embryos exist which allow a range of developmental stages to be grown *in vitro* for extended periods of time (Stafford and Davies, 1979; Wang *et al.*, 1987). They have been exploited to study, for example, the initiation and pattern of accumulation of storage proteins, the temporal expression of genes regulating these proteins and their spatial distribution in the embryonic tissues (Domoney *et al.*, 1980; Corke *et al.*, 1990; Yang *et al.*, 1990). Although at a gross macroscopic level the growth of the embryos *in vitro* can simulate the growth of those *in vivo*, they differ markedly when studied at a cellular level. Their primary experimental value is that they allow an analysis of certain developmental processes that are difficult to study *in vivo* and under environmental conditions that can be manipulated at will.

Meristem culture has proved to be of value in the elimination of many viruses in a range of plants including peas. Pea seed-borne mosaic virus

(PSBMV) has become one of the primary problems facing pea growers in North America and Europe (see Chapter 8). At one stage it was thought that radical measures to remove infected stocks from breeders' material might contain the disease; in the event this proved not to be possible. However Kartha and Gamborg (1978) successfully eliminated PSBMV from over 100 breeding lines of pea by the use of meristem culture.

Somaclonal variants have been considered to be a valuable source of genetic variation (Larkin and Scowcroft, 1981) but, while this is true for some horticultural species, there is still little evidence to support the claim for most crop plants and none for peas; neither have there been reports of the selection of useful mutants from cell cultures of peas grown *in vitro*. Haploid plants provide a rapid means of producing homozygotes from segregating populations, and have been exploited to this end in some breeding programmes. One of the methods of generating haploids is by anther or pollen culture; attempts to do so in peas have resulted in the production of haploid callus, and embryo-like structures were observed, but no haploid plants could be recovered (Gupta, 1975; Gosal and Bajaj, 1988).

Protoplasts

Landgren and Torrey (1973) were the first to isolate protoplasts from root cells and Constabel *et al.* (1973) from leaf mesophyll tissue. The latter succeeded in regenerating a proportion into microcalli by culturing them in microdrops. Stem apices of 3-day-old seedlings were used as a source material by Gamborg *et al.* (1975) but again no progress beyond the generation of callus was possible and no plants were regenerated. Landgren (1976a,b, 1981), von Arnold and Eriksson (1976, 1977), Crowder *et al.* (1979) and Jia (1982) examined the influence of different tissue sources, isolation and culture conditions as well as media but with no significant improvements in terms of survival and differentiation beyond those observed by earlier workers. The first reports of the successful regeneration of shoots from protoplast-derived calli were those of Puonti-Kaerlas and Eriksson (1988) and Lehminger-Mertens and Jacobsen (1989a). Both used a number of genotypes but the frequency of calli giving rise to shoots was low and no rooted plants were produced. Successful embryogenesis and regeneration of whole plants from protoplast-derived callus were finally achieved by Lehminger-Mertens and Jacobsen (1989b). Of six genotypes tested, two produced somatic embryos, from which fertile plants were derived.

Given this success, then, there are two ways in which protoplasts can be exploited in plant biotechnology: for cell fusion, or as vehicles for transformation. The latter topic will be considered later. Since the genus *Pisum* has only two species and these are sexually compatible to an extent that allows the exchange of genetic information, there is no incentive to exploit protoplasts for this reason. Any attempt at protoplast fusion must therefore be predicated on the assumption that intergeneric or interfamily fusions are likely to be successful, either in producing viable hybrid progeny or in allowing transfer of a limited amount of useful genetic information; neither assumption appears justified now. Soyabean-pea protoplast fusion products are capable of cell division (Kao et al., 1974; Constabel et al., 1975, 1976; Praznovsky et al., 1981), although the fusion products survive for a very limited period. Comparable levels of cell division and survival were obtained with the fusion products of pea and *Vicia hajastana* protoplasts (Kao and Michayluk, 1974; Kao et al., 1974). A more realistic approach as far as protoplast fusion is concerned might be therefore to expose the 'alien' parental line to ionizing radiation prior to fusion, in an attempt to introduce small segments from another legume into pea. However, even this is an extremely speculative approach in the absence of any clearly identifiable monogenic characters to be transferred and of a selection system for the incorporated fragments. There is no information on attempts to achieve organelle transfer in peas by protoplast fusion either. In summary, protoplast fusion offers little prospect in the genus *Pisum*.

Transformation

For reasons that are not obvious, transformation of grain legumes has proved a difficult, though not intractable, problem. The components of a transformation system are an appropriate cell or tissue to act as recipient and from which whole plants can be recovered, an efficient method of delivering constructs, and agents that allow selection and isolation of transformed cells, tissues and ultimately plants.

No success has been reported using pea protoplasts for transformation; the very low plating efficiencies of those derived from roots has been reported to be a primary mitigating factor by Schaerer and Pilet (1991). Leaf- and root-derived protoplasts showed transient expression of β-glucuronidase when electroporated with plasmid DNA containing the *uidA* gene (Hobbs et al., 1990). The influence of different promoters and of the presence or absence of a leader sequence under different electroporation conditions was investigated by these last authors, but the relevance of information derived from studies of transient expression to that required for transformation is still a matter of debate.

There are a number of cells and tissues other than protoplasts which in theory could be used as the recipient for transfer of exogenous DNA to pea. Callus cultures have been derived from a very wide range of pea tissues including apices, epicotyls, embryos, leaves, hypocotyls, roots, cotyledons, flower buds and stems, and plants have been regenerated from the first four [details of procedures and results can be found in reviews by Gantotti and Kartha (1986) and Griga and Novak (1990) and in the more recent papers of Tétu et al. (1990), Nielsen et al. (1991) and Ozcan et al. (1992)]. Given the ability to regenerate plants from callus and the well-established fact that most strains of peas are susceptible to infection by *Agrobacterium* (Robbs et al., 1991), the transformation of peas should be an attainable objective; the last 3 years have seen some progress in this respect. Puonti-Kaerlas et al. (1989) tested the response of five pea cultivars to three wild-type strains of *A. tumefaciens* and to two shooty, two rooty and one disarmed strain. Their first conclusion was that there was no difference in the response of the cultivars in terms of tumour induction but that susceptibility *in vivo* was not completely correlated with the response *in vitro*. The wild-type strain A281 was found to be more virulent than the wild-type C58 and the rooty strain GV3101 (pGV304) in the production of tumours both *in vivo* and *in vitro*. Epicotyl segments from 2-week-old seedlings and explants of stems and leaves were co-cultivated for 2 days with *Agrobacterium* and the calli generated from those infected with the disarmed strain were subsequently challenged on a medium containing 75 mg l^{-1} kanamycin. Two varieties – Filby and Petra – gave rise to resistant calli, most of which were derived from stem and leaf segments rather than the epicotyl. Of those tested, 91% showed nopaline synthase activity and, of the 12 transformants tested, all showed the presence on Southern blots of DNA fragments from the gene coding for kanamycin resistance, *nptII*, of the expected size. The two shooty and two rooty strains were ineffective (but see later). These were encouraging results in the sense that evidence of the integration of plasmid sequences from a disarmed strain into the pea genome was provided, but differentiation of plants from the callus could not be achieved.

Lulsdorf et al. (1991) tested the effect of *Agrobacterium* strain, different plasmids, period of co-cultivation (2–4 days) and the relative efficiencies of hygromycin (25 mg l^{-1}) and kanamycin (50 mg l^{-1}) as selective agents. The explants were stem sections from 5-day-old seedlings. The strain EHA101 was more effective than three C58 derivatives, LBA4404 and WR3095 in generating resistant calli, and a longer co-cultivation period was optimal. Beyond this it appeared that there were complex interactions of strain, genotype and co-cultivation protocols, and a precise definition of optimal conditions was not feasible. A caveat in the interpretation of these results, however, is that the only criterion of transformation was survival on the selection media.

The inability to differentiate and regenerate plants from callus in the experiments described so far was a severe limitation to progress, although as stated earlier there are many examples of successful differentiation of callus and even of the production of somatic embryos from certain tissues (Kysely et al., 1987; Kysely and Jacobsen, 1990; Tétu et al., 1990).

De Kathen and Jacobsen (1990) inoculated epicotyls and node explants from 7-day-old etiolated seedlings of the cultivar Madria with *A. tumefaciens* strains GV2260 or GV3850 carrying kanamycin- or hygromycin-resistance genes. A low frequency of plantlets showing resistance to either selectable marker was obtained, but none of the plants set seed.

A different approach was adopted by Zubko et al. (1990), who co-cultivated shoot and leaf segments with a shooty mutant of *A. tumefaciens* pGV2206. The callus derived from these retained its ability to produce shoots for at least 18 months and Southern blot analyses confirmed the presence of integrated *nptII* sequences in the genome of some of the plants. The problem with this approach is the inability to root the transgenic shoots and the need therefore to graft on to untransformed plants; however, this has been done and a further sexual generation produced.

Nauerby *et al.* (1991) inoculated thin cell layers of node segments from 10–12-day-old seedlings of the cultivar Bodil with *A. tumefaciens* C58 C1 containing one of two binary vectors; these were pGV2260::pGSGLUC1 and pGV3850::pGSGLUC1 with the *nptII* gene as the selectable marker and the *uidA (gus)* gene as reporter. Either the segments were co-cultivated with the bacteria or the nodes were injected with a needle coated with *Agrobacterium* before excision and plating on nutrient media. From 148 explants, eight shoots showing GUS activity were obtained. The first report of the generation of transformed rooted plants was that of Puonti-Kaerlas *et al.* (1990, 1992). Epicotyls from 10-day-old seedlings or shoot cultures derived from the epicotyls of five cultivars were co-cultivated with *A. tumefaciens* GV3101 (pGV2260::pGV1503) with GV3101 (pGV3850::pCAP212) or with GV3101 (pGV3850::pLD1). The three strains carried genes coding for resistance to hygromycin, kanamycin, or kanamycin plus methotrexate, respectively. The subsequent growth of callus was evaluated on a wide range of media and the comparative frequencies of transformation of the resulting callus established. In all instances Puget was the most responsive cultivar and the transformation rate with GV3101 (pGV2260::pGV1503), which carried the *hpt* gene for hygromycin resistance, was up to 3.7 times higher than that with the two other constructs. The best callus culture protocol involved sequential transfer through five different media; 1–15% of the calli in the variety Stivo and 0.2% in the cultivar Puget produced shoots, but they appeared only from hygromycin-resistant calli. The time required for shoot induction was 4–9 months. Seeds were produced on three of the six regenerated transformed lines of Stivo; the first generation progeny of one plant (S2) segregated 23 resistant : 7 sensitive to hygromycin, and the progeny of another plant (S3) segregated 3 resistant : 1 sensitive. By testing second-generation progeny from the resistant plants of S2, hygromycin resistance was shown to be stably transmitted through a sexual generation and to be due to a single dominant gene. Fragments of the appropriate size hybridizing to an *hpt* probe were detected in Southern blots of DNA from first- and second-generation resistant plants; the S2 plants had a single copy and the S3 had two copies of the *hpt* sequence. Northern blots confirmed the presence of a transcript of *hpt* in the first-generation plants from S3 and in the second-generation plants from S2. Unfortunately, when the first- and second-generation plants were subjected to cytological analysis, all were found to be tetraploid, which could be due to the extended period of callus culture before the plants differentiated.

To try and avoid the problems of regeneration and of cytological changes associated with an extended period of callus culture, Davies et al. (1993) adopted a different procedure for transforming peas. There have been several claims that inoculation of the meristems of seed can result in transformed plants. For example, Chee et al. (1989) successfully inoculated the plumule and cotyledonary node of germinating seeds of soyabean (*Glycine max*) with an avirulent strain of *A. tumefaciens* to produce transgenic plants. A similar approach was adopted by Penza et al. (1991), who successfully co-cultivated detached embryos from mature seed of cowpea (*Vigna unguiculata*) with *A. tumefaciens* to produce transformed shoots. Isolated meristems derived from germinating seeds of maize (Gould et al., 1991) and of petunia (Ulian et al., 1988) have been exploited successfully for transformation, using *Agrobacterium*. In all these reports it has been stressed that the primary motivation has been to avoid the problems associated with the differentiation of normal plants from transformed callus in these particular species. Davies et al. (1993) adopted a comparable method of inoculating the meristems in peas. Seeds of the cultivar Puget were germinated for 3 days, the testa, primary shoot and root were then removed and, using glass needles, the two cotyledonary buds were inoculated with *A. tumefaciens* C58/3 containing the binary vector SLJ1911. The vector contains *uidA* (*gus*) and *aph*3II genes driven by 35S promoters. The embryos were co-cultivated with the bacteria for 3–4 days, the cotyledons removed and the embryonic axis, with its two inoculated lateral buds, challenged with 60–75 mg l^{-1} aqueous solution of kanamycin for 24 h prior to plating on medium containing 100 mg l^{-1} kanamycin. The shoots that were positive for GUS activity were proliferated and then rooted prior to transfer to soil. Approximately 4 months elapsed between inoculation and the production of rooted plants. In one series of experiments four transformed plants were produced from 277 seeds. In three of these a meristematic callus was associated with the inoculated meristems. Of the four plants, one was partially and three fully pollen- and seed-fertile. Second-generation plants were all fully fertile and third- and fourth-generation fertile plants have been generated. The presence of *uidA* gene sequences in the genomic DNA of the first-generation plants was confirmed by a polymerase chain reaction procedure and by Southern blots of the DNA. Second-generation progeny were also tested for GUS activity and for the presence of integrated *uidA* sequences; the two were completely correlated. This method, while showing the feasibility of producing fertile diploid transgenic peas, is still difficult to reproduce consistently and needs to be further developed. The *bar* gene coding for resistance to phosphinothricin has also been used as a selectable marker, and plants showing resistance to phosphinothricin-containing herbicides have been generated.

Schroeder et al. (1993) have made the most significant progress in developing a reproducible transformation system for peas. The embryonic axis of developing seeds that were 2–5 days beyond the stage of maximum fresh weight was cut into longitudinal sections, after removing the root end, and immersed in a suspension of *A. tumefaciens* prior to plating on B5h (Brown and Atanassov, 1985) medium. The *Agrobacterium* strain used was critical and was the hypervirulent AGLI with the binary plasmid pSLJ1561 containing the *bar* and *npt*II genes driven by 35S promoters. The axis sections were co-cultivated for 4 days at 26°C and then plated on a Murashige and Skoog (1962) medium containing 10 mg l^{-1} Phosphinothricin (PPT) as the selective agent, for callus induction. Fifteen days later, and every subsequent 20 days, they were replated on a shoot induction medium containing 15 mg l^{-1} PPT. Rooted plants were subsequently produced which flowered and set seed. Putative transformants were tested for the presence of phosphinothricin acetyl transferase (PAT) activity, indicating the presence of a functional *bar* gene, and for neomycin phosphotransferase activity (NPT), for the *npt*II gene. Confirmation of their transgenic nature was obtained by testing leaves for their resistance to the PPT-containing herbicide Basta. The time between excision of the initial axis segments and production of mature transformed plants was c. 9 months. Between 1.5% and 2.5% of the starting explants gave rise to transformed plants in the two varieties used – Rondo and Greenfeast. PAT and NPT activities were detectable in the next generation of plants and segregated 3 : 1 for presence : absence.

The transformation of peas is thus an attainable goal. This achievement can be placed in perspective by comparing it with those obtained with other grain legumes. The most intensively studied has been soyabean (*Glycine max*); in this species transgenic plants have been produced by inoculating germinating seeds (Chee *et al.*, 1989) or cotyledonary explants (Hinchee *et al.*, 1988; Zhou and Atherly, 1990) with *A. tumefaciens* or by accelerating DNA-coated particles into the meristems of immature seeds (McCabe *et al.*, 1988) – the last being the current method of choice. Cowpea (*Vigna unguiculata*) has been transformed, using *A. tumefaciens* (Penza *et al.*, 1991), and in a recent paper (Russell *et al.*, 1993) the generation of transformed plants of *Phaseolus vulgaris* by electric discharge-induced particle acceleration of DNA into seed meristems was reported. Of the other grain legumes, moth bean (*Vigna aconitifolia*) has been transformed by direct uptake of DNA into protoplasts (Kohler *et al.*, 1987) or by co-cultivating protoplasts with *A. tumefaciens* (Eapen *et al.*, 1987). While there are many reports of transformed cells and callus tissues in these, as well as other, grain legume species, the production of transformed plants has still to be achieved in the majority.

On the basis of the results obtained with soyabean and *Phaseolus vulgaris* it will be important to test the efficacy of using microprojectiles to transform peas. In these experiments and those that have already proved successful in peas, seed meristems have been the target tissue and there is every indication that this should remain the material of choice for transformation experiments in legumes; the supply of material is virtually unlimited and the problems of somaclonal variation, which are often associated with callus culture, are minimized. Impressive progress has been achieved over the last few years in transforming peas and the prospects for achieving a routine method of transformation are excellent. The goals in the exploitation of transformation of peas are manifold and will be discussed in Chapter 11.

References

Bonner, J. and Addicott, F.T. (1937) Culture *in vitro* of excised pea roots. *Botanical Gazette* 99, 144–170.

Brown, D.C.W. and Atanassov, A. (1985) Role of genetic background in somatic embryogenesis in *Medicago*. *Plant Cell, Tissue and Organ Culture* 4, 111–122.

Chee, P.P., Fober, K.A. and Slightom, J.L. (1989) Transformation of soybean (*Glycine max*) by infecting germinating seeds with *Agrobacterium tumefaciens*. *Plant Physiology* 91, 1212–1218.

Constabel, F., Kirkpatrick, J.W., and Gamborg, O.L. (1973) Callus formation from mesophyll protoplasts of *Pisum sativum*. *Canadian Journal of Botany* 51, 2105–2106.

Constabel, F., Dudits, D., Gamborg, O.L. and Kao, K.N. (1975) Nuclear fusion in intergeneric herokaryons: a note. *Canadian Journal of Botany* 53, 2092–2095.

Constabel, F., Weber, G., Kirkpatrick, J.W. and Pahl, K. (1976) Cell division of intergeneric protoplast fusion products. *Zeitschrift Pflanzenphysiologie* 79, 1–7.

Corke, F.M.K., Hedley, C.L. and Wang, T.L. (1990) An analysis of seed development in *Pisum sativum*, XI. Cellular development and the deposition of storage protein in immature embryos grown *in vivo* and *in vitro*. *Protoplasma* 155, 127–135.

Crowder, A.J., Landgren, C.R. and Rockwood, L.L. (1979) Cultivar differences in starch content and protoplast yields from root cortical explants of *Pisum sativum*. *Physiologie Plantarum* 46, 85–88.

Davies, D.R., Hamilton, J. and Mullineaux, P. (1993) Transformation of peas. *Plant Cell Reports* 12, 180–183.

de Kathen, A. and Jacobsen, H.-J. (1990) *Agrobacterium tumefaciens*-mediated transformation of *Pisum sativum* L. using binary and cointegrate vectors. *Plant Cell Reports* 9, 276–279.

Domoney, C., Davies, D.R. and Casey, R. (1980) The initiation of legumin synthesis in immature embryos of *Pisum sativum* L. grown *in vivo* and *in vitro*. *Planta* 149, 454–460.

Eapen, S., Kohler, F., Gerdemann, M. and Shieder, O. (1987) Cultivar dependence of transformation rates in moth bean after co-cultivation of protoplasts with *Agrobacterium tumefaciens*. *Theoretical and Applied Genetics* 75, 207–210.

Gamborg, O.L., Shyluk, J.P. and Kartha, K.K. (1975) Factors affecting the isolation and callus formation in protoplasts from shoot apices of *Pisum sativum* L. *Plant Science Letters* 4, 285–292.

Gantotti, B.V. and Kartha, K.K. (1986) The pea. In: Evans, D.A., Sharp, W.R. and Ammirato, P.V. (eds), *Handbook of Plant Cell Culture – Techniques and Applications*, vol. 4. Macmillan, New York, pp. 370–418.

Gosal, S.S. and Bajaj, Y.P.S. (1988) Pollen embryogenesis and chromosomal variation in anther culture of three food legumes – *Cicer arientinum, Pisum sativum* and *Vigna mungo*. *Sabrao Journal* 20, 51–58.

Gould, J., Devey, M., Hasegawa, O., Ulian, E.C., Peterson, G. and Smith, R.H. (1991)

Transformation of *Zea mays* L. using *Agrobacterium tumefaciens* and the shoot apex. *Plant Physiology* 95, 426–434.

Griga, M. and Novak, F.J. (1990) Pea. In: Bajaj, Y.P.S., (ed.), *Biotechnology in Agriculture and Forestry*, vol. 10, *Legumes and Oil Seed Crops*. Springer Verlag, Berlin, pp. 65–99.

Gupta, S. (1975) Morphogenetic response of haploid callus tissue of *Pisum sativum* var B22. *Indian Agriculturist* 19, 11–21.

Hinchee, M.A.W., Cannor-Ward, D.V., Newell, C.A., McDonnell, R.E., Sato, S.J., Gasser, C.S., Fischhoff, D.A., Re, D.B., Fraley, R.T. and Horsch, R.B. (1988) Production of transgenic soybean plants using *Agrobacterium*-mediated DNA transfer. *Bio/Technology* 6, 915–922.

Hobbs, S.L.A., Jackson, J.A., Baliski, D.S., DeLong, C.M.O. and Mahon, J.D. (1990) Genotype- and promoter-induced variability in transient β-glucuronidase expression in pea protoplasts. *Plant Cell Reports* 9, 17–20.

Jia, S.-R. (1982) Factors affecting the division frequency of pea mesophyll protoplasts. *Canadian Journal of Botany* 60, 2192–2196.

Kao, K.N. and Michayluk, M.R. (1974) A method for high-frequency intergeneric fusion of plant protoplasts. *Planta* 115, 355–367.

Kao, K.N., Constabel, F., Michayluk, M.R. and Gamborg, O.L. (1974) Plant protoplast fusion and growth of intergeneric hybrid cells. *Planta* 120, 215–227.

Kartha, K.K. and Gamborg, O.L. (1978) Meristem culture techniques in the production of disease-free plants and freeze-preservation of germplasm of tropical tuber crops and grain legumes. In: Maraite, H. and Meyer, J.A. (eds), *Disease of Tropical Food Crops*. Université Catholique, Louvain, Belgium, pp. 267–283.

Kohler, F., Golz, C., Eapen, S., Kohn, H. and Schnieder, O. (1987) Stable transformation of moth bean *Vigna aconitifolia* via direct gene transfer. *Plant Cell Reports* 6, 313–317.

Kysely, W. and Jacobsen, H.-J. (1990) Somatic embryogenesis from pea embryos and shoot apices. *Plant Cell, Tissue and Organ Culture* 20, 7–14.

Kysely, W., Myers, J.R., Lazzeri, P.A., Collins, G.B. and Jacobsen, H.-J. (1987) Plant regeneration via somatic embryogenesis in pea (*Pisum sativum* L.). *Plant Cell Reports* 6, 305–308.

Landgren, C.R. (1976a) Patterns of mitosis and differentiation in cells derived from pea root protoplasts. *American Journal of Botany* 63, 473–480.

Landgren, C.R. (1976b) The influence of culture conditions on mitotic activity in protoplasts derived from *Pisum* root cortical explants. *Protoplasma* 87, 49–69.

Landgren, C.R. (1981) Gibberellin enhancement of the enzymic release of *Pisum* root cell protoplasts. *Physiologia Plantarum* 52, 349–352.

Landgren, C.R. and Torrey, J.G. (1973) The culture of protoplasts derived from explants of seedling pea roots. *Colloques Internationaux CNRS* no. 212, 281–289.

Larkin, P.J. and Scowcroft, W.R. (1981) Somaclonal variation – a novel source of variability from cell culture for plant improvement. *Theoretical and Applied Genetics* 10, 197–214.

Lehminger-Mertens, R. and Jacobsen, H.J. (1989a) Plant regeneration from pea protoplasts via somatic embryogenesis. *Plant Cell Reports* 8, 379–382.

Lehminger-Mertens, R. and Jacobsen, H.J. (1989b) Protoplast regeneration and organogenesis for pea protoplasts. *In Vitro* 25, 571–574.

Lulsdorf, M.M., Rempel, H., Jackson, J.A., Baliski, D.S. and Hobbs, S.L.A. (1991)

Optimizing the production of transformed pea (*Pisum sativum* L.) callus using disarmed *Agrobacterium tumefaciens* strains. *Plant Cell Reports* 9, 479–483.

McCabe, D.E., Swain, W.F., Martinell, B.J. and Christou, P. (1988) Stable transformation of soybean (*Glycine max*) by particle acceleration. *Bio/Technology* 6, 923–926.

Murashige, T. and Skoog, F. (1962) A revised medium for rapid growth and bioassays with tobacco tissue cultures. *Physiologia Plantarum* 15, 473–497.

Nauerby, B., Madsen, M., Christiansen, J. and Wyndaele, R. (1991) A rapid and efficient regeneration system for pea (*Pisum sativum*), suitable for transformation. *Plant Cell Reports* 9, 676–679.

Nielsen, S.V.S., Poulsen, G.B. and Larsen, M.E. (1991) Regeneration of shoots from pea (*Pisum sativum*) hypocotyl explants. *Physiologia Plantarum* 82, 99–102.

Ozcan, S., Barghchi, M. Firek, S. and Draper, J. (1992) High frequency adventitious shoot regeneration from immature cotyledons of pea (*Pisum sativum* L.). *Plant Cell Reports* 11, 44–47.

Penza, R., Lurquin, P.F. and Filippone, E. (1991) Gene transfer by cocultivation of mature embryos with *Agrobacterium tumefaciens*: application to cowpea (*Vigna unguiculata* Walp). *Journal of Plant Physiology* 138, 39–43.

Praznovsky, T., Kurnik, E., Paal, H. and Dudits, D. (1981) Fusion between soybean (*Glycine max*) and pea (*Pisum sativum*): protoplast division in heterokaryons. *Biologia* 29, 213–218.

Puonti-Kaerlas, J. and Eriksson, T. (1988) Improved protoplast culture and regeneration of shoots in pea (*Pisum sativum* L). *Plant Cell Reports* 7, 242–245.

Puonti-Kaerlas, J., Stabel, P. and Eriksson, T. (1989) Transformation of pea (*Pisum sativum* L.) by *Agrobacterium tumefaciens*. *Plant Cell Reports* 8, 321–324.

Puonti-Kaerlas, J., Eriksson, T. and Engström, P. (1990) Production of transgenic pea (*Pisum sativum* L.) plants by *Agrobacterium tumefaciens*-mediated gene transfer. *Theoretical and Applied Genetics* 80, 246–252.

Puonti-Kaerlas, J., Eriksson, T. and Engström, P. (1992) Inheritance of a bacterial hygromycin phosphotransferase gene in the progeny of primary transgenic pea plants. *Theoretical and Applied Genetics* 84, 443–450.

Robbs, S.L., Hawes, M.C., Lin, H.-J., Pueppke, S.G. and Smith, L.Y. (1991) Inheritance or resistance to crown gall in *Pisum sativum*. *Plant Physiology* 95, 52–57.

Russell, D.R., Wallace, K.M., Bathe, J.H., Martinell, B.J. and McCabe, D.E. (1993) Stable transformation of *Phaseolus vulgaris* via electric-discharge mediated particle acceleration. *Plant Cell Reports* 12, 165–169.

Schaerer, S. and Pilet, P.E. (1991) Roots, explants and protoplasts from pea transformed with strains of *Agrobacterium tumefaciens* and *A. rhizogenes*. *Plant Science* 78, 247–258.

Scheitterer, H. (1931) Versuche zur Kultur von Pflanzengeweben. *Archiv exp. Zellforschung* 12, 141–176.

Schroeder, H.E., Schotz, A.H., Wardley-Richardson, T., Spencer, D. and Higgins, T.J.V. (1993) Transformation and regeneration of two cultivars of pea (*Pisum sativum* L.) *Plant Physiology* 101, 751–757.

Stafford, A. and Davies, D.R. (1979) The culture of immature pea embryos. *Annals of Botany* 44, 315–321.

Tétu, T., Sangwan, R.S. and Sangwan-Norreel, B.S. (1990) Direct somatic embryogenesis and organogenesis in cultured immature zygotic embryos of *Pisum sativum* L. *Journal of Plant Physiology* 137, 102–109.

Ulian, E.C., Smith, R.H., Gould, J.H. and McKnight, T.D. (1988) Transformation of plants via the shoot apex. *In Vitro Cellular and Developmental Biology* 24, 951-954.
von Arnold, S. and Eriksson, T. (1976) Factors affecting the growth and division of pea mesophyll protoplasts. *Physiologia Plantarum* 36, 193-196.
von Arnold, S. and Eriksson, T. (1977) A revised medium for growth of pea mesophyll protoplasts. *Physiologia Plantarum* 39, 257-260.
Wang, T.L., Smith, C.M., Cook, S.K., Ambrose, M.J. and Hedley, C.L. (1987) An analysis of seed development in *Pisum sativum*. III. The relationship between the r locus, the water content and the osmotic potential of seed tissues *in vivo* and *in vitro*. *Annals of Botany* 59, 73-80.
Yang, L.J., Barratt, D.H.P., Domoney, C., Hedley, C.L. and Wang, T.L. (1990) An analysis of seed development in *Pisum sativum* X. Expression of storage protein genes in cultured embryos. *Journal of Experimental Botany* 41, 283-288.
Zhou, J.H. and Atherly, A.G. (1990) In situ detection of transposition of the maize controlling element (Ac) in transgenic soybean tissues. *Plant Cell Reports* 8, 542-545.
Zubko, E.I., Kuchuk, N.V., Tumanova, L.G., Vikonskaya, N.A. and Gleba, Y.Y. (1990) Genetic transformation of pea plants mediated by *Agrobacterium tumefaciens*. *Biopolymers and Cells* 3, 80-84.

Opportunities in Relation to Developments in Plant Biotechnology

11

D.R. DAVIES
John Innes Institute, Colney Lane, Norwich, NR4 7UH, UK

The opportunities that exist for improving any crop relate to the limitations to its productivity, to its current and potential future uses, and to the possibility of growing the crop in new areas. These in turn depend on there being the methods, as well as the genetic resources, available for achieving the improvements sought; the methods can include those of conventional breeding and/or those of biotechnology. In the past, the genetic resources have referred to the germplasm available in the crop or its near relatives, but these limitations no longer apply now that DNA sequences from any source can in theory be introduced into any other organism.

The current limitations of the pea crop differ with geographical region, farming system and end use. However a primary one in all instances is that imposed by pathogens and pests. The susceptibility to pathogens has not been an insuperable limitation up to now, because sources of resistance or tolerance to particular races of fungal, viral and bacterial pathogens have been identified in most of those instances where they have been sought (Hagedorn, 1985, 1989), although in some instances resistance is associated with undesirable qualities, such as the presence of polyphenols. Resistance has not been reported to those insect pests which attack peas. Thus conventional breeding can in theory solve many of the challenges posed by pathogens, but not by pests.

Yield of seed, and consistency of yield over seasons, is the next limiting factor. There is no doubt that progress has been achieved in terms of yield per hectare (see Chapter 1), but it is salutary to note that the average yield of dry seed in many regions of the world is less than 1 tonne per hectare, whereas in some European countries it is over 4, with yields of over 7 tonnes per hectare having been reported in Holland (Dantuma, 1983). While much of this difference is attributable to different agronomic practices and inputs, it also indi cates the scope for improvement. Total yield reflects many plant characteristics, including response to pathogens and pests, standing ability, harvest index, efficiency of partitioning of carbon, sink strength and a number of other physiological parameters, many of which have yet to be defined. Attempts have been made to relate yield to numbers of pods per node, seed per pod and seed size (Krarup and Davis, 1970; Pandey and Gritton, 1975) but no definitive conclusions have been drawn. Hedley and Ambrose (1981) have defined an ideotype for semileafless peas grown for their mature dry seed. It would have the genetic constitution *afaf* (i.e. with leaflets converted to tendrils) and be non-branching, early-flowering, small-seeded and with one pod at each flowering node. Conventional breeding programmes should be able to generate the necessary combination of attributes defined in such an idiotype. Although those genotypes that have their leaflets converted to tendrils have improved standing ability in comparison with those with normal leaflets, standing ability is still inadequate in wetter regions; a solution to this problem remains to be found.

The problem of inconsistency of yield reflects the interaction of a genotype with biotic or abiotic stresses. The former have been mentioned briefly earlier; the latter include cold and heat stress, drought and waterlogging (Davies *et al.*, 1985), but solutions to this latter category of problems are not yet available.

Limitations that relate to end use can be categorized into those that relate to the nutritional quality of dry seed used for animal feed, those that affect the consumer preferences of immature seed used for freezing, canning or dehydrating, and those that relate to the use of mature dry seed in human food. The value of peas as animal feed depends on the amount of protein, the protein composition, the total energy value and the amount and nature of antinutritional factors. Breeding programmes have not succeeded in influencing protein contents, although a sustained recurrent selection programme for the character has not been undertaken; there is the added problem in such work that protein content is markedly influenced by environment (Jermyn and Slinkard, 1977; Matthews and Arthur, 1985), as also is sulphur-amino acid content (P. Matthews, personal communication). Improvement in protein quantity and quality will have to depend on novel approaches, some of which were considered in Chapter 6. Antinutritional factors in peas include protease inhibitors, lectins and saponins, and genetic variation for all exists within peas (Huisman, 1991; Chapter 6). Tannins are only of concern in coloured-seeded varieties. The extent to which particular carbohydrates such as α-galactosides (Savage and Deo, 1989) may affect nutritional value and the extent to which the storage proteins themselves may be antigenic and affect their absorption by animals (Le Guen *et al.*, 1991) remains to be definitively established.

The improvements sought for seed harvested in an immature state relate to consumer preferences, which in turn depend on seed size and colour, rate of maturation, sugar and starch content and cell-wall composition (in so far as it affects texture). There is no evidence that existing germplasm stocks have been exhausted of sources of variation for many of these characters. The same is true of seed harvested in a mature dry state for traditional uses as a vegetable.

Many of the limitations to the productivity and current uses of the crop which have been described can be overcome, therefore, given an appropriate input of research and development. The germplasm resources in pea are extensive (see Chapter 1) but there remains, with this as with all crops, an incentive both to exploit existing resources more thoroughly and to continue to foster further plant collections.

Given this situation, what role is there and what opportunities exist for the emerging techniques of biotechnology? In Chapter 10 it was argued that tissue culture *per se*, embryo culture of interspecific and intergeneric hybrids and meristem culture for virus elimination and protoplast fusion were all likely to have little impact on crop improvement in peas. The induction of mutants, where clearly defined objectives and selection criteria are available, has proved very successful (see Chapter 5), and 24 mutant pea cultivars have been produced, in which the characters improved have included maturity date, determinate habit and lodging resistance (IAEA, 1991). However, it is the ability to introduce novel or modified DNA sequences through transformation that is opening up new opportunities for a great number of crops, and peas are no exception.

Among the primary targets for the genetic engineering of plants has been resistance to pests and pathogens. The introduction of DNA sequences that code for insecticidal compounds has been successfully achieved in a number of plants (Gatehouse *et al.*, 1991). These compounds include the *Bacillus thuringiensis* toxin (Peferoen, 1992), trypsin inhibitor from cowpea (Hilder *et al.*, 1987) and lectins (Boulter *et al.*, 1990).

This group represents only a sample of compounds that will ultimately be found to be effective insecticides and are not toxic to mammals; many of them are derived from those plants that have natural mechanisms for resisting insect damage. The pests of primary importance to pea vary with the region in which they are grown. van Emden *et al.* (1988) identify 18 species that are significant pests, attacking the vegetative parts, flowers, pods or seeds and causing substantial losses of yield. They include pea moth, bollworm, thrips, seed weevils, pod borer and cutworms. As resistance to these insect pests is at a premium, the opportunity offered by a genetic engineering route to resistance is an attractive one for the pea crop.

The genetic engineering of virus resistance is now an attainable goal in a number of instances and, although resistance to particular strains of many of the pea viruses is available, there are good reasons for exploiting this technology in peas. In the first place, the strain of a given virus that causes crop loss may change rapidly and new sources of resistance be unavailable; secondly, even if resistance is available in existing germplasm the time-scale for its introduction may be lengthy. Resistant transgenic plants have been obtained by introducing genes for viral coat protein or satellite RNA. The former simulates the natural phenomenon of cross-protection which is sometimes observed when a plant infected by a mild strain of virus is subsequently infected by a more virulent strain. The presence of satellite RNA modifies the severity of symptoms of some viruses. The first demonstration that transforming a plant with a viral coat protein gene conferred protection against that virus was that of Powell-Abel et al. (1986), but since then there have been many examples of its successful exploitation (Reavy and Mayo, 1992). Information coding for satellite RNA was first introduced into a transgenic plant by Baulcombe et al. (1986) with a consequent suppression of symptoms (Harrison et al., 1987). A different strategy involves exploiting antisense RNA. By inverting a cDNA, or part of a cDNA, for a given mRNA with respect to its promoter, a complement to the normal RNA or part of the RNA is produced; this can result in reduced amounts of the RNA and hence of its product. Antisense RNA has been used in an analogous manner to block RNA and DNA viruses and shown to reduce symptom severity (Bejarano and Lichtenstein, 1992). Another approach to the control of viruses has involved ribozymes, short RNA sequences that can cleave RNA molecules (Haseloff and Gerlach, 1988). Ribozymes are targeted to a specific RNA by flanking the ribozyme sequence with sequences homologous to the target RNA. Resistance to tobacco mosaic virus infection of tobacco has been achieved by this route (Young and Gerlach, 1990). Viral replicase sequences incorporated into plants are yet another source of resistance (see Longstaff et al., 1993). Given this variety of approaches to developing virus-resistant plants and the success of field trials (Sanders et al., 1992), there is every reason for optimism that one of these strategies will be successful with peas. A priority in this respect will be to generate resistance to pea seed-borne mosaic virus, a pathogen causing substantial problems in northern Europe. The coat protein gene of this virus has been cloned (M. Albrechtsen, personal communication) and its cDNA sequence will be incorporated into an *Agrobacterium* plasmid for introduction into pea. A similar approach is planned for pea enation mosaic virus, where again the coat protein sequence is available (R. Hull, personal communication).

Fungal pathogens pose a different set of challenges, but a multiplicity of approaches are currently being considered or evaluated. There are two that have progressed to a stage where heterologous genes have been incorporated into plants and positive results obtained. Chitinase genes from *Phaseolus vulgaris* or the bacterium *Serratia marcescens* incorporated into tobacco and into *Brassica napus* resulted in a reduction of damping off due to *Rhizoctonia solani* (Broglie et al., 1991; Howie et al., 1992). A gene for ribosome-inactivating protein from barley introduced into tobacco conferred protection against the same fungus (Logemann et al., 1992). Pathogenesis-related proteins and lectins are other candidate compounds for transfer, and yet other approaches involve enhancing natural resistance mechanisms (Lamb et al., 1992). In Chapter 8, the opportunities for modification or transfer of natural phytoalexins was discussed; the transfer of avirulence genes between pathogens was also described and this latter work has led to the suggestion that interspecific or intergeneric transfer of 'non-host'-resistance genes may be feasible (Lamb *et al.*, 1992).

The generation of resistance to herbicides by introducing DNA sequences from other plant species or bacteria is well documented (Mullineaux, 1992); whether there is the agronomic or economic justification for doing so in peas remains to be seen. That it is feasible has been demonstrated in our own work, in which bialophos-resistant peas have been produced by introducing the *bar* gene from the bacterium *Streptomyces* (Schroeder *et al.*, 1993; D.R. Davies, unpublished observation).

The composition of pea seeds can be drastically altered by mutations at particular loci; the lines incorporating mutations at the *r, rb, rug3, rug4* and *rug5* loci described in Chapter 5 illustrate the extent to which the amounts of starch, sugar and lipid can be changed, as well as the composition of the starch and protein. Natural variation in lipid composition also exists in the pea germplasm (Jones *et al.*, 1991; Chapter 5). The biochemical steps in starch biosynthesis in the pea seed and the enzymes involved have been characterized and have allowed the DNA sequences coding for the enzymes to be isolated (Chapter 6). This has already allowed some of the sequences to be introduced into potato, but the possibilities that can be envisaged for the manipulation of pea starch are our concern here. In the first place, antisense sequences (see above) for these enzymes can be introduced to analyse the effects on starch quality of blocking or partially inhibiting particular steps. The second option is to modify the coding sequences of these enzymes, in order to achieve a change in specificity, and then reintroduce them into pea; again this could lead to modified starches. The third alternative is to modify the promoter sequences regulating the activity of the genes. Since the frequency and extent of branching of the amylopectin and the proportion of amylose to amylopectin in starch are determinants of its properties in foods and in industrial uses, the ability to produce a range of starch forms in peas is an attractive one. It offers the possibility of giving added value to, and introducing new uses for, the crop.

The ability to modify the storage proteins is currently limited (see Chapter 6), and the extent to which there will be an incentive to exploit the quantitative and qualitative variation in lipid (Chapter 5) or to induce further variation remains to be seen. The feeding value of peas can certainly be improved by selecting for the low values of trypsin inhibitors (TI) that we now recognize in some genotypes of peas (Chapter 6), and opportunities for yet further reductions by introducing antisense sequences to the mRNA for TI will emerge. Another facet that has emerged in the work on TI is the possibility of developing an immunological assay for the levels of these substances in batches of feed and in breeding lines; this would be a considerable improvement on current methods of estimating amounts of TI.

In Chapter 6 the role of lipoxygenases in determining flavour was discussed, as was the existence of a genotype in which one of the main seed lipoxygenases is absent. The extent to which flavour will be improved in such genotypes is being tested. Another aspect of consumer preference is that of sugar content of seed; from the ability to perturb starch biosynthesis comes the opportunity to enhance the sucrose content of seed, and this again offers the prospect of some market niches for the frozen and fresh products, as well as possibly for the dry seed.

Genes that affect flowering time, patterns of development and morphology have been isolated in some plants, and the availability of these DNA sequences will aid the isolation of equivalent genes in other crops. One of the first such projects in peas is to identify the genes that affect leaf and stipule morphology (N. Ellis, personal communication). It is through such studies that information and techniques will emerge that will allow the identification of genes affecting morphological and physiological characters of direct agronomic importance, such as stem stiffness and maturation time. Now that the transformation of peas is possible, another route to the identification and isolation of genes can be considered. This is through the introduction of a heterologous transposable element, which, if mobilized, can insert into and inactivate genes. Such a strategy is being used to isolate genes in other crops, and the transposable element of choice has been *Ac* from maize. The analysis of the control of gene activity (see Chapter 4) is progressing apace; the controlled regulation in time and space of the action of specific genes is now attainable, and is going to be of particular significance in future crop production.

In Chapter 2, the generation of DNA markers such as restriction fragment length polymorphisms (RFLP) was described; these are among many such markers that now allow geneticists to provide identifiers at close intervals along all the chromosomes. Given such markers, plant breeders can follow a trait or traits in segregating generations by following a tightly linked DNA marker rather than the trait itself. This can be exploited for quantitative traits, for major gene attributes whose phenotype is difficult to assay, or simply to enable selection in a seedling rather than in a mature plant. The value of these DNA markers has been described by Tanksley *et al.* (1989), and the RFLP map of peas is already being exploited by plant breeders (N. Ellis, personal communication). One of the many uses of molecular maps of the chromosome is in establishing the extent of similarity (synteny) of the DNA sequences in chromosomes of different species and genera. There is evidence of synteny between peas and lentils, and peas and chickpea (Muehlbauer *et al.*, 1989), and this information can be used to identify and isolate useful genes that exist in other legumes, which can then be introduced into peas by transformation techniques.

In summary there are going to be substantial opportunities in the short to medium term for the genetic manipulation of peas, to enhance their resistance to pests and pathogens, to alter the composition of the seed, to improve its value as a feed and to broaden the range of uses. This last option is particularly relevant for the starch within the seed; the air classification of dry peas is a well-established commercial procedure to produce extracts that are enriched not only for starch, but also for protein and for fibre. In the longer term, opportunities for changing the morphology of the crop to enhance its standing capacity and to alter physiological characters such as flowering time and synchrony of pod development will undoubtedly emerge. The limitations will not be technological, but will be defined by the extent of investment in research and plant breeding, which in turn are related to the total value of the crop. While peas are an internationally important legume, the scale of investment required for developing and testing genetically engineered plants is such that commercial organizations can justify such investment only for major or high-value crops. The alternative for crops such as peas is to foster an internationally coordinated effort by governments and international agencies to identify priority targets and partition responsibilities for the research required to solve the problems. In the absence of such an approach, peas and probably most legumes other than soyabean will not benefit from the opportunities that are emerging from biotechnology for some time. This would be unfortunate because the grain legumes are an important component of agriculture worldwide and are an important component of the diet of many people in developing countries as well as having a role in maintaining the nitrogen status of impoverished soils. For many of these grain legumes, the pea is a model, and its manipulation will serve as an indicator of what it is possible to achieve through genetic engineering.

References

Baulcombe, D.C., Saunders, G.R., Bevan, M.W., Mayo, M.A. and Harrison, B.D. (1986) Expression of biologically active viral satellite RNA from the nuclear genome of transformed plants. *Nature (London)* 321, 446–449.

Bejarano, E.R. and Lichtenstein, C.P. (1992) Prospects for engineering virus resistance in plants with antisense RNA. *Trends in Biotechnology* 10, 383–387.

Boulter, D., Edwards, G.A., Gatehouse, A.M.R., Gatehouse, J.A. and Hilder, V.A. (1990) Additive protective effects of incorporating two different higher plant derived insect resistance genes in transgenic tobacco plants. *Crop Protection* 9, 351–354.

Broglie, K., Chet, I., Holliday, M., Cressman, R. Biddle, P., Knowlton, C., Mauvais, C.J. and Broglie, R. (1991) Transgenic plants with enhanced resistance to the fungal pathogen *Rhizoctania solani*. *Science* 254, 1191–1197.

Dantuma, G. (1983) Potential and actual yields of dried peas in north western Europe. In: Thompson, R. and Casey, R. (eds), *Perspectives for Peas and Lupins as Protein Crops*. World Crops: Production, Utilization, Description, vol. 8. Martinus Nijhoff Publishers, The Hague, pp. 165–168.

Davies, D.R., Berry, G.J., Heath, M.C. and Dawkins, T.C.K. (1985) Pea (*Pisum sativum* L.). In: Summerfield, R.J. and Roberts, E.H. (eds), *Grain Legume Crops*. Collins, London, pp. 147–198.

Gatehouse, J.A., Hilder, V.A. and Gatehouse, A.M.R. (1991) Genetic engineering of plants for insect resistance. In: Grierson, D. (ed.), *Plant Genetic Engineering*, Plant Biotechnology Series, vol. 1. Blackie & Son Ltd., London/Chapman and Hall, New York, pp. 105-135.

Hagedorn, D.J. (1985) Diseases of peas: their importance and opportunities for breeding for disease resistance. In: Hebblethwaite, P.D., Heath, M.C. and Dawkins, T.C.K. (eds), *The Pea Crop*. Butterworths, London, pp. 205-214.

Hagedorn, D.J. (ed.) (1989) *Compendium of Pea Diseases*. American Phytopathological Society, St Paul, Minnesota, 57pp.

Harrison, B.D., Mayo, M.A. and Baulcombe, D.C. (1987) Virus resistance in transgenic plants that express cucumber mosaic virus satellite RNA. *Nature* 328, 799-802.

Haseloff, J. and Gerlach, W.L. (1988) Simple RNA enzymes with new and highly specific endoribonuclease activities. *Nature* 334, 585-591.

Hedley, C.L. and Ambrose, M.J. (1981) Designing 'leafless' plants for improving yields of dried pea crops. *Advances in Agronomy* 34, 225-227.

Hilder, V.A., Gatehouse, A.M.R., Sheerman, S.E., Barker, R.F. and Boulter, D. (1987) A novel mechanism of insect resistance engineered into tobacco. *Nature* 330, 160-163.

Howie, W., Newbigin, E., Joe, L., Penzcs, E., Suslow, I. and Dunsmuir, P. (1992) Resistance to *Rhizoctania solani* in transgenic tobacco. *Sixth International Symposium of Molecular Plant Microbe Interaction, Seattle*, abstract.

Huisman, J. (1991) Quality and use of legume seeds for animal nutrition. In: Froud-Williams, R.J., Gladders, P., Heath, M.C., Jenkyn, J.F., Knott, C.M., Lane, A. and Pink, D. (eds), *Production and Protection of Legumes*. Aspects of Applied Biology 27, Association of Applied Biologists, Wellesbourne, UK, pp. 11-22.

IAEA (1991) Mutant varieties database. *Mutation Breeding Newsletter* (International Atomic Energy Agency, Vienna) 38, 22-49.

Jermyn, W.A. and Slinkard, A.E. (1977) Variability of percent protein and its relationship to seed yield and seed shape in peas. *Legume Research* 1, 33-37.

Jones, D.A., Barber, L.M. and Hedley, C.L. (1991) The genetics of fatty acid composition in pea seeds. In: Froud-Williams, R.J., Gladders, P., Heath, M.C., Jenkyn, J.F., Knott, C.M., Lane, A. and Pink, D. (eds), *Production and Protection of Legumes*. Aspects of Applied Biology 27, Association of Applied Biologists, Wellesbourne, UK, pp. 275-278.

Krarup, A. and Davis, D.W. (1970) Inheritance of seed yield and its components in a six parent diallel cross in peas. *Journal of the American Society for Horticultural Science* 15, 795-797.

Lamb, C.J., Ryals, J.A., Ward, E.R. and Dixon, R.A. (1992) Emerging strategies for enhancing crop resistance to microbial pathogens. *Bio/Technology* 10, 1436-1444.

Le Guen, M.P., Tolman, G.H. and Huisman, J. (1991) Antibody formation against pea proteins in piglets. In: Verstegen, M.W.A., Huisman, J. and den Hartog, L.A. (eds), *Digestive Physiology of the Pig*. Pudoc, Wageningen, pp. 99-103.

Logemann, J., Jach, G., Tommerup, H., Mundy, J. and Schell, J. (1992) Expression of a barley ribosome-inactivating protein leads to increased fungal protection in transgenic tobacco plants. *Bio/Technology* 10, 305-308.

Longstaff, M., Brigneti, G., Boccard, F., Chapman, S. and Baulcombe, D. (1993) Extreme resistance to potato virus X infection in plants expressing a modified component of the putative viral replicase. *EMBO Journal* 12, 379-386.

Matthews, P. and Arthur, A.E. (1985) Genetic and environmental components of variation in protein content of peas. In: Hebblethwaite, P.D., Heath, M.C. and Dawkins, T.C.K. (eds), *The Pea Crop*. Butterworths, London, pp. 369-382.

Muehlbauer, F.J., Weeden, N.F. and Hofman, D.L. (1989) Inheritance and linkage relationships of morphological and isozyme loci in lentil (*Lens* Miller). *Journal of Heredity* 80, 298-303.

Mullineaux, P.M. (1992) Genetically engineered plants for herbicide resistance. In: Gatehouse, A.M.R., Hilder, V.A. and Boulter, D. (eds), *Plant Genetic Manipulation for Crop Protection*. Biotechnology in Agriculture Series No. 7, CAB International, Wallingford, pp. 75-108.

Pandey, S. and Gritton, E.T. (1975) Genotypic and phenotypic variances and correlations in peas. *Crop Science* 15, 353-356.

Peferoen, M. (1992) Engineering of insect-resistant plants with *Bacillus thuringiensis* crystal protein genes. In: Gatehouse, A.M.R., Hilder, V.A. and Boulter, D. (eds), *Plant Genetic Manipulation for Crop Protection*. Biotechnology in Agriculture Series No. 7, CAB International, Wallingford, pp. 135-153.

Powell-Abel, P., Nelson, R.S., De, B., Hoffman, N., Rogers, S.G., Fraley, R.T. and Beachy, R.N. (1986) Delay of disease development in transgenic plants that express the tobacco mosaic virus coat protein gene. *Science* 232, 738-743.

Reavy, B. and Mayo, M.A. (1992) Genetic engineering of virus resistance. In: Gatehouse, A.M.R., Hilder, V.A. and Boulter, D. (eds), *Plant Genetic Manipulation for Crop Protection*. Biotechnology in Agriculture Series No. 7, CAB International, Wallingford, pp. 183-214.

Sanders, P.R., Sammons, B., Kaniewski, W., Haley, L., Layton, J., LaVallee, B.J., Delannay, X. and Tumer, N.E. (1992) Field resistance of transgenic tomatoes expressing the tobacco mosaic virus or tomato mosaic virus coat protein genes. *Molecular Plant Pathology* 82, 683-685.

Savage, G.P. and Deo, S. (1989) The nutritional value of peas (*Pisum sativum*). A literature review. *Nutrition Abstracts and Reviews (Series A)*, (2), 66-68.

Schroeder, H.E., Schotz, A.H., Wardley-Richardson, T., Spencer, D. and Higgins, T.J.V. (1993) Transformation and regeneration of two cultivars of pea (*Pisum sativum* L.). *Plant Physiology* 101, 751-757.

Tanksley, S.D., Young, N.D., Paterson, A.H. and Bonierbale, M.W. (1989) RFLP mapping in plant breeding: new tools for an old science. *Bio/Technology* 7, 257-264.

van Emden, H.F., Ball, S.L. and Rao, M.R. (1988) Pest, disease and weed problems in pea, lentil, faba bean and chickpea. In: Summerfield, R.J. (ed.), *World Crops: Cool Season Food Legumes*. Kluwer Academic, Dordrecht, pp. 519-534.

Young, M. and Gerlach, W.L. (1990) Ribozyme activity against plant pathogen RNAs. In: *Abstracts of Workshop on Genome Expression and Pathogenesis of Plant RNA Viruses*, Serie Universitaria 253, Fundacion Juan March, Madrid, p. 31.

Index

Abscisic acid
    and lipoxygenases 146
    and wilting 204
Abscisic acid-responsive proteins 148
ADP glucose pyrophosphorylase 105, 108, 109, 125, 126, 128, 129
Agrobacterium 294–297
Albumins 132, 142
    PA1 and PA2 143, 144
    structure 143, 144
    synthesis 143
    variation 144
Alleles *see* Appendices A and B
Amylopectin 121, 125, 131
Amylose 121, 123, 125, 131
    and *rugosus* loci 105–107
Antinutritional factors 146, 149, 304
Antisense RNA 306, 307
aphII see kanamycin resistance
Asparagine synthase 246
Aspergillus nidulans 219
ATP, hydrolysis in N_2 fixation 244
Avirulence genes and *Pseudomonas* infection 225–227

Bacillus thuringiensis toxin 305
Bacterial blight 7
Bacteroids 262–266
Baker's yeast, lipoxygenase expression in 145, 146
Banding *see* chromosomes, banding patterns
bar see phosphinothricin resistance
Basta *see* phosphinothricin resistance
β-glucuronidase 66, 73, 293, 296
Bowman-Birk inhibitor *see* trypsin inhibitors
Branching
    and environment 195
    genes 194
    influence of flowering and internode length genes on 195, 197–200
    patterns 193–200
    and photoperiod 195
Breeding, methods 5

Callus 292
Carbon
    flux in starch synthesis 128–131
    supply to cotyledons 124
    translocation to nodules 247, 248
Cauliflower mosaic virus 35S promoter 68, 77, 78, 296, 297

Index

Chalcone synthase 16, 24, 29, 30, 31, 220, 258
Chaperone proteins in seeds 138
Chloramphenicol acetyl transferase 66, 271
Chlorophyll a/b binding protein 27
  gene expression 175
  genes 75, 76
Chromosomes
  arm-length ratios 14
  banding patterns 14
  duplications 31
  location in nuclei 38
  numbers 14, 25
  polytene 15
  rearrangements 14, 18, 23
  satellites 14
  secondary constrictions 26
  translocations 14, 17, 18, 19, 23
  trisomics 14
Convicilin
  genes 17, 19, 137, 139, 140
  structure 137
  synthesis 137
Cotyledons *see* seed, cotyledons
Crop
  as break crop 3
  leafless 8, 9
  production 4
  semileafless 7, 8
  uses 1, 3, 114, 303
  world distribution 4
  yields 4, 6, 7, 8, 303
Cytoplasmic markers 22

Disease
  bacterial 7, 222–227
  fungal 6, 7, 219–221
  resistance response genes 218, 219–220
  viral 6, 7, 31, 227–229
DNA *see* genome, chloroplast, DNA; genome, nuclear, DNA
DNA-binding proteins 62, 69–76, 143, 271

DNA-protein-binding motifs 69–76, 142–143

11S proteins 133, 150
Embryo
  culture 100–102, 291
  mutants, morphology 109–111
  see also seed, embryo
Emulsification and seed protein isolates 150
Endoreduplication, in cotyledons 96, 102
Endosperm *see* seed, endosperm
Enhancers 62, 63, 68
Epidermis, air spaces 170
Erysyphe pisi 7
Evolution 5, 6, 15, 22
  of plastid genome 50, 54

Fasciation 205
5S RNA genes 16, 31, 32, 35
Flavonoids and *nod* gene expression 253
Flowering 180–193
  and branching 195
  genes 182, 183
Fusarium solani 7, 217
Fusarium oxysporum 7, 221

Gelling and seed protein isolates 150
Gene expression
  control of 62–67, 77, 78, 142
  differential 61, 130–131, 140–142
  glutamine synthase 245, 249
  in response to pathogens 219, 220
  leghaemoglobin 247
  light-regulated 65–76
  nodulins 269–271
  Rubisco 65–76, 175
  seed storage protein 140–143
Genes *see* Appendices A and B
Genetic distance 22, 23

Index

Genome
  chloroplast 49
    DNA 22, 49–55
    deletions 53
    evolution 50–54
    inheritance 50, 55
    inverted repeats 50, 51, 53
    mutants 51
    rearrangements 51, 52
    recombination 53
    replication 50, 54, 55
  mitochondrial 49
  nuclear
    DNA
      base composition 26
      content 25
      methylation 27, 28
      replication 36–37, 38
      interspersion patterns 26
      repeated sequences 25, 27, 29
      size 25, 26
Gibberellic acid 166, 173–180, 199, 200
Globulins 132, 133
Glucose-1-phosphate 124
Glucose-6-phosphate 124
Glutamate synthase 245, 249
Glutamine synthase 245
Glycosylation, of vicilin 135, 136
GOGAT *see* Glutamate synthase
Granule-bound starch synthase 127
gus see β-glucuronidase

Heatshock, promoter 78
Height, plant 6
Herbicide resistance 307
Homoeotic genes
  leaf morphology 166–173
  flowering 189
Host range, *Rhizobium* 238
Hybrids, intergeneric 6
Hydrogenase 248, 249
Hydrotropism 205
Hygromycin resistance 295
Hypersensitive reaction 222

Ideotypes 304
In situ hybridization 15, 16
Indole-3-acetic acid 180, 185, 200
Infection thread 261, 262
Inocula, rhizobia 271
Internode length 173, 175, 177, 178, 179, 180, 187
  and branching 195
  and flowering 193
Introns, splicing 63
Isochores 27

Kanamycin resistance 294, 295, 296, 297
Karyotype 14, 22, 39, 40

lacZ 250
Late flowering, locus 184–185
Leaf morphology 166–173
Leaflets 169–172
Lectins 148–149, 305
  and rhizobial attachment 259–260
  and *Rhizobium* symbiosis 149
  structure 149
  synthesis 148, 149
Leghaemoglobin 246, 247, 271
Legumin
  genes 17, 19, 21, 34, 97, 134, 135, 138, 139
  mRNA 97, 99, 101, 105, 106, 140, 141
  structure 133–135
  synthesis 101, 106, 135, 137
Linkage 16–25
Lipid
  rugosus loci 105, 114
  variation in composition and content 114
Lipoxygenases 142, 145
  and abscisic acid 146
  and bread making 145
  genes 17, 19, 21
  genetics 145
  and methyl jasmonate 146
  mRNAs 145
  off-flavours 145
  and vining peas 145

Loci *see* Appendices A and B
Luciferase 66

Meiosis, mutants 14
Meristem
    culture 291–292
    inoculation 296
Methyl jasmonate and
        lipoxygenases 146
Mitochondrial genes 49
Mutants
    branching 193–200
    fertility 166
    flowering 180–193
    fruit development 202–204
    internode 173–180
    leaf 166–173
    nodulation 266–268
    pod 6, 202
    senescence 200–201
    starch 6, 103–108, 112, 114, 123, 128–132
    stipules 8, 167–172
    testa 89

Nectria haematococca 219
Nitrogen, contribution to soil by fixation 240–241
Nitrogen fixation 237–273
    genes 264, 265
Nitrogenase 243–244
Nod-factors 257–266
Nodulation 249–266
    host specificity 250–253
    mutants, peas 266–268
    Rhizobium genes 252
Nodule, root
    anatomy 238–243
    physiology 243–249
Nodulins 269–271
nptII see kanamycin resistance
Nucleolus 38
Nucleolus organizers 35

Oxygen, and nitrogen fixation 246

Pathogenesis-related proteins 218
Pathogens 217–231
Pea early browning virus 229
Pea mosaic virus, resistance 229
Pea seed-borne mosaic virus 6, 31, 227–229, 291–292
    resistance 228
    sequence 228
    transmission 228
Peronospora pisi 7
Peronospora viciae 221
Phaseolin, engineering of 150
Phosphinothricin resistance 296, 297
Phosphoglucomutase 108, 109, 125
Photomorphogenesis 173–176, 178
Photoperiod
    and apical senescence 201
    and pod development 203
    response 185
Phytoalexins 217, 218–220
Phytochrome 74
    and photomorphogenesis 173–176, 178, 199
Pisatin *see* phytoalexins
Plastid
    DNA *see* genome, chloroplast, DNA
    starch synthesis 124–125
Pod
    development 202–204
    shattering 6, 202
poly(A) tail 63
Polysaccharides and rhizobial infection 261
Polytene *see* chromosomes, polytene
PR proteins *see* pathogenesis-related proteins
Processing of seed protein precursors 135, 136, 137, 138, 150
Promoter deletions
    Rubisco genes 68–73
    seed protein genes 142–143
Protease inhibitors 143, 146–148
Protein
    bodies 135, 137, 150
    DNA-binding *see* DNA-binding proteins

Protein *Contd*
  seed 132–152
    biotechnology 150
    content 9
      effect of environment on 304
    cysteine content 132, 143, 148, 150, 152
    methionine content 132, 150
    quality 9
    rugosus loci 105
    storage 9, 31, 96, 113
      gene expression 140–143
      gene map positions 138–140
      genetics 138–140
      variation 139
Protoplasts 292
  fusion 293
PsbMV *see* Pea seed-borne mosaic virus
Pseudomonas syringae 7, 218, 222–227
Pyrophosphatase 125

r locus *see* Appendix A
r_b locus *see* Appendix A
Rearrangements *see* chromosomes, rearrangements
Recombinant inbreds 20
Recombination 18, 32–36, 37
Regeneration
  from callus 293
  from protoplasts 292
Reporter genes 66
Resistance genes, mapping 229
Resistance to pathogens 217
RFLP 18–22, 31, 128, 145, 221, 308
Rhicadhesin 258
Rhizobium 237–273
Ribozymes 306
Ribulose bisphosphate carboxylase *see* Rubisco
RNA, stability 63, 142
Rogue 8, 23
Root
  development 204–205
  hair deformation and nodulation 254, 260
  nodules 237–273

rot 7, 217
rRNA genes 16, 24, 26, 31, 32, 34, 35, 36, 37, 38, 39, 95
Rubisco 27
  gene expression 65–76, 175
rugosus loci *see* Appendix A

Satellites *see* chromosomes, satellites
Scanning tunnelling electron microscopy and vicilin structure 136
Seed-borne mosaic virus *see* pea seed-borne mosaic virus
Seed
  cellular development 92–102
  coat 6
  cotyledons
    cell population 93
    development 92
    epidermal cells 93
    transfer cells 93
  development 83–89
    r loci and 103–109
  embryo
    cell division 94, 95, 97, 101
    cell expansion 94, 95, 101
    globular 91
    growth, development 87, 88, 89–109
    heart-stage 91
    nutrient flow 87, 88, 90
    osmotic pressure 104, 105, 106
    sac 87
  endosperm 86
    development 86, 92
    formation 92
  growth analysis 83–87
  induced mutants 106–109
  'lags' in growth 84, 85, 86
  mutants 103–109
  round 103–109
  rugosus loci and 89
  size 3, 84, 87, 95
    in reciprocal crosses 87, 88
  sugar 6
  suspensor 90

Seed *Contd*
  testa
    characteristics, mutants 89
    development 86, 88
    and gibberellic acid 178, 179
    sieve tubes 87
    wrinkled 103–109
Semileafless *see* crop, semileafless
Senescence 200–201
7S proteins 133, 150
Signalling in nodule formation 249, 254–256
Somatic embryos 292
Splicing *see* introns, splicing
Standing ability 6
Starch
  branching enzyme 105, 107, 108, 125, 127, 128, 131
  composition and seed development 123
  control of synthesis 131
  grain morphology and *rugosus* loci 103, 104
  modification 131–132
  and *rugosus* loci 105, 107, 108, 112, 114, 123, 128–132
  structure 121–123
  synthase 125, 126–127, 129
  synthesis 123–131
  uses 2
Stem elongation 177, 178
Stipules 167–172
Sucrose, seed, *rugosus* loci 104, 113
Sulphur
  content of protein *see* protein, seed, cysteine/methionine content
  deficiency 142
Susceptibility to pathogens 217
Symbiosis
  mutants, peas 266–268
  rhizobia 266–271
Synteny 308

Targeting, storage proteins 137, 150, 151

TATA box 62, 63, 74, 271
Telomeres 16, 29, 38
Tendrils 166–172
Testa *see* seed, testa
35S promoter *see* cauliflower mosaic virus 355 promoter
Tissue culture 291–292
Transcription 62, 63, 68–76
  factors *see* DNA-binding proteins
  initiation 36
Transformation 293
Transgenic plants
  and nodulin promoters 271
  and phytochrome 175
  and Rubisco 67–69, 77
  and seed protein genes 142–143, 151
  and virus resistance 306
Translation, control of expression 63
Translocation
  plastid glucose-6-phosphate 124
  see also chromosomes, translocations
Transposons 16, 29–30, 39, 107, 128, 308
Trisomics *see* chromosomes, trisomics
Trypsin inhibitors 146–148, 305, 307
  physiological role 147
  sequences 147
  variation in amount 146

uidA *see* β-glucuronidase

Variegation 29
Vernalization 186, 187, 188
Vicilin
  genes 17, 19, 21, 24, 97, 137, 139, 140
  mRNA 97, 99, 101, 105, 140
  structure 135–137
  synthesis 96, 97, 101, 135–137
Viruses 7, 227–230, 305–306
  detection 230

Wax, leaf surface 170–171

waxy genes 126
Wilting 204
Winter hardiness 8, 204

Yield *see* crop, yields

Zn-finger protein 75

Appendix A: Pea genes, alleles, loci

a 6, 16, 19, 29, 31, 138, 140, 165
a2 31
A3 226
Adh 17, 19, 21
Adh-1 19
af 6, 17, 19, 165, 167, 169, 171, 304, 172, 173
age 205
alb 165
alt 165
apo 6, 202
apu 167, 169, 171
ar 89, 95, 165, 201, 203
arg 29, 167, 170
art-1 205
Asc 194, 200

b 165
brz 264–265
bt 21, 203
bulf 165

cab 17, 19, 21, 72
ce 21, 165
chi-6 165

coch 21, 167, 170
coch^{het} 167, 170
cov 165
cp 203
cr 165
cri 167, 170, 173
cry 179
Cvc 17, 19

def 89, 203
det 181–184, 189, 192, 198
dgl 165
di 89
dm 170, 181–184, 192
dne 170, 181–187, 193, 195, 197, 201

e 181–184, 187, 189, 198, 201
efr 185
en 166, 204
ep1 89
ep2 89
er 166

fa 16, 19, 190, 193, 205

fds 181–184, 192
fn 202
fna 202
fo 171
foe 89
fr 194, 196
fru 194, 196
Fs 17, 19, 21
fsd 181–184, 188, 189, 198

gi 181–184, 187, 188, 189, 195, 198
gp 16, 17, 19, 21, 165
GS 17, 19, 21
gty 6, 17, 19, 21

Hin 89
ho 185, 194, 200
hr 170, 186, 189
Hr 181–184, 195, 201, 202

i 16, 17, 19
ins 171

k 138

la 179
la cry^s 174, 195, 199
lat 171
le 6, 16, 17, 19, 145, 174, 176, 177, 179, 193, 195, 199
legA 134, 138, 139, 142
legD 138
legJ 134, 138
legK 138
len 165
len-2 165
lf 170, 181–185, 187, 192, 195, 198, 202, 204
lg-1 17, 19, 21
lg-J 17, 19
lgr 174, 179, 195, 200
lh 176, 177, 179, 193, 195, 199, 201
lip1 174, 175, 176
lk 170, 174, 180, 195, 199

lka 174, 180, 193, 195, 199, 202
lkb 170, 174, 180, 195, 199, 200, 202
lkc 174, 180, 195, 200
lkd 174, 180, 195
lm 173, 174, 204
Lox 17, 19, 21, 145
lr 166
ls 174, 176, 177, 179, 180, 193, 195, 199, 200
lv 165, 173, 174, 175, 193, 195, 199
lw 174, 175, 193

mifo 89

n 6, 95, 201, 202, 203
na 174, 175, 176, 178, 179, 193, 195, 199
nod-1 264–265, 268
nod-2 264–265
nod-3 264–265, 268
Np 203

p 6, 202, 203
pim 191, 192, 202
pl 204
Pl 17, 19
ppd 170, 181–187, 193, 195, 197
pra 185
pro 194, 200
pu 165
pur 29

r 6, 16, 17, 19, 21, 84, 88, 89, 103–107, 112, 114, 123, 128, 130–132, 138, 141, 142
R2 218, 225
ram 194, 196
r_b 6, 17, 19, 21, 84, 88, 89, 103–107, 112, 114, 126, 128, 130–132, 138
rbcS 17, 19, 21, 61, 67–79
rbcS-3A 67–79
rbcS-E9 68
rbcS3.6 75

red 171
rms-1 195, 196, 197
rms-2 194, 196, 197, 200
rms-3 194, 196, 197
rms-4 194, 196, 197
rms-5 194, 196, 197
Rrn1 16, 17, 19
Rrn2 17, 19, 21
rug3 107–109, 113, 131
rug4 107, 131
rug5 107, 131
rugosus see r, r_b, *rug*

sbm 6, 166
ser 171
sil 167, 170, 172, 173
sin 6, 203
sln 174, 176, 178, 179
sn 165, 170, 181–189, 193, 195, 197, 198, 201, 202, 203, 204
st 17, 19, 21, 165, 169, 171
st^{bs} 167, 169
sym-1 to sym-28 264–265
sym-1 268
sym-2 17, 19, 31, 264–265, 268, 272
sym-5 268
sym-18 268
sym-19 268

tac 173
Td 167, 170
te 203
tl 17, 19, 21, 165, 166, 167, 169, 171, 172, 173
twp 203

uni 167, 169
uni^{tac} 167, 169, 171
up 171

v 6, 16, 19, 202, 203
Vc-2 17, 19, 21
Vc-3 17, 19, 21
Vc-5 17, 19, 21
veg-1 181–184, 189, 191, 195, 201–202
veg-2 181, 184, 189–192, 195, 198–202

wa 165, 170
was 170
wb 165, 170, 171
wel 165, 170
wil 204
wlo 165, 170, 171, 172
wsp 165, 170

Appendix B: Bacterial genes, alleles, loci

aphII 293–297
avrPpiA1 225, 226, 227
avrPpaA1 225
avrPmaA2 225

bar 296, 297

fts z 262

hpt 295

hup 248

lacZ 250

nif-A 245
nod A to *nod X* 250–256
nod D see also 262
nod X see also 272
nptII 294, 295

uidA 293